Cooling of Microelectronic and Nanoelectronic Equipment

Advances and Emerging Research

WSPC Series in Advanced Integration and Packaging

Series Editors: Avram Bar-Cohen *(University of Maryland, USA)*
Shi-Wei Ricky Lee *(Hong Kong University of Science and Technology, ROC)*

Published

Vol. 1: Cost Analysis of Electronic Systems
by Peter Sandborn

Vol. 2: Design and Modeling for 3D ICs and Interposers
by Madhavan Swaminathan and Ki Jin Han

Vol. 3: Cooling of Microelectronic and Nanoelectronic Equipment: Advances and Emerging Research
edited by Madhusudan Iyengar, Karl J. L. Geisler and Bahgat Sammakia

WSPC Series in Advanced Integration and Packaging | Vol. 3

Cooling of Microelectronic and Nanoelectronic Equipment

Advances and Emerging Research

Editors

Madhusudan Iyengar
Facebook, USA

Karl J L Geisler
3M, USA

Bahgat Sammakia
Binghamton University, State University of New York, USA

World Scientific

NEW JERSEY • LONDON • SINGAPORE • BEIJING • SHANGHAI • HONG KONG • TAIPEI • CHENNAI

Published by

World Scientific Publishing Co. Pte. Ltd.
5 Toh Tuck Link, Singapore 596224
USA office: 27 Warren Street, Suite 401-402, Hackensack, NJ 07601
UK office: 57 Shelton Street, Covent Garden, London WC2H 9HE

Library of Congress Cataloging-in-Publication Data
Iyengar, Madhusudan.
Cooling of microelectronic and nanoelectronic equipment : advances and emerging research / Madhusudan Iyengar (Facebook, USA), Karl J.L. Geisler (3M, USA), & Bahgat Sammakia (Binghamton University, USA).
pages cm. -- (WSPC series in advanced integration and packaging ; vol. 3)
Includes bibliographical references and index.
ISBN 978-9814579780 (hardcover : alk. paper)
1. Electric machinery--Cooling. 2. Microelectronics. 3. Nanoelectronics. I. Geisler, Karl J. L. II. Sammakia, Bahgat, 1952– III. Title.
TK2189.I94 2014
621.38--dc23

2014026665

British Library Cataloguing-in-Publication Data
A catalogue record for this book is available from the British Library.

Printed in Singapore by B & Jo Enterprise Pte Ltd

DEDICATION TO DR. AVI BAR-COHEN IN CELEBRATION OF HIS 65TH BIRTHDAY

We are delighted to dedicate this book to celebrate the 65th birthday of Professor Avi Bar-Cohen. Dr. Avram Bar-Cohen is an internationally recognized leader in thermal science and technology, an Honorary Member of ASME, and Fellow of IEEE, and Distinguished University Professor in the Department of Mechanical Engineering at the University of Maryland. His publications, lectures, short courses, and research outcomes, in conjunction with professional service in ASME and IEEE, have helped to create a scientific foundation for the thermal management of electronic components and systems, while pioneering techniques for energy-efficient sustainable design of manufactured products. His current research focuses on two-phase microgap and microchannel coolers for high heat flux electronic components, on-chip thermoelectrics for hot spot remediation, thermal control of power amplifiers and photonics, and polymer-fiber composite heat exchangers. Bar-Cohen was the general chair for the 2010 International Heat Transfer Conference in Washington DC and is the President of the Assembly of International Heat Transfer Conferences. He is currently on assignment as a Program Manager in the Microsystem Technology Office at the U.S. Department of Defense's Defense Advanced Projects Agency (DARPA) in Virginia.

In addition to Honorary membership in ASME, Bar-Cohen's honors include the Luikov Medal from the International Center for Heat and Mass Transfer in Turkey (2008), ASME's Heat Transfer Memorial Award (1999), Curriculum Innovation Award (1999), Edwin F. Church Medal (1994) and Worcester Reed Warner Medal (2000), and the Electronic and Electrical Packaging Division's Outstanding Contribution Award (1994) as well as the InterPack Achievement Award (2007). Bar-Cohen was the founding chair of the IEEE Intersociety Conference on Thermal Management in Electronic Equipment (ITHERM) in 1988 and was recognized with the IEEE CPMT Society's Outstanding Sustained Technical Contributions Award (2002), the ASME/IEEE ITHERM Achievement Award (1998), and the THERMI Award from the IEEE/Semi-Therm Conference (1997). Most recently, Professor Bar-Cohen received the 2014 IEEE Components, Packaging and Manufacturing Technology Field Award. This annual CPMT award is the highest honor given in the area of electronics packaging, and considered by many to be the "Nobel Prize" in the field.

Professor Bar-Cohen has co-authored Design and Analysis of Heat Sinks (Wiley, 1995) and Thermal Analysis and Control of Electronic Equipment (McGraw-Hill, 1983). As editor of The Encyclopedia of Thermal Packaging (World Scientific, 2013), he also contributed as co-author of Volume 3: Dielectric Liquid Cooling of Immersed Components. He has co-edited 16 books in this field and authored/co-authored over 400 journal papers, refereed proceedings papers, and book chapters. He has delivered 70 keynote, plenary and invited lectures at major technical conferences and institutions, and he holds 8 US and 3 Japanese patents. He has advised to completion 64 PhD and master's students at the University of Maryland, the University of Minnesota and the Ben Gurion University (Beer Sheva, Israel), where he began his academic career in 1972. He was on the faculty in the Mechanical Engineering department at the University of Minnesota from 1989 to 2001, serving as Director of the Thermodynamics and Heat Transfer Division from 1992 to 1997. Starting in 1997, he directed the University of Minnesota Center for the Development of Technological Leadership and beginning in 2000 held the Sweatt Chair in Technological

Leadership. From 2001 to 2010, he served as the Chair of the Mechanical Engineering department at the University Maryland.

Avram Bar-Cohen began his professional career at the Raytheon Company in Massachusetts in 1968 and for the past 44 years has been involved in the design, analysis, and optimization of thermal systems, with an emphasis on the thermal packaging of electronic equipment. He has lectured widely, published extensively in the archival heat transfer and packaging literature, and taught many short courses on this subject, at both universities and major conferences in the US and abroad. He served as General Manager and Executive Consultant for packaging and physical modeling at Control Data Corporation, 1984-1989, held a succession of academic appointments, from Lecturer to Professor, in the Department of Mechanical Engineering at the Ben Gurion University of the Negev (Israel), 1973-1988, and was on the faculty at the Massachusetts Institute of Technology, 1977-1978, and the Naval Postgraduate School, 1982.

Many of Prof. Bar Cohen's students are currently in leadership positions around the nation and around the globe. Some of who contributed to chapters in this book covering a wide range of topics related to thermal management and thermal packaging of micro- and nanoelectronics, including enhanced heat transfer, heat sinks, liquid cooling, phase change materials, synthetic jets, computational heat transfer, electronics reliability, 3D packaging, thermoelectrics, data centers, and solid state lighting. This broad range of topics as well as the outstanding nature of the research and publications is a direct measure of the impact that Prof. Bar-Cohen has had and continues to have on research and development in the field of electronic systems integration and packaging.

We are truly delighted to dedicate this work to Avi, on behalf of the electronics packaging community, as a thank you for his continuing scientific and engineering contributions but even more so his visionary leadership, thoughtful mentoring, and heartfelt friendship.

Madhu Iyengar, Bahgat Sammakia, and Karl Geisler

Contents

CHAPTER 1

A REVIEW OF COOLING ROAD MAPS FOR 3D CHIP PACKAGES

Dereje Agonafer

Jenkins Garrett Professor
Site Director, NSF IUCRC Center in Energy Efficient Systems
Director, Electronics, MEMS & Nanoelectronics Systems Packaging Center
University of Texas at Arlington, U.S.A.
agonafer@uta.edu

The microelectronics industry has thrived through dimensional scaling and corresponding reduction in cost and increase in performance. It has been reported that the average selling price of a transistor has reduced from a few dollars in the early 50's to a billionth of a dollar in the early 2000. It has, however, become more difficult to sustain reduction in cost by scaling. Also, while new technology nodes results in reduced gate delay, it also effects an increase in the interconnect delay. One approach to delaying new technology node and improving performance is through reduction in interconnect delay through packaging. In particular, 3-D Through-Silicon-Via (3D TSV) technology is emerging as a powerful technology to reduce package footprint, decrease interconnection power, higher frequencies, and provide efficient integration of heterogeneous devices. TSVs provide high speed signal propagation due to reduced interconnect lengths as compared to wire-bonding and SOC (system-on-chip). However, with many advantages of 3D ICs over conventional 2D counterpart, there are some inherent thermal-mechanical-electrical challenges that need to be addressed before 3D ICs could become mainstream. This chapter talks about a few of the 3D TSV IC challenges from the thermal, mechanical and the performance standpoint of view. It also discusses a novel technique for high powered 3D IC cooling to sub-ambient temperatures using thermo-electric cooler (TEC).

Introduction

The convergence and miniaturization of computing and communications dictates building up rather than out. As the consumers demand more functions on their hand-held electronic devices, the need for more devices such as memory, CPU and GPU in hand-held type footprints is increasing. This results in high package density. Chip-stacking (3-D) is emerging as a powerful tool that satiates such Integrated Circuit (IC) package requirements. 3-D looks to be the future of hand-held electronics; hence, making it an important research area. Access to the third dimension has significantly simplified chip-level communications and transfer of information among the processing elements and has provided rapid access to memory and configurable logic. The active devices are confined to a plane in the upper surface of a semiconductor crystal with several layers of fixed interconnects separated by dielectrics above. As planar device miniaturization continues to its ultimate limits, the complexity of circuit interconnections for 2-D devices becomes a limitation for performance and drives up power dissipation [1]. 3-D technology would enable extremely dense solid-state memory to be arrayed within a few microns of the processing elements, which reduces access times. The 3-D arrangement also provides opportunities for new circuit architectures based on the geometrical ability to have greater numbers of interconnections among multi-layer active circuits. A 3-D FPGA would overcome the interconnect limitations, resulting in greater silicon efficiency per function (number of used gates/total number of gates), faster signal/data throughput, and faster switching of the gate-level configuration. True 3-D integrated circuits can operate at higher clock rates and can consume less power over their 2-D implementations, as the 3-D arrangement minimizes the length of circuit interconnects [1]. Stacking for low-device-interconnects (LDI) has been well developed with current stacks exceeding 15 devices enabled by wire-bonded interconnects. High-density-interconnects (HDI), however, necessitates area array packaging with much higher number of interconnects as well as reduced footprints. Three dimensional (3-D) stacking of the processor and memory components in high computing applications reduces the communication delay in a multi-core system owing to reduced system size and shorter interconnects [2]. The shorter interconnection length between the processing and memory components in a multi-core system lowers the overall system access latencies and boosts the system performance. "*Comparing with CSP (chip scale package), which has a silicon efficiency of about 80%; 3-D stacked packages are around 300% without increasing the thickness or the footprint of the package*" [3]. Integration in the Z direction is achieved by stacking dies or packages and interconnecting them with WB (wire bonding), FC (flip chip) or TSV (through silicon via) [3]. For a conventional single chip package, heat can be dissipated through the top (spreader and eventually through the heat sink), bottom (through the substrate and eventually PCB). But for a 3-D package commonly used in handheld products, heat dissipation is even more difficult, as the luxury of having a heat sink is not there due to space constraints. However, this 3-D integration of the processors and memory exacerbates the reliability and thermal problems due to high thermal resistance of the stacked designs. To overcome these thermal challenges and provide faster inter-chip electrical interconnection, Through Silicon Via (TSV) technology is being implemented in 3-D electronics. TSV technology for die-on-die interface has the potential of further improving the IC performance and package assembly efficiency. TSVs allow 3-D chips to be interconnected directly and prove to be critical components for enabling high-speed communication between circuits on stacked die. TSVs provide both the thermal and electrical interconnection between the stacked dies. 3-D TSV technology offers many advantages such as shorter interconnect length, higher silicon efficiency, ease of reliability testing and heterogeneous integration of different chips leading to a highly miniaturized footprint [1]. Miniature CMOS image sensor used in mobile phones is one of the early applications of the TSV technology and currently TSV application is emerging in DRAMs, flash memories, network devices, logic, etc., [3,4].

Although the 3-D TSV technology is being termed as the "next big thing" in the semiconductor arena and has the potential of revolutionizing the packaging industry, it has some inherent issues that need to be addressed before it could be implemented in the mainstream electronics industry. TSV fabrication process, thermal management of 3-D TSV packages, chip real estate (CRE) compensation to accommodate TSVs and thermo-mechanical TSV/chip package interaction (CPI) are some of the key issues in this technology. Furthermore, there are a number of processes and methodologies that are considered proprietary and may require licensing agreements and additional fees for their use. Industry roadmaps, however, continue to point toward the eventual use of TSV in developing new generations of high performance system-in-package products [4]. Progress in this area has accelerated through the cooperation and joint development programs between a number of government, industry and technical universities. Figure 1 and 2 show the chip stacking trend and 2-D vs 3-D interconnection, respectively [5, 10]. In this chapter, a review of thermal-mechanical-electrical challenges for 3-D packaging is presented.

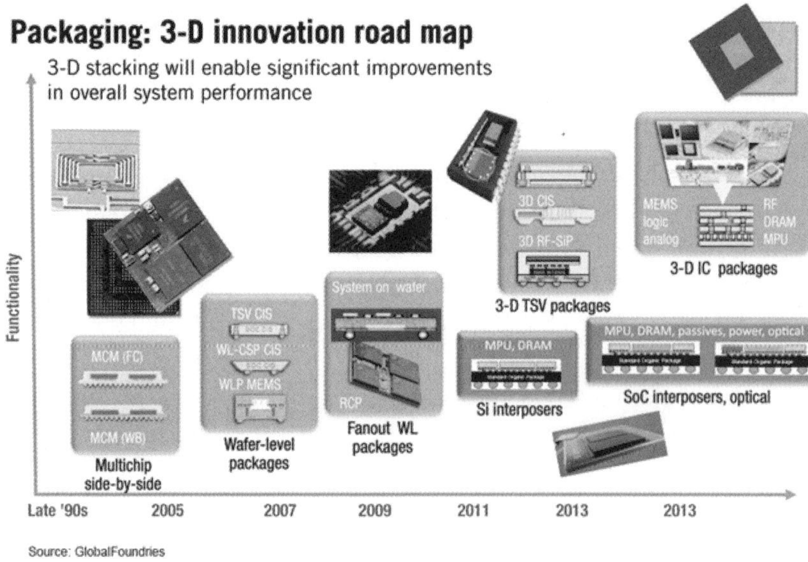

Fig. 1. 3D Innovation Roadmap [5].

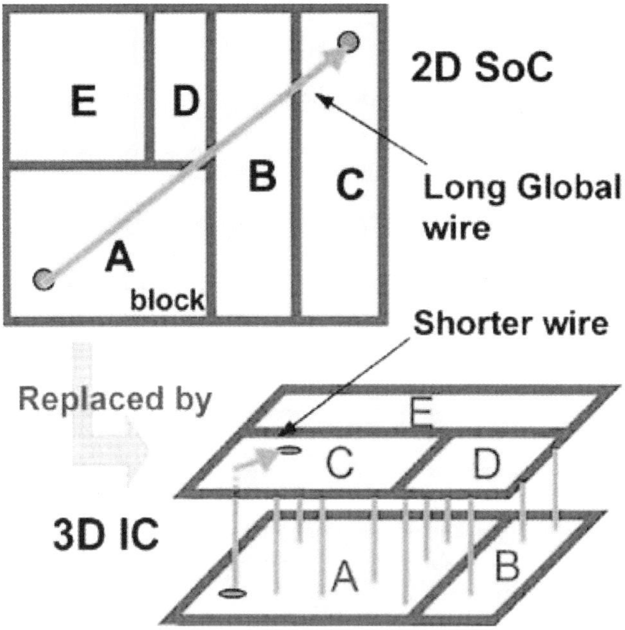

Fig. 2. 2D vs 3D Interconnect Length Variation [9].

TSV Fabrication

Emerging 3D chips stacking and MEMS/Sensors packaging technologies are using DRIE (Deep Reactive Ion Etching) to etch through-silicon via (TSV) for advanced interconnections. The interconnection step can be done prior to or post CMOS manufacturing, each requiring different etch process performances. Unlike other techniques, DRIE has the capability to etch feature sizes ranging from sub-micron to millimeter width. The main specificity of the DRIE is that etch rate is sensitive to the total exposed area and the aspect ratio. For the TSV applications, where the total exposed area is lower than 10%, high etch rates are achievable. The fabrication process is shown in Figure 3 [4]. There will be significant investment required in the equipment and process for via ablation, mask and coating systems, precision imaging systems, alloy plating and chemical etching systems.

Through Silicon Via ablation process variations include:

- Via-first integration forms TSV holes in the basic silicon wafer prior front end semiconductor processing
- Via-middle integration forms the small via holes in the wafer following front-end transistor formation and local interconnect processes
- Via-last hole information and plating processes are performed from the backside surface following wafer thinning

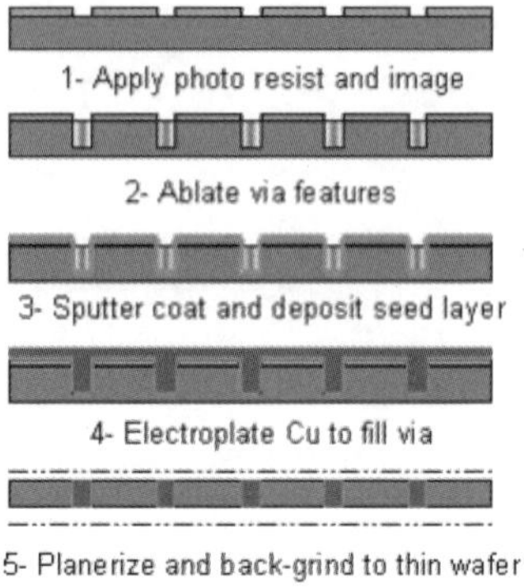

Fig. 3. TSV Fabrication [4].

After the formation of holes they are filled with a conductive polymer or plated closed with copper using an electroplating process. The copper via filling process is achieved through a number of alloy deposition steps that begin with applying a thin adhesion layer to the wafers surface and via features using an RF magnetron sputtering process. This is followed by a metal-organic compound deposition to provide a conformal, continuous, and low resistivity Cu seed layer. Electroplating is finally performed using a copper sulfa mate plating solution. Mechanical grinding systems typically use a two-step process that includes coarse grinding with a thinning rate of about 5 microns per second followed by fine-grinding at a rate of only 1 micron per second). When completed, the plated copper blind via features that were initially processed from the active top-side surface of the wafer will be exposed on the backside for subsequent wafer-to-wafer joining. Another alternative process is Wet chemical Etching Wet chem. thinning is accomplished by directing a thin stream of an etching agent over the surface of the rotating wafer. The etching chemistry developed for silicon wafer thinning is a combination of Hydrofluoric acid (HF) and Nitric acid (HNO3). Etching rates can be modified by adjusting the percentages of the chemical elements. The etch rate for silicon wafers will depend on the spin rate and the flow of the etching agent across the wafer surface. The target etching rate for wet chemical spin etching of silicon is about 10 μm per minute. In preparation for wet etching, the active surface of the wafer is protected, either by applying a mask coating or by use of special chucks designed to protect the surface during the process [4].

Thermal-Mechanical-Electrical Challenges in 3-D

Substantial research is being conducted to understand and characterize the thermal- mechanical-electrical behavior of 3-D packages with TSVs. As mentioned above, TSVs serve as the thermal and electrical interconnects within the stack. Thermal management and the thermo-mechanical interaction of TSVs and multiple stacked die with the package is the key. Since the filler metal in TSVs usually has higher thermal conductivity compared to silicon, it may potentially be used as a heat conduction path [6, 7]. Thermal resistance reduces with increasing TSV counts [8] for low-power, sub-Watt 3D ICs. A coupled thermal and mechanical analysis was completed to accurately understand the role of TSVs thermal management and structural integrity of the package [8]. As expected, the overall thermal resistance of the package decreased owing to the inclusion of Cu TSVs, as shown in Figure 4. However, this is accompanied by a penalty in terms of silicon efficiency. Also, TSV placement requires a transistor keep-out zone (KOZ) due to mechanical stresses around it, thereby significantly reducing the silicon efficiency, as shown in Figure 2. Moreover, the placement of each TSV requires sacrifice of space on the transistor plane where transistors could otherwise have been placed. To study the effect of TSV footprint on the interconnect delay; a flip chip package is modeled with 2 stacked chips dissipating 1W each (low powered) and the diameter, count and pitch of the TSV is varied within the desired range (constraining the area occupied by TSVs). As shown in Figures 6a and 6b, the TSV footprint substantially affects the RC (resistance capacitance) delay [9].

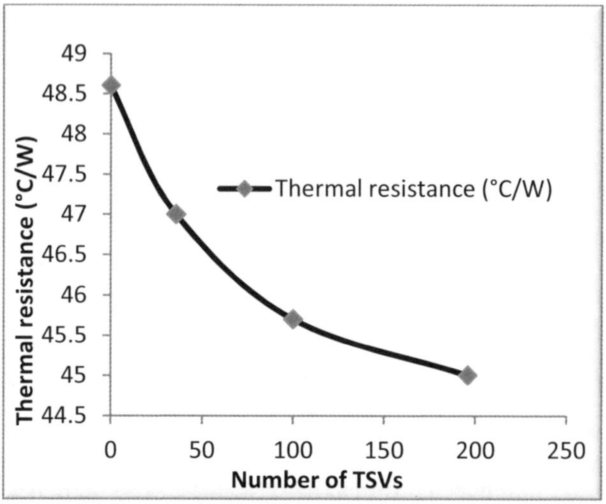

Fig. 4. Low powered device [8].

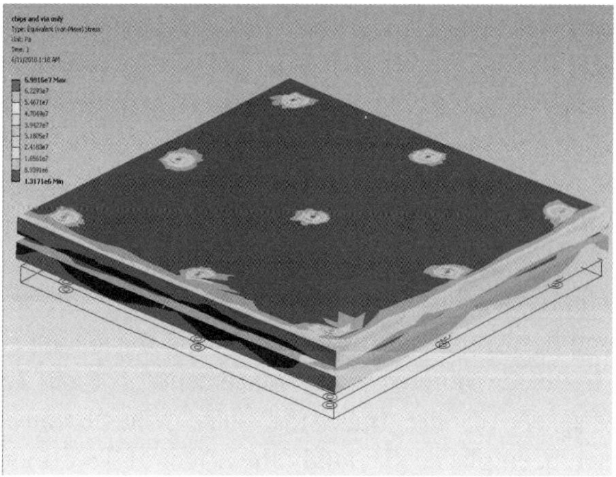

Fig. 5. Stress spots at the TSV/Chip interface (low powered) [8].

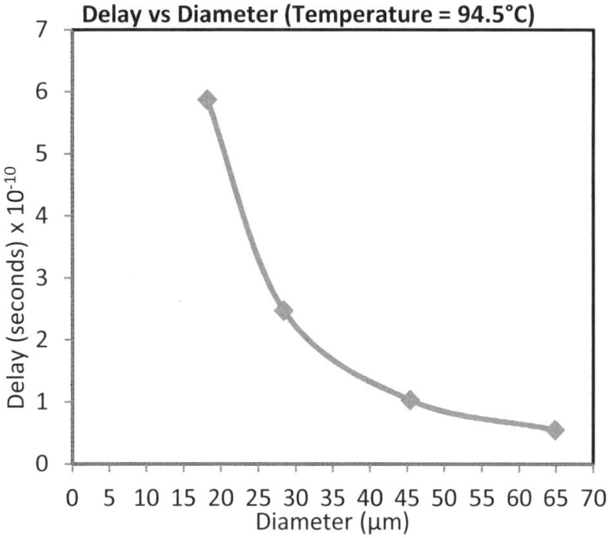

Fig. 6. a) Delay variation with TSV footprint [9].

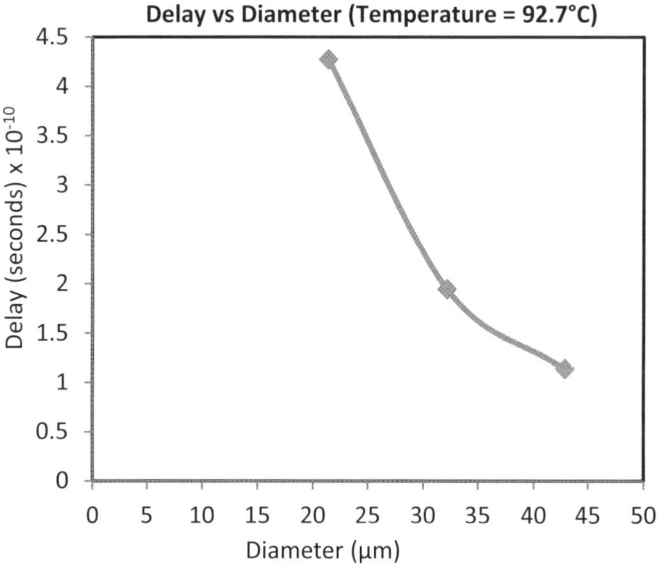

Fig. 6. b) Delay variations with TSV footprint [9].

This illustrates the complex interplay of various objectives in TSV design, and the need for a multi-disciplinary approach for resolving the various inherent trade-offs, in order to come up with an optimal TSV solution for the future generation electronics. The critical role played by TSVs in enabling 3D integration warrants a thorough investigation of mechanical stress effects caused by thermal expansion mismatch between silicon and the usual metal fillers constituting the TSV. Of particular interest, would be the role of the filler metal, dielectric thickness and geometry of the shallow trench isolation (STI) around a TSV. Parametric analysis of the TSV footprint to optimize thermal transport and minimize KOZ is a key area that needs to be investigated. Furthermore, another important issue is the heat generation due to electrical current (joule heating) in the TSV and its effect on the resultant thermo-mechanical response. Effect of the thermal stresses induced due to the TSV filler material on the transistor performance should also be quantified to design a robust package.

In a 3-D package, chips can be arranged in different architectures. To understand the effect of various die architectures on the thermal and mechanical response of a stacked package, Agonafer et al. [1] performed a comprehensive thermal-mechanical analysis on different architecture packages, as shown in Figure 7.

Thermal analysis was done for various parameters such as PCB thermal conductivity, higher "heat transfer coefficient" value, underfill, and copper spreader case. It is clearly evident from Figure 8 that the chip layout or architecture has negligible effect on the junction temperature. Of the three architectures evaluated, the spacer architecture was the easiest to improve. However, none of the thermal management strategies evaluated offered a significant improvement in performance. This implies that the root cause for the elevated temperature profiles is the high power buildup in the package die stack (i.e., memory and logic dice). The three stack architectures evaluated assume that all the dice are functioning simultaneously—the worst case scenario. However, in practice, the various die do not function simultaneously. The degree of functionality of each die, especially the memory die, is determined by the bus architecture. The thermal management strategies evaluated have demonstrated the significance of system design. Efforts must be concentrated on the system design processes (placement of components on the PCB, bus architecture, etc.) to improve the overall performance of the package. Thus it is clear from the study that thermal issues are not of a concern for low power applications. As shown by thermo-mechanical analysis, mechanical issues have a significant effect on package architecture for low-density interconnects, as shown by the study. Accelerated thermal cycling was performed on the modeled CSP using finite element analysis. Temperature variation was from −40 to 125°C with 15 minutes ramps and 15 minutes dwells (one-hour cycle).

(a)

(b)

(c)

Fig. 7. a) Rotated Stack, b) Staggered, c) Spacered [1].

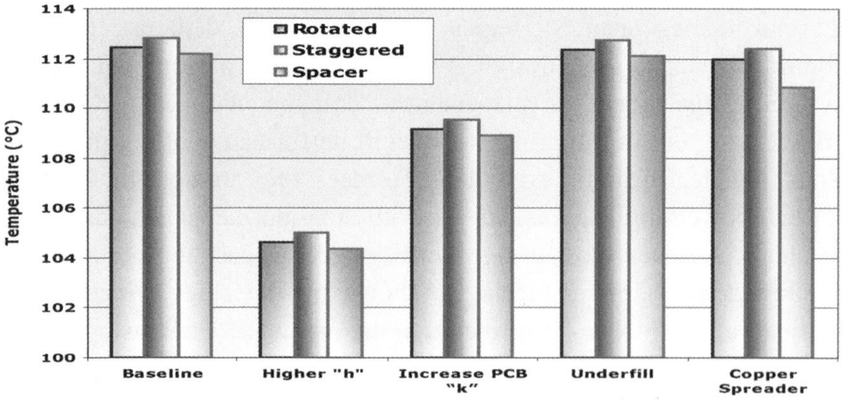

Fig. 8. Temperature results for different architectures [1].

The plastic work increases with the number of die, which indicates risk for solder joint reliability. As we decrease the die count, which means more mold compound material (higher CTE than die) and a higher mean package CTE (e.g., higher mold compound CTE) enhances the solder joint reliability. For 3-die and 5-die stacked packages, spacer-die architecture shows better characteristic life over the other two packages. For a 7-die stacked package, rotated-die architecture shows better characteristic life over the other two packages. Plastic strain accumulation and the worst solder joint location vary with various die stacking configurations. Die stresses in various die stacking configuration also vary, and the worst case is found with the pyramid 7 die stacking. This means that die stacking in stack packages surely has an effect on solder joint reliability and on ball architecture.

Presently, stack dice are used widely as low-power memory applications because thermal management of 3D architecture such as high-power processors inherits many thermal challenges. Inadequate thermal management of three-dimensional integrated circuits (3D-ICs) leads to reduction in performance, reliability, and ultimately system catastrophic failure. Heat dissipation of 3D systems is highly nonuniform and nonunidirectional due to many factors such as power architectures, transistors packing density, and real estate available on the chip. The potential problem with local sub-ambient cooling 3DICs is a slim probability of packaging a small subambient cooling device or system in a tight real estate available on the motherboard dedicated for the processor and its cooling apparatus such as heat sink and fan, as shown in Figure 9.

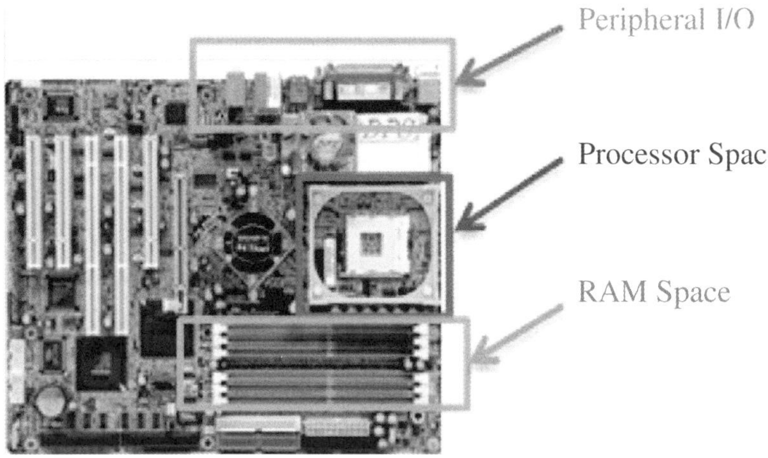

Fig. 9. Intel Pentium 4 Mother Board [11].

The higher the processor power, the larger the thermoelectric module must be, which leads to larger heat sink footprint requirement. The growth in heat sink size will overshadow other area on the motherboard such as RAM space and external peripheral I/O connections as illustrated in Figure 9. If more than one thermoelectric were used to achieve the desired cooling target, the temperature of the processor might be lower, but conventionally only three thermoelectric modules can be stacked at once in the cascade-style structure, as shown in Figure 10 below. Cascading thermoelectric modules might create a structural stability problem for high vibration applications. The primary objective of this research was to explore the possibility of sub-ambient cooling future for 3D-ICs as processing power and physical dimensions of 3DICs continue to grow vertically. Figure 10 and 11 show the conventional techniques that have been used to cool the 3-D IC, however the Multi-Dimenional-Heat-Transfer-System (MHTS) is a far superior thermal solution for high powered 3-D ICs. Figure 12 shows the novel MHTS method of cooling a 3-D IC (25W power) to sub-ambient temperatures using Thermo-Electric Cooler (TEC).

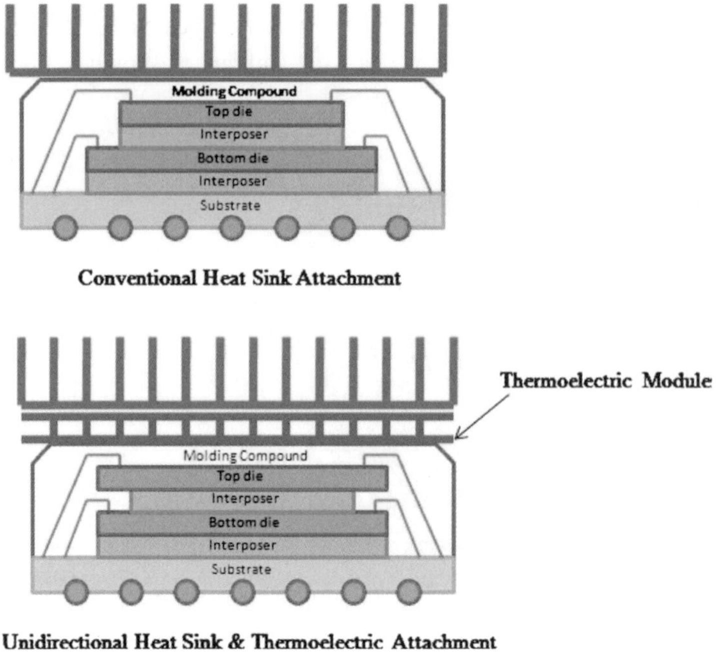

Fig. 10. Cooling Techniques for 3-D ICs [11].

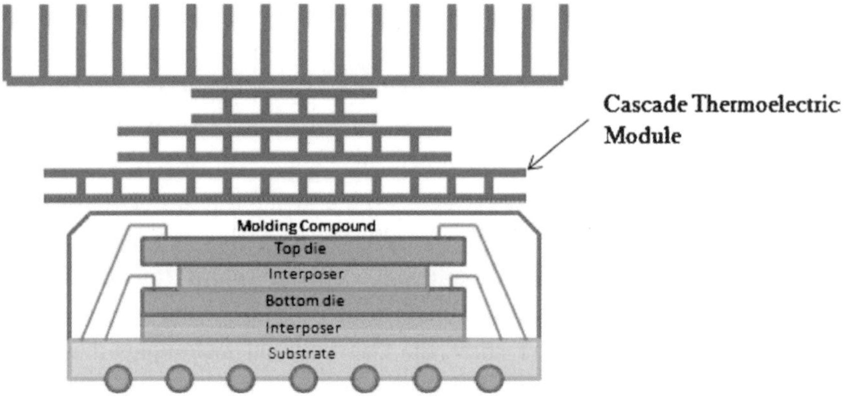

Fig. 11. Cascade TEC Module [11].

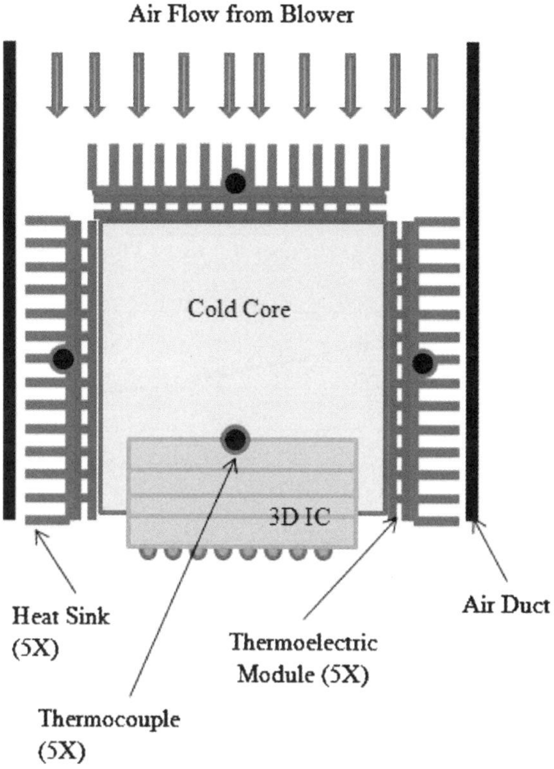

Fig. 12. MHTS [11].

It is clearly evident from Figures 13 and 14 that the Multi-Dimenional-Heat-Transfer-System (MHTS) can deliver sub-ambient cooling with less air flow rate and TEM power compared with the conventional 1D passive cooling and 1D active cooling cases presented above. At only 22 CFM and 90 WTEM power input, the MHTS can achieve a temperature of approximately 13°C. The advantage of the MHTS is the ability to direct the cold TEM toward the root of the cold core. The cold core in this case acted as an extended fin, which was being cooled by enclosed thermoelectric modules. In many other design cases and patents, the inventors concentrated on adding many TEM to achieve the desire cooling temperature, but the heat transfer and leakage surrounding the TEM were completely ignored, which severally penalized the cooling efficiency. In this case, adding more TEM does not guaranteed lower chip temperature if the real estate within the motherboard could not accommodate larger heat sink footprint. The unique design of the MHTS can accommodate four or more TEMs without violating the real estate allocated and still achieve the desire sub-ambient temperature.

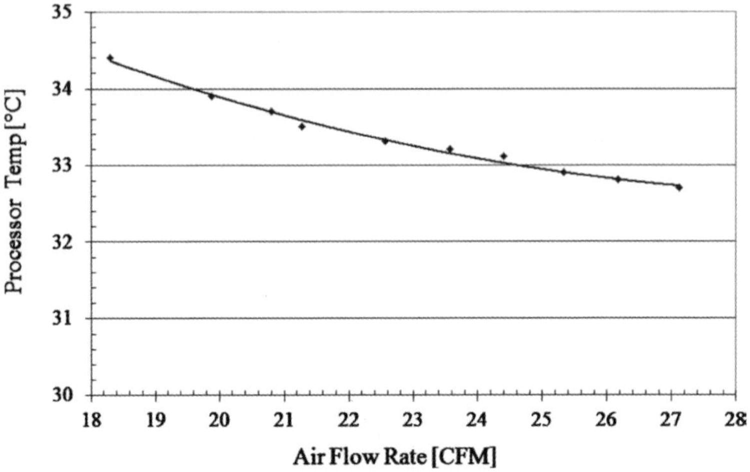

Fig. 13. Processor Temperature vs CFM (without MHTS) [11].

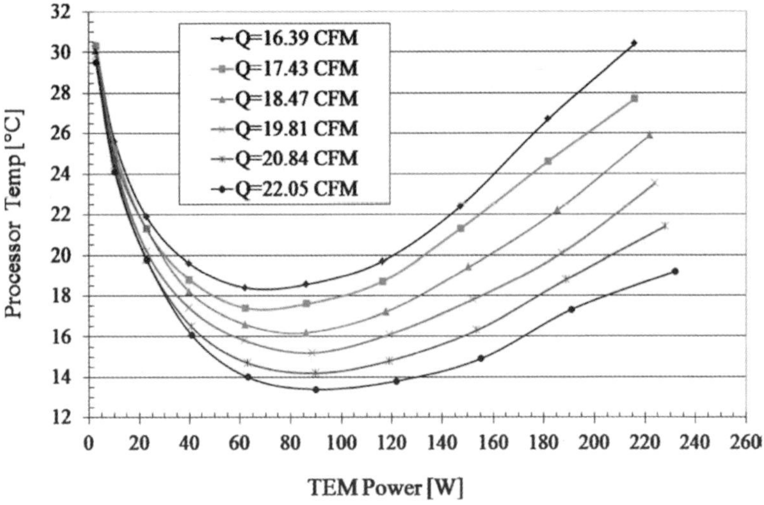

Fig. 14. Processor Temperature vs TEM power (MHTS) [11].

Acknowledgement

The author wants to acknowledge current and former students who have participated in this work: Dr. Abhijit Kaisare, Dr. Mohammad Hossain, Dr. Saket Karajgikar, Baek Sung, Yong Lee Roksana Akhter, Thiagarajan Raman, and Nuwan Rodrigo,. In particular, the author wants to thank Fahad Mirza, PhD student, for help in preparing the manuscript.

References

1. Agonafer, D., Kaisare, A., Hossain, M., Lee, Y., Sandur, B, Dishongh, T., Pekin , S., "Thermo-Mechanical Challenges in Stacked Packaging", Heat Transfer Engineering, 29 (2), pp. 134–148.
2. Venkatadri, Sammakia, B., Srihari, K., Santos, D., "A Review of Recent Advances in Thermal Management in Three-Dimensional Chip Stacks in Electronic Systems", Journal OF Electronics Packaging, Vol. 133, 041011, pp. 1–15.
3. Zheng, J., Zhang, Z., Chen, Y., Shi, J., "3D Stacked Package Technology and its Application Prospects," 2009 International Conference on New Trends in Information and Service Sciences.
4. Solberg, V., "Implementing TSV for 3D Semiconductor Packaging", Surface Mount Technology Association International Conference, 2011, Fort Worth, TX.
5. http://www.eetimes.com/document.asp?doc_id=1278829
6. Goplen, B., and Sapatnekar, S.: 'Thermal via placement in 3D ICs', *Proc. Intl. Symp. on Physical Design*, 2005, pp. 167–174.
7. Cong, J., and Zhang, Y.: 'Thermal via planning for 3-D Ics', *Proc. IEEE/ACM Intl. Conf. Computer Aided Design*, 2005, pp. 745–752.
8. Mirza. F, Muralidaran, B., Mynampati, P., Karajgikar, S., Agonafer, D., "Parametric Coupled Thermal and Structural Analysis of Through Silicon Vias in 3-D Packaging," IMECE 2010.
9. Mirza, F., Raman, T., Iftakher, A., Iqbal, S., Agonafer, D., "Parametric Thermal Analysis of TSVs in a 3-D Module based on Interconnect Delay and Silicon Efficiency", ITHERM 2012, San Diego, CA.
10. Private Communications, AMD ITRI 2011.
11. Phan, H., Agonafer, D., "Experimental Analysis Model of an Active Cooling Method for 3DICs Utilizing Multidimensional Configured Thermoelectric Coolers", Journal of Electronics Packaging, Vol. 132, June 2010, pp. 025401–4.

CHAPTER 2

THERMAL PERFORMANCE MAPPING OF DIRECT LIQUID COOLED 3D CHIP STACKS

Karl J.L. Geisler

3M Electronics & Energy Group Laboratory
Saint Paul Saint Paul, MN 55144, U.S.A.
kjgeisler@mmm.com

Avram Bar-Cohen

University of Maryland, Department of Mechanical Engineering,
College Park, MD 20742, U.S.A.
abc@umd.edu

Chip stacks are a crucial building block in advanced 3D microsystem architectures and can accommodate shorter interconnect distances between devices, leading to reduced power dissipation and improved electrical performance. Although enhanced conduction can serve to transfer the dissipated heat to the top and sides of the package and/or down to the underlying PCB, effective thermal management of stacked chips remains a most difficult challenge. Immersion cooling techniques, which provide convective and/or ebullient heat transfer, along with buoyant fluid flow, in the narrow gaps separating adjacent chips, are a most promising alternative to conduction cooling of three dimensional chip stacks. Application of the available theories, correlations, and experimental data are shown to reveal that passive immersion cooling—relying on natural convection and/or pool boiling—could provide the requisite thermal management capability for 3D chip stacks anticipated for use in much of the portable equipment category. Alternatively, pumped flow of dielectric liquids through the microgaps in 3D stacks, providing single phase and/or flow boiling heat absorption, could meet many of the most extreme thermal management requirements for high-performance 3D microsystems.

1. Introduction

While homogeneous and heterogeneous chip stacks are a crucial building block in advanced microsystem architectures, effective thermal management of stacked chips remains a most difficult challenge. In addition to providing increased packaging density, 3D die stacking accommodates shorter interconnect distances between devices, drastically improving electrical performance [1]. Current applications of 3D stacking are focused primarily on wirebonded memory chips for increased memory densities. Emerging 3D packaging developments are being driven by through-silicon via (TSV) interconnects, die thinning, and wafer-to-wafer bonding technologies [2].

A first-order approach for cooling 3D devices is to rely on conduction through various solid layers up to the top of the package – where a heat sink or coldplate can be attached - and/or down to the underlying PCB. More advanced thermal packaging approaches might include between-die solid conductors (interposers) to improve lateral heat spreading to the package periphery [3], but contact resistance between layers and the relatively low thermal conductivity of the electrical insulators typically considered for this role would be expected to yield large temperature gradients. Further, this approach may well be limited by one's ability to dissipate that heat into the surrounding air, given space limitations outside the package. On the other hand, immersion cooling techniques, which provide fluid flow in the narrow channels formed between stacked dies and high heat transfer coefficients directly on the die surfaces, could be very efficient in dissipating high volumetric heat generation rates, isothermalizing the silicon layers, and accommodating Joule heating in high current-density interconnects.

While some researchers, *e.g.* [4] [5], are considering cooling 3D "porous" structures with arrays of microchannels with small hydraulic diameters, integrated 3D silicon structures will likely be fabricated in layers. In this case, contact resistance between the layers may be significant and must be considered. By simply allowing coolant to flow across the entire gap between silicon layers and not relying on microchannel sidewall conduction, this complication may be avoided. Further benefits include elimination of detrimental flow instabilities (flow maldistributions, oscillations, and reversals) caused by thermal coupling between parallel microchannels [6].

Single phase natural convection in dielectric liquids provides a quantitative baseline for direct liquid cooling of 3D chip stacks. Nucleate pool boiling also includes the advantages of passive, buoyancy-driven flow, but with significantly increased heat transfer rates above natural convection. Further enhancement in the dissipated heat flux and volumetric heat transfer rates can be achieved by the introduction of forced flow through the narrow channels in the chip stack, both single phase forced convection and flow boiling. Using available heat transfer coefficient correlations, the maximum achievable heat transfer rates from individual chips, optimal gap dimensions, and upper performance limits for immersion cooling of 3D chip stacks can be determined. Sample results for representative 3D chip stacks in the dielectric electronics cooling liquid FC-72 are generated, and performance maps summarizing the upper limits for various cooling approaches are presented and discussed.

2. Passive Immersion Cooling

Passive techniques, which require no input power, are often desired for improved reliability and ease of implementation when compared to active, forced flow approaches. While heat transfer performance may be relatively limited, capabilities may be sufficient for certain application classes, in particular stacked memory and low-power logic devices.

2.1. *Single Phase Natural Convection*

Single phase natural convection is the easiest form of direct liquid cooling to implement and provides a lower bound for pool boiling and a benchmark solution for other advanced cooling approaches. An illustration of the basic geometry of a die stack appears in Figure 1. The gravity vector is taken to be parallel to the die length, *L*, and perpendicular to the stack height, *H*.

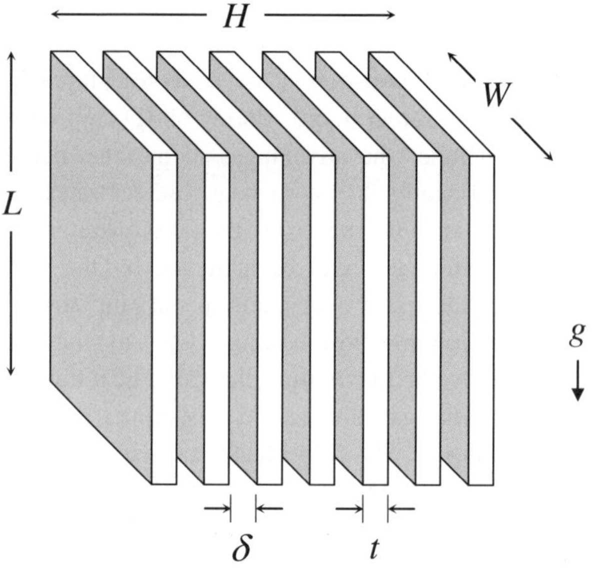

Fig. 1. Die stack geometry with nomenclature.

Leveraging the composite Nusselt number correlation derivations of Bar-Cohen and Rohsenow [7] and including the Prandtl number correction of Bar-Cohen and Schweitzer [8], Geisler [9] validated a composite Nusselt number correlation for symmetrically-heated, vertical microchannels immersed in saturated FC-72 at atmospheric pressure (101.3 kPa), based on the use of the fully-developed and isolated plate asymptotes, as

$$\mathrm{Nu}_\delta = \frac{h\delta}{k} = \left[\left(\frac{24}{\mathrm{El}}\right)^2 + \frac{0.705^{-2}}{\sqrt{\mathrm{El}}}\right]^{-1/2} \tag{1}$$

where the Elenbass number, El, is defined as,

$$\mathrm{El} \equiv \mathrm{Ra}_L \frac{\delta^4}{L^4} = \frac{c_p \rho^2 g \beta (T_{\mathrm{die}} - T_{\mathrm{amb}}) \delta^4}{\mu k L} \tag{2}$$

Following Elenbaas [10], the heat transfer coefficient defined by equation (1) is based on the temperature difference between the channel wall and the fluid at the channel inlet. Using the definitions of the

Nusselt number and heat transfer coefficient, the total volumetric heat transfer rate from an array of vertical dies can be written as

$$\frac{q}{V} = 2k\Delta T\left(\frac{\mathrm{Nu}_{\delta}}{\delta(\delta + t)}\right) \tag{3}$$

where the dies are $L \times W$ in size with thickness t, and H is the height of the entire stack—as illustrated in Figure 1. Optimum die spacings, which maximize the volumetric cooling rate from a chip stack may be found by substituting the channel Nusselt number expression, equation (1), into equation (3), taking the derivative of the resulting expression with respect to δ, and setting the result equal to zero. This process yields the following transcendental equation.

$$\left(2\delta + 3t - a^{-2}\mathrm{El}^{3/2}\delta\right)_{\mathrm{opt}} = 0 \tag{4}$$

where

$$a = 24 \cdot 0.705 = 16.92 \tag{5}$$

The optimum Elenbaas number in the thin die limit, equation (4) with $t = 0$, is equal to El = 69.0 (Nu = 1.66) and occurs in the transition region between the fully developed channel and isolated plate asymptotes. The effect of die thickness on optimum spacing is relatively subtle [9]. As die thickness increases, fewer dies can be accommodated in a fixed stack height, and larger gaps are needed to maximize the stack volumetric heat transfer rate. Unfortunately, since the heat transfer coefficient increase is modest, the total chip stack heat transfer rate decreases as the die thickness increases, even at optimum spacings.

Figure 2 shows the volumetric heat dissipation rates, as a function of die thickness for optimally-spaced die stacks, obtained with equation (4). Over the geometric parameter ranges of interest ($t < 0.5$ mm, $L = 10$–30 mm, and $\Delta T = 10$–30°C), optimum die spacings are found to range from 0.3 mm to 0.8 mm, while volumetric heat dissipation rates reach several tens of MW/m^3. While volumetric heat transfer rates for vanishingly thin die immersed in FC-72 can be expected to reach as high as 50 MW/m^3, or 50 W/cm^3, for more typical configurations where die thickness is approximately equal to these optimum spacings, volumetric dissipation

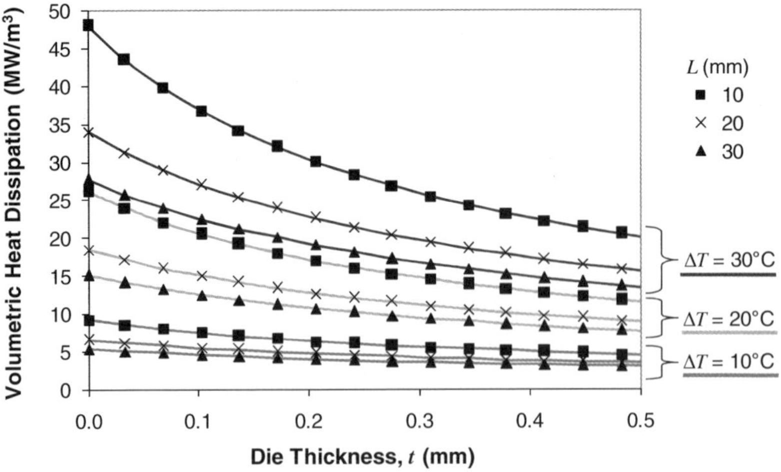

Fig. 2. Natural convection volumetric heat dissipation rates for optimally-spaced die stacks in saturated FC-72 at atmospheric pressure.

rates are reduced to approximately half of the thin die values. Further, single phase natural convection heat transfer rates drop sharply at sub-optimum spacings. For the conditions of Figure 2, reducing spacings to a third of the optimum values results in volumetric heat dissipation reductions of 80% to 90%.

2.2. *Pool Boiling*

Unfortunately, the desirable chemical inertness and electrically insulating characteristics of candidate immersion fluids also tend to be accompanied by relatively poor thermal properties. As a consequence, even the optimum natural convection heat transfer coefficients result in cooling rates that are insufficient to meet the demanding requirements of high performance microprocessors. Fortunately, nucleate boiling, which provides significant increases in heat flux with very modest increases in surface temperature, may be employed to substantially enhance heat transfer rates from microelectronics immersed in dielectric liquids.

The upper limit of nucleate boiling is the Critical Heat Flux (CHF), representing the point where the large volume of generated vapor prevents fresh liquid from replenishing the supply at the surface. As a result, the surface becomes blanketed by vapor and increases greatly in temperature. As this temperature excursion can be on the order of 100°C or more, electronics cooling applications are most often restricted to the single phase and nucleate boiling regimes, with CHF as an absolute upper limit. The well known Kutateladze-Zuber CHF relation [11] was developed for saturated pool boiling on large, thick, upward-facing horizontal plates.

$$q''_{\text{K-Z}} = \frac{\pi}{24} C_{\text{lg}} \sqrt{\rho_{\text{g}}} [\sigma_{\text{l}} g(\rho_{\text{l}} - \rho_{\text{g}})]^{1/4} \tag{6}$$

In confined chip stack geometries, boiling activity can be expected to produce axial quality and vapor fraction gradients, leading to a distinct progression of two phase flow regimes and an increase in fluid buoyancy in the inter-chip channel. However, as boiling activity increases, the growing vapor fraction may eventually lead to dry out (premature CHF) at the heated surfaces, beginning at or near the channel exit, and causing severe heat transfer degradation. Based on earlier work by Monde *et al.* [12] and Bonjour and Lallemand [13], Geisler and Bar-Cohen [14] correlated CHF for symmetrically heated pool boiling channels as

$$q''_{\text{chan}} = q''_{\text{K-Z}} \left[1 + \gamma \left(\frac{L}{\delta} \right)^{1.517} \right]^{-1} \tag{7}$$

where

$$\gamma = 1.08 \times 10^{-4} \left(\rho_{\text{l}} / \rho_{\text{g}} \right)^{1.343 (p/p_c)^{0.252}} \tag{8}$$

The volumetric cooling rate attained at critical heat flux (CHF) may be expressed as

$$\left(\frac{q}{V} \right)_{\text{CHF}} = \frac{A q''_{\text{CHF}}}{V} = \frac{2nLW q''_{\text{CHF}}}{nLW(\delta + t)} = q''_{\text{CHF}} \frac{2}{(\delta + t)} \tag{9}$$

where n is the number of chips , $H/(\delta+t)$, and q''_{CHF} is the pool boiling critical heat flux.

Geisler and Bar-Cohen [15] combined equations (6) and (7) with a factor from the literature based on heated surface properties that captures the reduction in CHF expected with decreasing die thickness [16], yielding

$$\left(\frac{q}{V}\right)_{\mathrm{CHF}} = \frac{\pi}{24}\frac{C_{\mathrm{lg}}\sqrt{\rho_{\mathrm{g}}}\left[\sigma_{\mathrm{l}} g(\rho_{\mathrm{l}}-\rho_{\mathrm{g}})\right]^{1/4}}{1+\gamma(L/\delta)^{1.517}} \times\left(\frac{t\sqrt{(\rho c_{p}k)_{\mathrm{h}}}}{t\sqrt{(\rho c_{\mathrm{p}}k)_{\mathrm{s}}}+0.1}\right)\left(\frac{2}{\delta+t}\right) \tag{10}$$

Fortunately, the negative effect of decreasing die thickness on CHF does not significantly counteract the volumetric benefits until $t < 10$ μm [9].

Figure 3 shows volumetric heat dissipation results for 20 mm long dies immersed in saturated FC-72 at atmospheric pressure as predicted by equation (10). At vanishingly small channel spacings, CHF goes to zero, while the array volume asymptotes to a constant $V = nLWt$. At large spacings, corresponding to $L/\delta < 10$, boiling channels achieve their large spacing CHF limit, and additional spacing increases provide no improvement in overall heat transfer but continue to increase the array volume. Thus, an optimum channel spacing exists in the range $0 < \delta <$

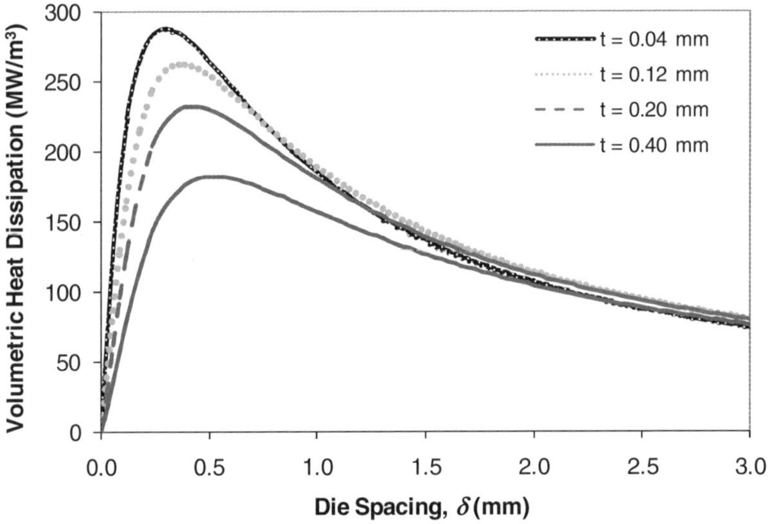

Fig. 3. Pool Boiling volumetric heat dissipation at CHF for 20 mm long stacked silicon dies immersed in saturated FC-72 at atmospheric pressure.

L/10 where volumetric heat dissipation is maximized. Maximum volumetric heat dissipation rates range from 200 MW/m^3 to 500 MW/m^3, or 200 W/cm^3 to 500 W/cm^3, for die lengths from 30 mm to 10 mm—an order of magnitude higher than can be achieved in single phase natural convection.

Optimum die spacings that maximize volumetric heat dissipation may be found by taking the derivative of equation (10) with respect to channel spacing and setting it equal to zero. Algebraic manipulation yields

$$\left(\frac{L}{\delta_{\text{opt}}}\right)=\left[1.517\gamma\left(1+\frac{t}{\delta_{\text{opt}}}\right)-\gamma\right]^{-1/1.517} \tag{11}$$

Figure 4 shows optimum spacing results, calculated using equation (11) over a range of die length and thickness for saturated FC-72 at atmospheric pressure (γ = 0.002412). As was the case in the natural convection optimization, as die thickness increases, the number of dies and, hence, gaps per unit volume decreases, thus decreasing the volumetric penalty of larger spacings, while providing modestly higher

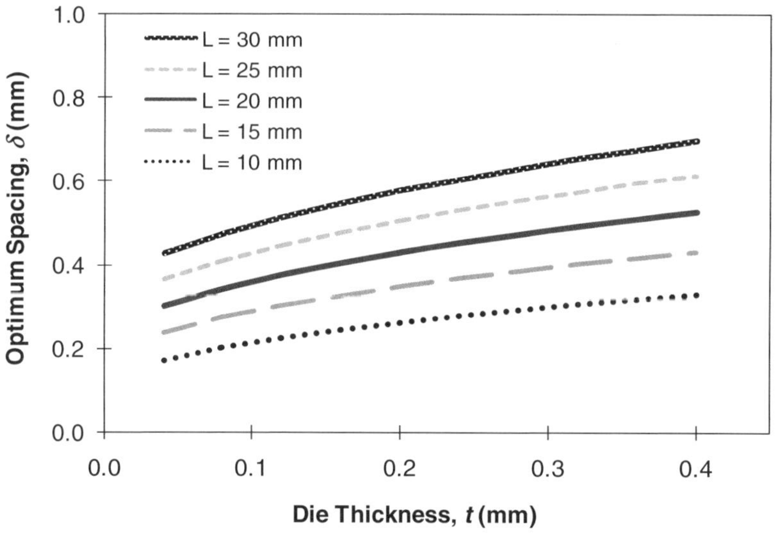

Fig. 4. Optimum pool boiling die spacing calculated using equation (11) for saturated FC-72 at atmospheric pressure.

heat transfer coefficients. It is interesting to note that optimum die spacings are approximately half a millimeter for both single phase natural convection and pool boiling. While these thermally-optimum spacings may be significantly larger than through-silicon interconnect structures currently in development, clear passive immersion cooling advantages would derive from implementing somewhat larger chip gaps that are in the range of more traditional solder bumps/columns or the interposers anticipated for use in heterogeneous chip stack applications. Moreover, even at sub-optimum chip gaps, pool boiling can offer 2 to 3 times the volumetric cooling rates achieved by natural convection.

3. Pumped Liquid Systems

Attention now shifts to direct liquid cooling approaches which require active pumping of the working fluid, starting with single phase forced convection. Microfabrication techniques have been used to create microchannel heat sinks and coldplates since the early 1980's [17]. Such liquid cooled microchannel coolers are characterized by sub-millimeter gaps and large channel aspect ratios. Flow in such channels results in very thin boundary layers and convective heat transfer coefficients that easily exceed the natural convection values cited above and rival the highest heat transfer rates achievable in pool boiling. Applying this technique to 3D chip stacks, the entire stack may be seen to be analogous to a microchannel cooler, with dielectric liquid forced between the tightly-spaced chips.

3.1. *Single Phase Forced Convection*

The thermal behavior of single phase flow in high aspect ratio microchannels formed between parallel plates or within a chip stack may be predicted with the Kakac *et al.* [18] correlations for heat transfer from isoflux surfaces in developing laminar flow. The average Nusselt number, based on the hydraulic diameter and divided into three flow length segments, is given by:

$$\mathrm{Nu}_{D_h} = 2.236x^{*-0.333} \quad \text{for} \quad x^* < 0.001 \tag{12}$$

$$\mathrm{Nu}_{D_h} = 2.236x^{*}-0.333+0.9 \quad \text{for} \quad 0.001 < x^{*} < 0.01 \tag{13}$$

$$\mathrm{Nu}_{D_h} = 8.235+0.0364/x^{*} \quad \text{for} \quad x^{*} > 0.01 \tag{14}$$

where

$$\mathrm{Nu}_{D_h} \equiv \frac{hD_h}{k} \tag{15}$$

$$x^{*} = \frac{L/D_h}{\mathrm{RePr}} \tag{16}$$

and

$$D_h = 2\delta \quad (\text{for } \delta << \mathrm{W}) \tag{17}$$

The Reynolds number, which is based on hydraulic diameter, may be expressed in terms of mass flux, G (kg/m^{2}s). Alternately, for the case of fixed volumetric flow rate, V, the Reynolds number is shown to be independent of chip spacing:

$$\mathrm{Re} = G\left(\frac{2\delta}{\mu}\right) = \left(\frac{\rho V}{\delta W}\right)\left(\frac{2\delta}{\mu}\right) = \frac{2\rho V}{\mu W} \tag{18}$$

With the Nusselt number correlations provided above, the wall heat transfer coefficient is based on the average temperature of fluid in the channel, $(T_{\mathrm{out}}+T_{\mathrm{in}})/2$, where

$$\left(T_{\mathrm{out}} - T_{\mathrm{in}}\right) = \frac{q}{c_p G\delta W} \tag{19}$$

Combining this equation with the definition of the heat transfer coefficient yields:

$$q = \left(T_{\mathrm{die}} - T_{\mathrm{in}}\right)W\left(\frac{1}{2Lh} + \frac{1}{2c_p G\delta}\right)^{-1} \tag{20}$$

Considering that the volume of a single channel is $LW(\delta+t)$, volumetric heat dissipation may be expressed as

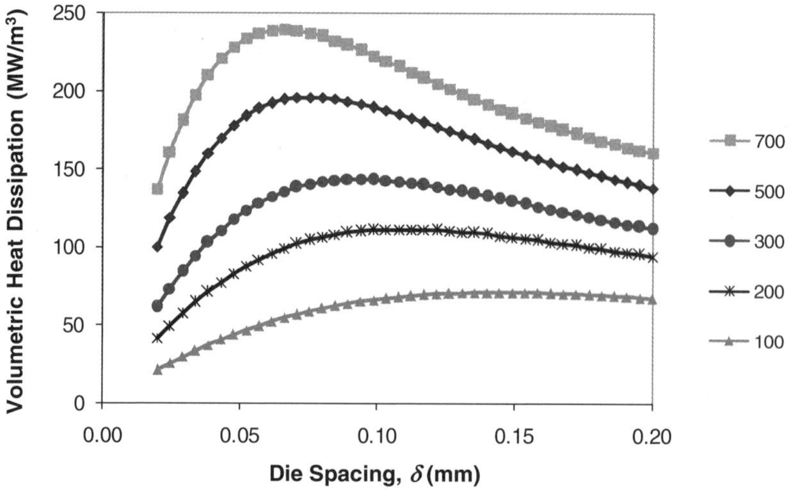

Fig. 5. Volumetric heat dissipation results for single phase forced convection cooled die stacks with $L = 20$ mm, $t = 0.3$ mm, and $\Delta T = 30°C$.

$$q = \left(T_{\text{die}} - T_{\text{in}}\right)W\left(\frac{1}{2Lh} + \frac{1}{2c_pG\delta}\right)^{-1} \tag{21}$$

where the heat transfer coefficient, h, is obtained from equations (12) through (16).

Sample results for the dependence of volumetric heat dissipation on die spacing are shown in Figure 5 for saturated FC-72 at atmospheric pressure, assuming a die length of 20 mm and thickness of 0.3 mm, with an average die temperature 30°C above the inlet, and for various channel mass fluxcs. Thcsc data demonstrate than when mass flux, G, (and, therefore, average channel velocity) is fixed, there exists an optimum spacing which maximizes the volumetric heat transfer. For the examples shown in Figure 5, the optimum spacings range from 0.05 mm to 0.1 mm, and volumetric heat dissipation rates approach 250 MW/m^3. Beyond these optima, spacing increases reduce volumetric density without improving heat transfer. At smaller spacings total liquid flow rate is reduced for fixed velocity, and the sensible heat rise in the channel affects heat transfer at a rate faster than die stack volume is reduced.

In order to find these optima explicitly, one can take the derivative of equation (21) with respect to channel spacing and set it equal to zero. This process yields the following fourth order polynomial in δ.

$$\begin{aligned} 0 &= \frac{8 \cdot 0.0364^2 + 4 \cdot 0.0364}{8.235} \Gamma \delta^4 \\ &- \frac{2}{t}\delta^3 + (4 \cdot 0.0364 - 1)\delta^2 + \frac{8.235}{2\Gamma} \end{aligned} \tag{22}$$

where

$$\Gamma = \frac{Gc_p}{kL} \tag{23}$$

Detailed expressions for the coefficients in equation (22) are shown to demonstrate their lineage back to the Nusselt number correlation, equation (14), which has been chosen based on the expectation that the optimum spacing will produce near fully-developed conditions at the duct exit.

Unfortunately, analytical solutions to the roots of equation (22) are prohibitively complex. Numerical solution of equation (22) must be pursued carefully, not only to select the proper root, but also to ensure convergence, as the values of the polynomial coefficients may differ by

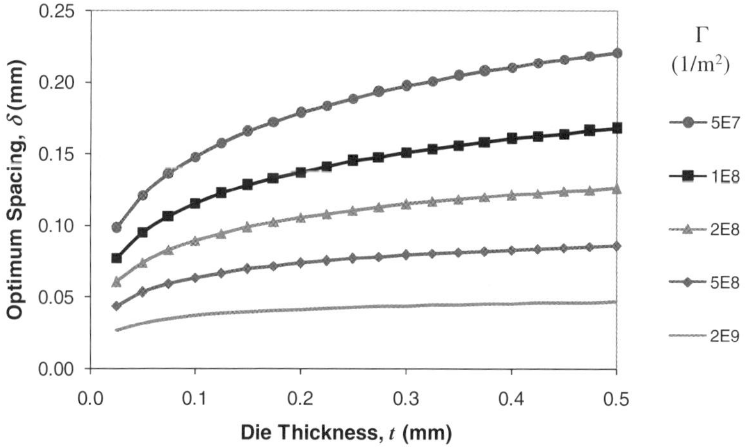

Fig. 6. Optimum spacing as a function of die thickness for single phase forced convection in saturated FC-72 at atmospheric pressure.

fifteen to twenty orders of magnitude. It is interesting to note that the optimum spacing is only dependent on temperature through the temperature dependence of the fluid thermal conductivity and specific heat. Further, the relationship between die thickness and optimum spacing depends only on Γ. Figure 6 shows the results of numerical solution of equation (22) and demonstrates that over a wide parametric range, optimum spacings fall between 0.02 mm and 0.2 mm.

Figure 7 shows volumetric heat dissipation rates as a function of die thickness for optimally-spaced stacks of 20 mm long dies in saturated FC-72 at atmospheric pressure, with an average die to inlet temperature rise of 30°C. Over a range of mass flux from 100 to 700 kg/m^{2}s and die thicknesses between 0.1 and 0.5 mm, heat transfer rates are shown to range from 100 to 500 MW/m^3—comparable to the results shown in the previous section for die stacks optimized for pool boiling. For a given set of conditions, the single phase forced convection provides less heat transfer per die than pool boiling, but the optimum spacing is smaller so the net result is comparable volumetric heat transfer. Over a wider parametric range (L = 10–30 mm, and ΔT = 10–30°C), optimum die spacings range from 40 to 190 μm, while volumetric heat dissipation rates span 13 to 850 MW/m^3. Corresponding Nusselt numbers are in a

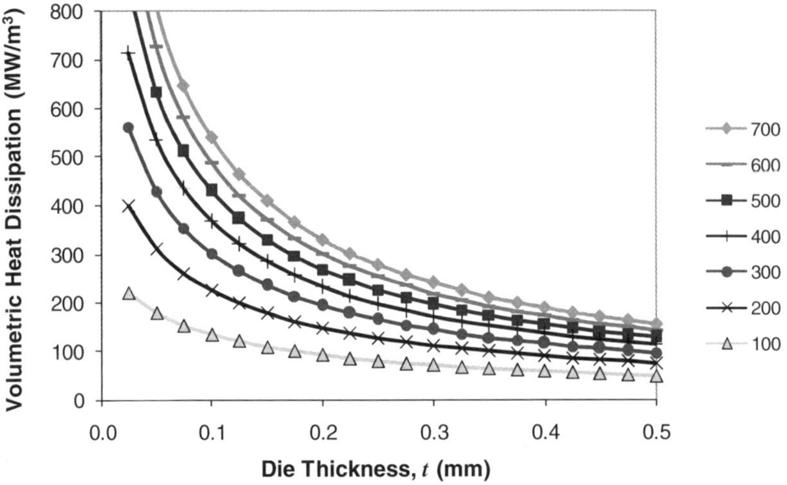

Fig. 7. Single phase forced convection volumetric heat dissipation rates for optimum die stacks in saturated FC-72 at atmospheric pressure, L = 20 mm, ΔT = 30°C.

very narrow range, from 8.3 to 8.9, signifying near fully developed channel flow (x^* = 0.06–0.2), and Reynolds numbers are deep into the laminar regime, less than 300, validating the appropriateness of the various correlations employed.

For off-optimum spacings, the transition to turbulence is expected to begin at a Reynolds number of 2300. Beyond this point, the McAdams equation for turbulent single phase forced convection may be employed:

$$\mathrm{Nu}_{D_h} = 0.023\mathrm{Re}_{D_h}^{0.8}\,\mathrm{Pr}^{0.333} \tag{24}$$

3.2. *Flow Boiling*

The relatively high heat transfer rates attainable with nucleate pool boiling and forced convection of dielectric liquids flowing through miniature channels makes the forced evaporative flow of such liquids in a heated microgap channel between chips a most promising candidate for the thermal management of advanced semiconductor devices. Interestingly, gas-driven flow of a dielectric liquid through miniaturized channels, formed by the microgaps between gallium arsenide chips and adjacent circuit boards, was used to provide thermal management for a compact, three-dimensional, high performance module in the Cray-3 computer developed in the 1990's [19]. Regrettably, few validated correlations exist for flow boiling in miniature channels, but understanding and identification of the prevailing flow regimes (i.e. stratified, intermittent, bubble, annular) can facilitate a regime-informed extrapolation of available data and traditional flow boiling correlations to the analysis and evaluation of the two-phase heat transfer coefficients in miniature channels.

The inherent complexity of two-phase thermal transport and its parametric sensitivities in miniature channel flow can be expected to lead to substantial variations in the heat transfer coefficient as the vapor quality increases from near-zero (saturated liquid) to near-unity (saturated vapor) conditions. Bar-Cohen and Rahim [20], examined the predictive accuracy of five classical two-phase heat transfer correlations, namely those due to Chen [21], Kandlikar [22], Gungor-Winterton [23], Gungor-Winterton Revised [24], and Shah [25]. The comparison of these correlations to the available miniature channel data for refrigerants and dielectric liquids was conducted by dividing the available data into three categories: Intermittent Flow, Annular Flow-low quality, and Annular Flow–moderate quality. The peak heat transfer coefficient reported for each set of conditions was used as the criteria for separating the "low quality" and "moderate quality" sub-regimes of annular flow. When the data are aggregated in this "regime-sorted" manner, Bar-Cohen and Rahim [20] showed that selecting the best fitting of the classical correlations for each of the flow regime categories, yields predictive agreement with regime-sorted heat transfer coefficients that does not depart significantly from the agreement found in large pipes and channels. In particular, the dominant, low quality Annular data was found to be correlated by the Chen correlation to within an average discrepancy of just 24% and standard deviation of 31%, while the Shah correlation provides agreement to within an average discrepancy of 32% and standard deviation of 40% for data in the Intermittent regime.

As a validation exercise, Bar-Cohen and Rahim [21] considered area-averaged experimental heat transfer coefficients from Kim *et al.* [22], spanning a range of 5 kW/m^{2}K to 7.5 kW/m^{2}K. In agreement with expectations, all but one of these heat transfer coefficient values are seen to be bounded by predictions based on the Shah [25] and Chen [23] correlations. Based on these and other results, it is expected that the highest heat transfer rates in two-phase flow through miniature channels will be attained in the annular regime, relying on the evaporation of vanishingly thin liquid films. The Chen correlation finds the two phase heat transfer coefficient as the sum of the nucleate boiling contribution (microscopic contribution, or h_{mic}) and the convective boiling contribution (macroscopic contribution, or h_{mac}).

$$h_{\text{TP}} = h_{\text{mic}} + h_{\text{mac}} \tag{25}$$

The microscopic contribution was found as

$$h_{\text{mic}} = 0.00122\left[\frac{k_{\text{l}}^{0.79} c_{p,\text{l}}^{0.45} \rho_{\text{l}}^{0.49}}{\sigma^{0.5} \mu_{\text{l}}^{0.29} h_{\text{lg}}^{0.24} \rho_{\text{g}}^{0.24}}\right] \times \left[T_{\text{w}} - T_{\text{sat}}\left(P_{\text{l}}\right)\right]^{0.24}\left[P_{\text{sat}}\left(T_{\text{w}}\right) - P_{\text{l}}\right]^{0.75} S \tag{26}$$

where S is the suppression factor which was originally presented by Chen in a graphical format. The empirical fit of the suppression factor as introduced by Collier was used in this study as

$$S(\mathrm{Re}_{tp}) = \left(1.25 + 2.56\times10^{-6}\,\mathrm{Re}_{tp}^{1.17}\right)^{-1} \tag{27}$$

The macroscopic contribution was found using the Dengler-Addoms correlation with a Prandtl number correction factor to generalize the correlation beyond water as a working fluid.

$$h_{mac} = h_l F(X_{tt}) \mathrm{Pr}_l^{0.296} \tag{28}$$

where the empirical fit of the enhancement factor, F, as a function of the Martinelli parameter, X_{tt}, provided by Collier was used in this study as

$$F(X_{tt}) = 1 \text{ for } X_{tt}^{-1} \le 0.1 \tag{29}$$

$$F(X_{tt}) = 2.35\left(0.213 + \frac{1}{X_{tt}}\right)^{0.736} \text{ for } X_{tt}^{-1} > 0.1 \tag{30}$$

The macroscopic contribution was found to be dominant in microchannel flow boiling as nucleate boiling is suppressed because of the very thin liquid film and insufficient superheat at the heated surface.

Given that the Chen correlation represents an upper limit and doesn't reflect full behavior for all flow regimes, *etc.*, no optimum can be calculated as with the other mechanisms. Predictions are shown in following section.

4. Performance Maps

In addition to the range of possible volumetric cooling rates discussed in the individual sections above, it is instructive to make a direct comparison of the achievable heat densities associated with various heat transfer modes through the use of performance maps. For purposes of the initial comparison, it will be assumed that saturated FC-72 at atmospheric pressure is the working fluid and that the chips are 20 mm square and 0.05 mm thick with a maximum allowed chip surface temperature of 85°C. Fluid flow rates and velocities for natural convection and pool boiling are determined by the physics of these

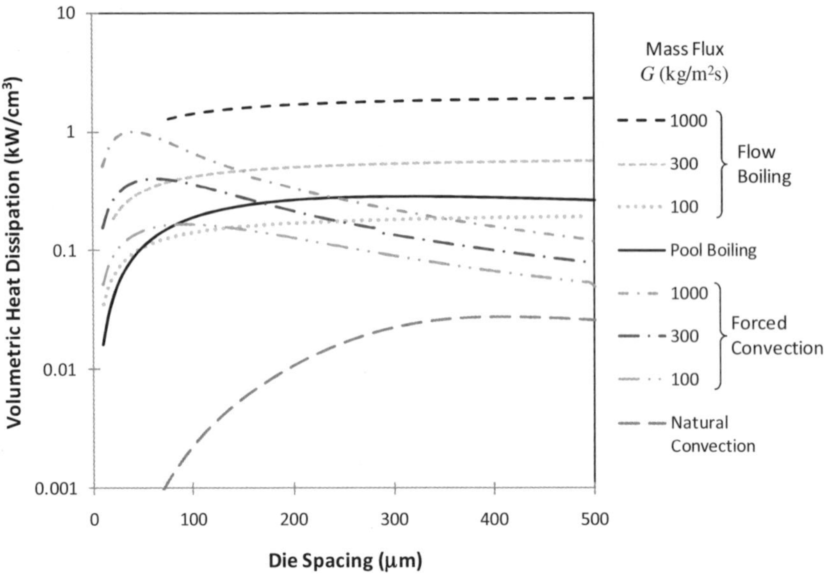

Fig. 8. Volumetric heat dissipation performance map for 20 mm × 20 mm × 0.050 mm dies.

buoyancy-driven flows, while for pumped systems, flow rates must be specified.

Figure 8 shows a 3D chip stack immersion cooling performance map, expressed in terms of volumetric heat dissipation versus gap dimension. In Figure 9, the data behind Figure 8 are displayed in terms of chip surface heat flux. The forced convection curves in Figures 8 and 9 are based on a condition of fixed mass flux, and Reynolds numbers are less than 2300 for all of the conditions presented, signifying laminar flow. Unsurprisingly, single phase natural convection provides the lowest heat transfer capability of the three heat transfer modes. For gap spacings in the 100 μm to 200 μm range, single phase forced convection and pool boiling provide roughly equivalent volumetric heat transfer performance, with pool boiling improving at larger spacings and forced convection improving with reductions in gap size. Maximum single phase forced convection volumetric heat dissipation is shown in Figure 14 to peak around 1 kW/cm^3 at die spacings near 40 μm with a forced flow rate of 1000 kg/m^{2}s. At small spacings, single phase heat transfer rates tend to

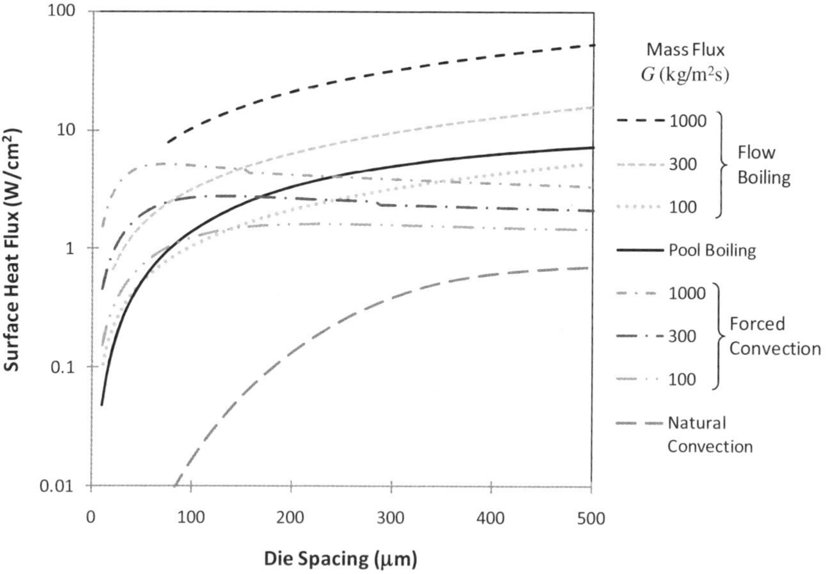

Fig. 9. Heat flux performance map for 20 mm × 20 mm × 0.050 mm dies.

be higher than achievable with flow boiling, suggesting that single phase forced convection will prevail at spacings less than 50 microns. However, flow boiling is able to maintain and improve upon the highest heat transfer rates at increasingly larger spacings. The flow boiling curves of Figure 14 are based on an assumption of 50% exit flow quality.

Passive techniques (natural convection and pool boiling) require no input power and may be easier to implement but provide somewhat limited heat transfer capability. While flow boiling suggests great promise for cooling of extreme high heat density devices, complexities inherent to multiphase flow systems make heat transfer predictions significantly more complicated and significantly less accurate than comparable single phase calculations. As a result, multiphase flow and heat transfer models and correlations require heavy experimental validation and evaluation of specific implementation details. Fundamental research is needed to solidify the physical understanding of two phase flow phenomena in the context of realistic device constraints and to develop robust predictive models for heat transfer, pressure drop, and critical heat flux that include the effects of geometry, surface modifications, pressure, saturation temperature variations, inlet liquid subcooling, and new fluids or combinations of fluids. It is expected that many of these effects can be exploited to push flow boiling heat transfer rates substantially higher than the results presented herein. In fact, all of these results could be improved using traditional single- and two-phase enhancement techniques, pushing these limits even further.

5. Conclusions

The foregoing analyses and discussion have explored the thermal management potential of direct liquid cooling of 3D chip stacks, a crucial building block in advanced microsystem architectures of the future. The introduction of miniature liquid passages between adjacent chips to provide convective and/or ebullient heat transfer was shown to offer a most promising alternative to conduction cooling of three-dimensional chip stacks. It was shown that cooling densities as high as 40 W/cm^3 for natural convection, 200 to 400 W/cm^3 for pool boiling, and approaching 1 kW/cm^3 for forced convection and flow boiling could be achieved using FC-72 as the dielectric coolant.

It may, thus, be surmised that passive immersion cooling—relying on natural convection and/or pool boiling—could provide the requisite thermal management capability for heterogeneous 3D chip stacks anticipated for use in much of the low-power equipment category (memory stacks, communications, etc.). Alternatively, pumped flow of dielectric liquids through microgaps could meet many of the most extreme thermal management requirements for high-performance 3D microsystems. Due to the thermofluid complexity of ebullient phenomena and the paucity of data for boiling of dielectric liquids in microgaps, much more fundamental research is needed to solidify the physical understanding of these phenomena and to underpin the basic relations needed for the design and optimization of 3D immersion cooled microsystems. Then, further impacts of realistic device implementations, such as non-uniform heat dissipation ("hot spots") and the effect of flow blockages due to die interconnect, can be explored.

Nomenclature

El	Elenbaas number
F	forced convection enhancement factor
g	gravitational acceleration (m/s2)
G	mass flux (kg/m2s)
h	heat transfer coefficient (W/m2K)
h_{fg}	latent heat of evaporation (J/kgK)
H	fin height (m)
k	thermal conductivity (W/mK)
L	fin/heat sink length (m)
n	number of dies
Nu	Nusselt number
P	pressure (Pa)
P_{r}	reduced system pressure, P/Pcrit
Pr	Prandtl number
q	heat flow (W)
q'	heat flow per unit length (W/m)
q''	heat flux (W/m2)
Ra	Rayleigh number
S	nucleate boiling suppression factor
t	die thickness (m)
T	temperature (ºC)
W	heat sink width (m)
x^*	dimensionless channel length
X_{tt}	Martinelli parameter

Greek Symbols

δ	die/channel spacing (m)
Γ	single phase optimization parameter
γ	fluid density parameter
ψ	two phase enhancement ratio
ρ	mass density (kg/m3)
σ	surface tension (N/m)

Subscripts

amb	ambient
CHF	critical heat flux
crit	critical
g	gas/vapor state
h	heater
in	inlet
l	liquid state
mac	macroscopic
mic	microscopic
opt	optimum
out	outlet
sat	saturated
tp	two phase

References

1. R. Zhang, K. Roy, C.-K. Koh, D.B. Janes, Power trends and performance characterization of 3-dimensional integration, *Proceedings of the 2001 IEEE International Symposium on Circuits and Systems*, 4 (2001) 414–417.
2. J.U. Knickerbocker, P.S. Andry, B. Dang, R.R. Horton, M.J. Interrante, C.S. Patel, R.J. Polastre, K. Sakuma, R. Sirdeshmukh, E.J. Sprogis, S.M. Sri-Jayantha, A.M. Stephens, A.W. Topol, C.K. Tsang, B.C. Webb, S.L. Wright, Three-dimensional silicon integration, *IBM Journal of Research & Development*, 52 (6) (2008) 553–569.
3. D. Gerlach, Y.K. Joshi, Parametric thermal modeling of 3D stacked chip electronics with interleaved solid heat spreaders, *Proceedings of the Tenth Intersociety Conference on Thermal and Thermomechanical Phenomena in Electronic Systems*, (2006) 1208–1212.
4. M. Rayasam, S. Chaparala, D. Farnam, B.G. Sammakia, G. Subbarayan, Thermal solution maps: A strategy for thermal design of three-dimensional packages, *Journal of Electronic Packaging*, 131 (1) (2009) 011015, 9 pages.
5. Y.J. Kim, Y.K. Joshi, A.G. Fedorov, Y. Lee, S.K. Lim, Thermal characterization of interlayer microfluidic cooling of three-dimensional IC with non-univorm heat flux, *Proceedings of the Seventh International ASME Conference on Nanochannels, Microchannels and Minichannels*, ICNMM2009-82133, (2009) 10 pages.
6. N. Thiagarajan, R.J. Jones, D.T. Pate, S.H. Bhavnani, Thermal characteristics of two-phase flow of a dielectric fluid in surface augmented microchannels,

Proceedings of the Eleventh Intersociety Conference on Thermal and Thermomechanical Phenomena and Emerging Technologies in Electronic Systems, ITHERM (2008) 189–196.

7. A. Bar-Cohen, W.M. Rohsenow, Thermal optimum spacing of vertical, natural convection cooled, vertical plates, *Journal of Heat Transfer*, 106 (1984) 116–122.
8. A. Bar-Cohen, H. Schweitzer, Convective immersion cooling of parallel vertical plates, *IEEE Transactions on Components, Hybrids, and Manufacturing Technology*, CHMT-8 (3) (1985) 343–351.
9. K.J.L. Geisler, *Buoyancy-Driven Two Phase Flow and Boiling Heat Transfer in Narrow Vertical Channels*, Ph.D. Thesis, Department of Mechanical Engineering, University of Minnesota, 2007.
10. W. Elenbaas, Heat dissipation of parallel plates by free convection, *Physica*, 9 (1) (1942) 665–671.
11. N. Zuber, On the stability of boiling heat transfer, *Transactions of the ASME*, 80 (1) (1958) 711–720.
12. M. Monde, H. Kusuda, H. Uehara, Critical heat flux during natural convective boiling in vertical rectangular channels submerged in saturated liquid, *Journal of Heat Transfer*, 104 (2) (1982) 300–303.
13. J. Bonjour, M. Lallemand, Effects of confinement and pressure on critical heat flux during natural convective boiling in vertical channels, *International Communications in Heat and Mass Transfer*, 24 (2) (1997) 191–200.
14. K.J.L. Geisler, A. Bar-Cohen, Confinement effects on nucleate boiling and critical heat flux in buoyancy-driven microchannels, *International Journal of Heat and Mass Transfer*, 52 (2009) 2427–2436.
15. K.J.L. Geisler, A. Bar-Cohen, Passive immersion cooling of 3-D stacked dies, *IEEE Transactions on Components and Packaging Technologies*, 32 (3) (2009) 557–565.
16. M. Arik, A. Bar-Cohen, Effusivity-based correlation of surface property effects in pool boiling CHF of dielectric liquids, *International Journal of Heat and Mass Transfer*, 46 (2003) 3755–3764.
17. R.J. Phillips, Microchannel heat sinks, in: A. Bar-Cohen, A.D. Kraus, (Eds.), *Advances in Thermal Modeling of Electronic Components and Systems*, Volume 2, ASME Press, New York, 1990.
18. S. Kakac, R.K. Shah, W. Aung, *Handbook of Single-Phase Convective Heat Transfer*, John Wiley & Sons, New York, 1987.
19. A. Bar-Cohen, G. Sherwood, M. Hodes, G. Solbrekken, Gas-assisted evaporative cooling of high density electronic modules, *IEEE Transactions on Components, Hybrids, and Manufacturing Technology*, 18 (3) (1995) 502–509.
20. A. Bar-Cohen, E. Rahim, Modeling and prediction of two-phase microgap channel heat transfer characteristics, *Heat Transfer Engineering*, 30 (8) (2009) 601–625.
21. J.C. Chen, Correlation for boiling heat transfer to saturated fluids in convective flow, industrial and engineering chemistry, *Process Design and Development*, 5 (3) (1966) 322–329.

22. S.G. Kandlikar, P. Balasubramanian, An extension of the flow boiling correlation to transition, laminar, and deep laminar flows in minichannels and microchannels, *Heat Transfer Engineering*, 25 (3) (2004) 86–93.
23. K.E. Gungor, R.H.S. Winterton, A general correlation for flow boiling in tubes and annuli, *International Journal of Heat and Mass Transfer*, 29 (3) (1986) 351–358.
24. K.E. Gungor, R.H.S. Winterton, Simplified general correlation for saturated flow boiling and comparisons of correlations with data, *Chemical Engineering Research and Design*, 65 (2) (1987) 148–156.
25. M.M. Shah, A new correlation for heat transfer during boiling flow through pipes, *ASHRAE Transactions*, 82 (2) (1976) 66–86.

CHAPTER 3

DYNAMIC THERMAL MANAGEMENT CONSIDERING ACCURATE TEMPERATURE-LEAKAGE INTERDEPENDENCY

Bing Shi and Ankur Srivastava

Department of Electrical and Computer Engineering
University of Maryland, College Park, USA,
{bingshi,ankurs}@umd.edu

In this paper, we develop an accurate dynamic thermal management DTM approach considering the interdependency of temperature and leakage. By modeling the leakage-thermal interdependence as a quadratic polynomial, we develop accurate analytical equations that capture the thermal transient. We also identify *all* the situations in which thermal runaway would occur, which should be avoided by DTM. We then present a discrete dynamic programming algorithm that performs thermal-aware task and speed scheduling using the model we derived. Compared to a linear leakage-thermal model,[1] owing to our more accurate model, our scheme resulted in 18.2% better performance while maintaining the temperature below constraint.

1. Introduction

1.1. *Thermal Issues in Modern Computer Chips*

In the high performance computer systems, the performance improvement can be achieved by many approaches, such as increased clock frequency and transistor integration density. These approaches usually lead to increased system power dissipation as well. Modern CPUs dissipate an average of 70−100W power while spatial and temporal power variations result in hotspots with even higher power density (up to $300W/cm^2$). The coming years will continue to witness a significant increase in CPU power dissipation due to advanced multi-core architectures and 3D integration technologies. Increase in CPU power density is usually accompanied by drastic increase in chip temperature. Nowadays the problems of increased chip power density, leakage power and system temperature have become major obstacles for further improvement in chip performance.

Recent data shows that more than 50% of all IC failures are related to thermal issues.[2] For instance, excessive temperature reduces the electron and hole mobilities which leads to increase in circuit propagation delay;[3,4] thermal variations and hotspots on chip cause reliability problems such as circuit mismatch and reduced chip lifetime (due to the cumulative damage caused by excessive temperature);[5–7] higher temperature increases the leakage power which in-turn further increases the temperature. Hence, loss of performance and reliability due to unpredictable thermal hotspots has become a major issue and limiting factor for further performance improvement in modern computer systems.

Many efforts have been made to mitigate the thermal issues in CPU chips. These efforts can be classified into three categories: dynamic thermal management schemes,[8–14] materials with better thermal property[15,16] and advanced cooling schemes.[17–20] In this paper, we focus on the first scheme: dynamic thermal management.

1.2. *Dynamic Thermal Management*

In dynamic thermal management (DTM), the chip operation is controlled during runtime for curtailing thermal emergencies.[1,21,22] Thermal management can be achieved by controlling processor knobs such as core frequency and supply voltage,[13,14,23–25] scheduling of tasks etc,[26] which in effect, control the power dissipation in different parts of the chip. These schemes basically manage the chip temperature through controlling the heat dissipation rate/pattern. For example, in dynamic voltage and frequency scaling (DVFS), the supply voltage and operating frequency of micro-processors are dynamically controlled to reduce the chip power consumption, thereby reducing the temperature as well.

1.3. *Temperature-leakage Interdependency*

With continued scaling, the impact of leakage power is growing as well. Today, up to 50% (or even more) of the total power consumption is leakage power.[27] It has been shown that leakage and temperature are highly interdependent: higher temperature increases the leakage power which in-turn further increases the temperature.[28–30] This cross-coupling effect increases the importance and difficulty of chip thermal management. The interdependence between temperature and leakage has been known for years and several attempts have been made during design time to better estimate/control the leakage and temperature through various design decisions.[31,32] For example[31] estimates the chip thermal and leakage profile while accounting for their interdependence.[32] estimates the chip leakage profile while accounting for thermal variability. Such approaches are mostly interested in analyzing the steady state behavior and use a combination of numerical and analytical techniques to capture this thermal-leakage relationship.

1.4. *Motivation of DTM Considering Temperature-leakage Interdependency*

DTM on the other hand is a dynamic approach where we need to be able to predict the fashion in which temperature changes in time for a set of processor control knob options. For example, for a given choice of frequency and initial temperature, we would like to know the time at which the temperature will violate the thermal constraint. The complex relationship between leakage and temperature can result in two different dynamic behavior of temperature. Either the temperature increases or decreases and eventually settles to a steady state value. Or, the temperature increases uncontrollably, thereby leading to thermal runaway. We would like to capture both transient behaviors. A steady state analysis approach will not suffice in such scenario. Recent work presented in[1,33] captures the leakage, thermal dependence using a linear model. Although their approach is interesting, the linear model has limited accuracy and may mispredict thermal runaway. The approach in[14] uses a piecewise linear approximation which is an improvement over the pure linear model. But, they use approximation schemes to solve the thermal differential equations thereby leading to limited accuracy estimates.

In this paper, we develop an accurate approach by modeling the leakage-thermal interdependence as a quadratic polynomial. Existing work presented in[32,34] has verified the accuracy of the quadratic model. We develop accurate *analytical* equations that capture the thermal transient. We also identify the situations in which thermal runaway would occur. Using our approach, any DTM scheme can estimate the temperature dynamics over time as a function of processor frequency and also if a specific choice of frequency will lead to thermal runaway. Our model is generic and can be applied in any DTM scheme regardless of whether a discrete or continuous mathematical optimization framework is used. We demonstrate the applicability of our model in a dynamic programming based DTM scheme for performing task speed assignment and scheduling. Specifically, the following contributions have been made. 1) An analytical model for the temperature dynamics in time parameterized w.r.t the initial temperature and the processor speed. 2) The model can predict the thermal behavior in time for any choice of initial temperature and frequency. 3) The model enumerates *all cases* where thermal runaway will definitely occur. These cases have been tabulated and could be used by any DTM scheme as system states that must be avoided. It is important to note that we have identified all the cases of thermal runaway. 4) Incorporated the model in a dynamic programming based DTM scheme for allocating task schedules and speed. Experimental results indicate that compared to a linear leakage-thermal model,[1] owing to our more accurate model, our scheme resulted in 18.2% better performance while maintaining the temperature below constraint. We also show that using the model which does not consider the leakage-thermal interdependence in DTM results in under-estimate of the real temperature and the solution will considerably violate the thermal constraint.

Although, the quadratic polynomial model is an approximation as well, using more accurate models (such as[35]) would force us to use numerical schemes for estimating the thermal dynamics. Such an approach would not be amenable to efficient optimization during DTM. Also existing work presented in[32,34] has already verified the accuracy of the quadratic model.

Rest of the paper is organized as follows. Section 2 develops the thermal model considering leakage-thermal interdependence. We then explore a dynamic programming based thermal-aware task speed scheduling problem using our model in section 3. The experimental result is given in section 4.

2. RC Thermal Model Considering Leakage

2.1. *Single Core RC Thermal Model*

The thermal behavior of a processor can be modeled by an RC pair. Here we adopt the lumped RC thermal model similar to the one used in.[21,36] As shown in Figure 1, voltage represents the temperature, and current represents the power consumption in the processor. The resistance represents a potential heat transfer path through the packaging, while the capacitance indicates the ability of the processor to store heat.

According to the RC thermal model in Figure 1, the thermal behavior of the processor can be described as the following differential equation:

$$\frac{dT}{dt} = -\frac{T - T_{amb}}{RC} + \frac{P_{total}}{C} \tag{1}$$

where $T(K)$ is the chip temperature, $T_{amb}(K)$ is the ambient temperature, $P(W)$ is the total power consumption of the chip, $R(K/W)$ is the thermal resistance, and $C(J/K)$ is the thermal capacitance. By assuming that the power does not change in time (just for the time being) and solving differential equation 1, the chip temperature T is formulated as equation 2, assuming the initial temperature is T_0.

$$T = (T_0 - T_{amb} - RP_{total})e^{-t/RC} + RP_{total} + T_{amb} \tag{2}$$

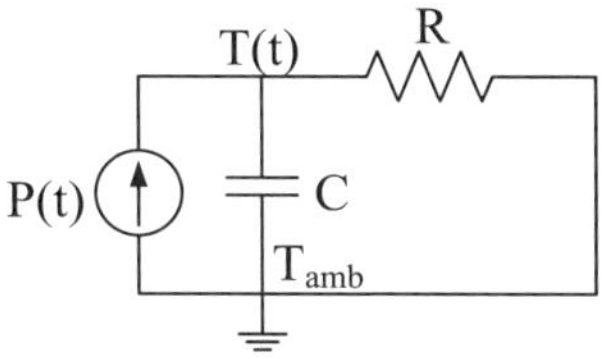

Fig. 1. RC Thermal Model

2.2. *Leakage and Temperature Dependency*

As the IC technology scales down to deep sub-micron domain, leakage becomes a large component of power consumption. Today, up to 50% (or even more) of the total power consumption is leakage power.[27] Therefore, simply approximating the total power consumption by dynamic power[37,38] will lead to an underestimation of the real temperature.

There are three components of leakage power: Band-To-Band-Tunneling (BTBT) leakage, sub-threshold leakage and gate leakage. The BTBT leakage and sub-threshold leakage increases with temperature while the gate leakage is rather insensitive to temperature.

The works in[32,34] model the thermal dependence of leakage as a polynomial function of temperature:

$$P_l = a \cdot T^2 - b \cdot T + d \tag{3}$$

where a, b and d are technology dependent parameters and could be generated by polynomial approximation of the accurate leakage model in.[35]

From equation 2 and 3, we can see that temperature and leakage power consumption influence each other. That is, higher leakage power leads to higher temperature while higher temperature will further increase the leakage power. Because of this dynamic cross coupling of temperature and leakage power, simply assuming that leakage power P_l does not change in time and substituting $P_{total} = P_d + P_l$ into equation 2 cannot capture the dynamic nature of the interdependence between temperature and leakage.

In this paper, we first develop a thermal model that take leakage power and its dynamic interdependence with temperature into consideration. Here we assume that dynamic power is linearly dependent on frequency, that is $P_d = kf$, where f is the frequency and k is a constant (although our method is trivially generalizable to more accurate models where $P_d = kf^{\alpha}$). We get the expression of total power consumption as shown in equation 4.

$$P_{total} = P_l + P_d = a \cdot T^2 - b \cdot T + d + kf \tag{4}$$

By substituting the power model in equation 4 into the thermal differential equation 1, we get the new thermal differential equation as shown in equation 5.

$$\begin{aligned}\frac{dT}{dt} &= -\frac{T - T_{amb}}{RC} + \frac{P_l + P_d}{C} \\ &= -\frac{T - T_{amb}}{RC} + \frac{a \cdot T^2 - b \cdot T + d + kf}{C} \\ &= \frac{a}{C}T^2 - (\frac{b}{C} + \frac{1}{RC})T + \frac{Rkf + Rd + T_{amb}}{RC} \\ &= A \cdot T^2 + B \cdot T + D(f)\end{aligned} \tag{5}$$

where

$$A = \frac{a}{C} \quad B = -(\frac{b}{C} + \frac{1}{RC}) \quad D(f) = \frac{Rkf + Rd + T_{amb}}{RC} \tag{6}$$

Assuming the frequency f does not change in time (for the time being), we would like to solve the differential equation to derive the underlying equation governing the thermal transient. The temperature profile described in differential equation 5 can be derived by solving the following integration:

$$\int_{T_0}^{T} \frac{dT}{AT^2 + BT + D(f)} = \int_0^t dt \tag{7}$$

The solution of equation 7 depends on the values of parameters A, B, $D(f)$ and initial temperature T_0.

2.3. *Thermal Model with Leakage Power*

In this section we present our analytical model that captures the temperature dynamic w.r.t time parameterized using initial temperature T_0 and processor frequency f. This model is created by solving differential equation 5 and is a complete model that captures all combinations of input parameters T_0 and f. Using our model, we also enumerate *all* the cases for which thermal runaway occurs. Our analytical model could be used by any DTM scheme. We first define the following:

$$f_m = \frac{1}{k}\left(\frac{(Rb+1)^2}{4aR^2} - \frac{T_{amb}}{R} - d\right)$$
$$T_{s1} = \frac{-B + \sqrt{B^2 - 4AD(f)}}{2A}, \; T_{s2} = \frac{-B - \sqrt{B^2 - 4AD(f)}}{2A} \tag{8}$$

Since the model is rather complex, we first enumerate the model and then prove its correctness by presenting the derivation. The different input parameter combinations T_0, f have been partitioned into the following cases.

(1) Case 1:
   (a) Case 1a: $f < f_m$ and $T_0 \leq T_{s1}$
   (b) Case 1b: $f < f_m$ and $T_0 > T_{s1}$
(2) Case 2:
   (a) Case 2a: $f = f_m$ and $T_0 \leq -B/2A$
   (b) Case 2b: $f = f_m$ and $T_0 > -B/2A$
(3) Case 3: $f > f_m$

Note that these cases enumerate all possible input combinations. Now we highlight the model for each of these cases.

Case 1a:

$$T(t) = \frac{\sqrt{B^2 - 4AD(f)}}{2A} \frac{x_0 + 1 + (x_0 - 1)e^{\sqrt{B^2-4AD(f)}t}}{x_0 + 1 - (x_0 - 1)e^{\sqrt{B^2-4AD(f)}t}} - \frac{B}{2A} \tag{9}$$

where

$$x_0 = \frac{2AT_0 + B}{\sqrt{B^2 - 4AD(f)}} \tag{10}$$

Case 1b: The temperature will subject to thermal runaway.

Case 2a:

$$T(t) = \frac{2AT_0 + B}{-A(2AT_0 + B)t + 2A} - \frac{B}{2A} \tag{11}$$

Case 2b: The temperature will subject to thermal runaway.

Case 3: The temperature will subject to thermal runaway.

We will derive all these cases subsequently. It is important to note that our analytical model has the following benefits.

a. Based on the initial temperature, one can choose the frequency such that thermal runaway will not occur;

b. One can choose the frequency such that the desired temperature profile can be achieved;

c. The equations of temperature profile are purely analytical;

d. These are optimal rules that consider all possible cases for the temperature profile. If we develop thermal-aware frequency policies based on these rules, we are guaranteed that the temperature will never violate the constraints and/or the runaway condition will never occur.

This model can be easily incorporated in many DTM schemes. For example a DTM scheme can use it to evaluate a set of possible frequency choices for a given initial temperature. This model can be used to estimate the amount of time it will take to reach a desired temperature for a given f, T_0 assignment. It can also be incorporated in a non-linear programming based optimization framework. Next we present the derivation.

2.3.1. *Case 1:*

Note that in this case $f < f_m$, that means $B^2 - 4AD(f) > 0$. So function $AT^2 + BT + D(f) = 0$ has two real roots T_{s1} and T_{s2} (described in equation 8) under this scenario. The integration in equation 7 can be solved as follows:

$$\frac{1}{A}\int_{T_0}^{T} \frac{dT}{(T - T_{s1})(T - T_{s2})} = \int_0^t dt \tag{12}$$

We define:

$$x(t) = \frac{2AT(t) + B}{\sqrt{B^2 - 4AD(f)}} \qquad x_0 = x(0) \tag{13}$$

we get

$$\begin{aligned}
&\frac{2}{\sqrt{B^2 - 4AD(f)}} \int_{x_0}^{x} \frac{1}{(x-1)(x+1)} dx = \int_0^t dt \\
&\Rightarrow \ln \frac{x-1}{x+1}\bigg|_{x_0}^{x} = \sqrt{B^2 - 4AD(f)}t\bigg|_0^t \\
&\Rightarrow x(t) = \frac{x_0 + 1 + (x_0 - 1)e^{\sqrt{B^2 - 4AD(f)}t}}{x_0 + 1 - (x_0 - 1)e^{\sqrt{B^2 - 4AD(f)}t}}
\end{aligned} \tag{14}$$

By substituting equation 14 into the expression of $x(t)$ in equation 13, we can get the expression for the temperature as shown in equation 9. Here x_0 is an initial condition which is a function of both the initial temperature T_0 and frequency f. It will decide the temperature profile.

There are two subcases associated with different initial condition T_0. We describe Case 1b first and then Case 1a.

Case 1b $f < f_m$ and $T_0 > T_{s1}$: It can be shown that for this case $x_0 > 1$. If we plot equation 9 with this as x_0, we find that the temperature $T(t)$ quickly diverges to infinity. This case is shown in Figure 2(c). As we can see from the figure, the temperature will diverge and increase to infinity in a very short time, which results in thermal runaway. So we should avoid this case, unless the runtime of the task is substantially short.

Case 1a $f < f_m$ and $T_0 \leq T_{s1}$: It can be shown that for this case $x_0 \leq 1$. When x_0 equals exactly to 1, the temperature will stay at $T = T_{s1}$ (this can be illustrated by simply substituting the appropriate values in equation 9). But the problem with this case is, as long as there is a tiny fluctuation which make the temperature increase a very small value, the processor will jump out of steady state and enter case 1b, which is thermal runaway. So in practical implementation, we should avoid this case as well.

When $x_0 < 1$ (which happens when $T_0 < T_{s1}$), equation 9 governs the thermal dynamics. In fact, it can be proved that temperature must always converge to T_{s2} as $t \to \infty$. For the sake of illustration, this case is plotted in Figures 2(a) and 2(b). For the same thermal equation 9, the actual thermal behavior is different for $T_{s2} < T_0 < T_{s1}$ (plotted in Figure 2(a)) than for $T_0 \leq T_{s2}$ (plotted in Figure 2(b)). Both cases converge to T_{s2}.

2.3.2. *Case 2:*

Here $f = f_m$, therefore $B^2 - 4AD(f) = 0$. So the function $AT^2 + BT + D(f) = 0$ has one real root $-B/2A$. By solving equation 7, we obtain the following:

$$
\begin{aligned}
&\frac{1}{A}\int_{T_0}^{T}\frac{dT}{(T+\frac{B}{2A})^2}=\int_0^t dt\\
&\Rightarrow -\frac{1}{A}\cdot\frac{1}{T+\frac{B}{2A}}\Big|_{T_0}^{T}=t\Big|_0^t\\
&\Rightarrow T(t)=\frac{2AT_0+B}{-A(2AT_0+B)t+2A}-\frac{B}{2A}
\end{aligned}
\tag{15}
$$

There are two subcases in this situation depending on the initial temperature T_0.

Case 2a $T_0 \leq -B/2A$: 1) If the initial temperature T_0 is lower than the value $-B/2A$, the temperature will slowly increase to $-B/2A$ as in Figure 3(a). 2) If the initial temperature equals exactly to $-B/2A$, the temperature will stay at the initial temperature $-B/2A$. So $-B/2A$ is the steady state temperature when $f = f_m$. Hence in this situation the thermal dynamics is governed by equation 15 (or equation 11).

Case 2b $T_0 > -B/2A$: When we incorporate this case in equation 15, we find that the temperature quickly diverges to infinity (as shown in Figure 3(b)), which results in thermal runaway. We should also avoid this case.

2.3.3. *Case 3:*

In this case, $B^2 - 4AD(f) < 0$ since $f > f_m$. We can solve the integration in equation 7 and get the temperature profile by following steps:

$$
\begin{aligned}
&\int_{T_0}^{T}\frac{dT}{AT^2+BT+D(f)}=\int_0^t dt\\
&\Rightarrow \frac{4A}{4AD(f)-B^2}\int_{T_0}^{T}\frac{dT}{(\frac{2A}{\sqrt{4AD(f)\quad B^2}}(T+\frac{B}{2A}))^2+1}=\int_0^t dt\\
&\Rightarrow \frac{2}{\sqrt{4AD(f)-B^2}}\arctan\frac{2A}{\sqrt{4AD(f)-B^2}}(T+\frac{B}{2A})\Big|_{T_0}^{T}=t\Big|_0^t\\
&\Rightarrow T(t)=\frac{\sqrt{4AD(f)-B^2}}{2A}\tan(y(t))-\frac{B}{2A}
\end{aligned}
\tag{16}
$$

$$
y(t)=\frac{\sqrt{4AD(f)-B^2}}{2}t+\arctan(\frac{2A}{\sqrt{4AD(f)-B^2}}(T_0+\frac{B}{2A}))
\tag{17}
$$

As we can see from equation 16 and Figure 3(c), the temperature profile in this case follows a tangent function $\tan(y(t))$ which increases very rapidly

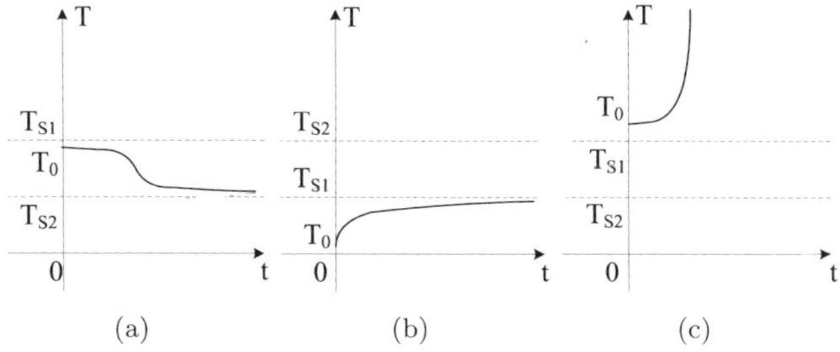

Fig. 2. (a)(b) represent case 1a, (c) represents case 1b

towards infinity when $y(t)$ approaches $\pi/2$. This happens regardless of the initial temperature T_0. So we should avoid this situation as well. As a result, we set the maximum frequency $f_{max} = f_m$, and the frequency should never exceed f_{max}.

3. Thermal Aware Speed and Task Scheduling

Following the discussion of previous section, we can use the rules derived there in any dynamic thermal management scheme to constraint the frequency so that thermal runaway condition does not occur. Also we can create temperature profiles such that thermal constraints are never violated. Our rules can be used in both continuous (mathematical) and discrete optimization frameworks.

In this section, we present a discrete dynamic programming algorithm that performs thermal-aware task and speed scheduling using the model we derived in the previous section, which accounts for the non-linear leakage and temperature dependence. Our primary objective is not to approach the DTM problem with a new algorithmic technique, but to empower a large class of DTM algorithms with a better way of predicting the thermal impact of their decisions.

3.1. *Problem Formulation*

Given:

(1) a periodic sequence of n tasks $\mathcal{J} = \{J_1, J_2, ..., J_n\}$, whose corresponding workload is $\mathcal{W} = \{w_1, w_2, ..., w_n\}$, respectively;

(2) m distinct frequency levels $\mathcal{F} = \{f_1, f_2, ..., f_m\}$, so the execution time of the i'th task using frequency f_j is w_i/f_j;

(3) l sleep state where the frequency $f = 0$, and the length of l sleep

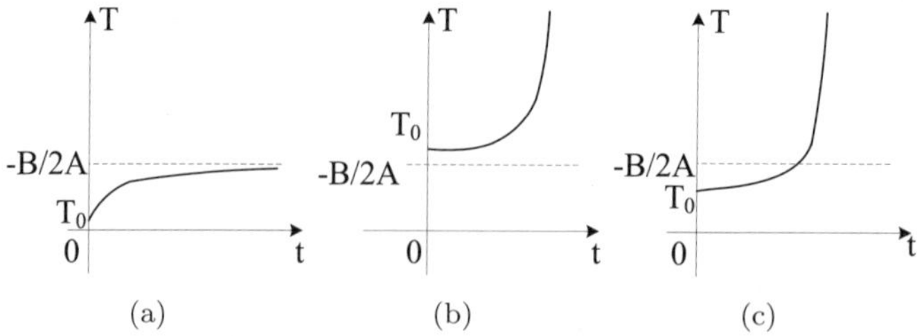

Fig. 3. (a): Case 2a, (b): Case 2b, (c): Case 3

states are $l_1, l_2, ..., l_l$;

(4) the initial temperature is T_0,

We would like to assign a frequency for each task and a potential sleep state after each task so that the total execution time of the n tasks is minimized. During the execution of each task, the frequency does not change. Also the temperature should never exceed the maximum temperature constraint T_{max} and after the execution of all tasks, the final temperature decreases below T_0.

This problem is NP-complete.[36] But we present a dynamic programming method to find solutions reasonably close to optimal. Many versions of this problem have been studied before.[21,36] Note that this problem is a demonstrator of the practicality of the model we developed in the previous section.

3.2. *Dynamic Programming with Fixed Task Order*

We first explore the optimization problem assuming that the task order is already given. In section 3.3, we will incorporate task order rescheduling to this problem. The steps to solve this speed scheduling optimization problem with dynamic programming is as follows:

(1) Sort the frequency in ascending order so that $f_1 < f_2 < ... < f_m$.

(2) Traverse the tasks from first to the last.

(3) For the first task, the initial processor temperature is T_0. For each of the frequency states, use the model to predict the thermal profile and/or potential runaway state. Note that the task will take time w_0/f_i to execute for each frequency. Prune out frequencies that result in runaway or violation of thermal constraint. It is noteworthy that if frequency f_j violates the constraints then frequencies $f_{j+1}...f_m$ will violate as well and do not need to be evaluated. This will give is a set of solutions with varying latency and final temperature. Each solution k is characterized by a latency and

temperature tuple (d_k, T_k). Here latency d_k is the time it takes to finish task 1 and final temperature T_k is the temperature upon completion of task 1 for this specific frequency assignment. Among these solutions we can prune out those which are clearly sub-optimal than others (higher latency and higher final temperature). This pruning process results in a left-over set of co-optimal solutions $\mathcal{C} = \{(d_1, T_1), (d_2, T_2), ..., (d_K, T_K)\}$ as illustrated in Figure 4.

(4) Now evaluate the possibility of adding a potential sleep state. For each co-optimal solution from the previous set $\mathcal{C}$, incorporate the possibility of adding each of the l sleep states. For a solution $(d_k, T_k) \in \mathcal{C}$, adding a sleep state l_i would make the overall latency become $d_k + l_i$. The final temperature at the end of the sleep state (l_i time units) could be computed using our model with T_k as the initial temperature and $f = 0$. Combining each co-optimal solution in $\mathcal{C}$ with each sleep state in l will give us a set of new solutions with total latency and final temperature as parameters. Once again we can generate a new co-optimal set $\mathcal{C}$ of solutions using the pruning criteria in Figure 4.

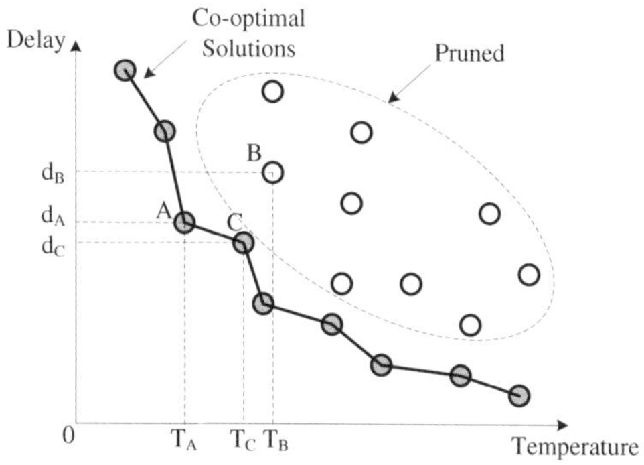

Fig. 4. Multi-objective optimization

(5) Now the set $\mathcal{C}$ represents the total latency vs final temperature tradeoff for finishing task 1. Note that none of these solutions violate the thermal constraint or result in thermal runaway.

(6) Now pick the next task J_2. For each solution $(d_k, T_k) \in \mathcal{C}$, use T_k as the initial temperature and compute the total latency of running J_2 at each of the available frequencies, plus the latency of previous task (and sleep states). Also compute the thermal profile using our model. Prune out solution that result in thermal runaway and also those that violate

the thermal constraint. Among the remaining set of solutions with total latency and final temperature, prune out sub-optimal solutions using the criteria of Figure 4. This will give us a new set of co-optimal solutions $\mathcal{C} = \{(d_1, T_1), (d_2, T_2), ..., (d_K, T_K)\}$ where d_k is the total latency of executing J_1 and J_2 with potential sleep states in between and T_k is the final temperature. Now repeat step (4) for adding new sleep states.

(7) Repeat step (6) for all subsequent tasks.

(8) After the last task pick the solution with smallest latency whose final temperature is less than T_0.

With these steps, we are guaranteed to find the optimal solution since during pruning, we only prune those solutions that violate the temperature constraints and the suboptimal solutions, and we did not prune any potential optimal solutions. Although this method is not proved mathematically with polynomial runtime, in practical implementation, as will be shown in the simulation result, it does prune many solutions and the runtime is fast. We can also use more strict pruning criteria to prune more co-optimal solutions to gain runtime. However, we will not explore it in this paper since it is not the focus of this paper.

3.3. *Considering Task Rescheduling*

Rescheduling the task order may further decrease the task execution time. So here we combine task rescheduling with speed scheduling in this problem to find better task order and the corresponding frequency assignment for each task. Once again our primary objective is to demonstrate the applicability of our model in different DTM paradigms.

3.3.1. *Dynamic Programming with Task Scheduling*

To consider task rescheduling, we also use the dynamic programming method, but add one more dimension of choice to the dynamic programming: task to be executed.

(1) Just as before, we start with initial temperature T_0. Instead of evaluating just the first task J_1, we evaluate the possibility of starting from any of the n tasks. For each of the tasks as the first task, we get a set of co-optimal solutions with (d_k, T_k) as parameters. Following this we can add potential sleep states similar to the previous algorithm. This set of solutions (for all tasks) represent that one task has been completed (although the specific task may be different for each solution).

(2) This set of co-optimal solutions may be too many. We can use several

heuristic based pruning techniques to reduce the potential solution space without potentially loosing out on optimality. The pruning mechanism we used in this paper quantifies each solution by the total workload that has been completed by it. Hence now each solution (d_k, T_k) also has an associated *total workload executed* parameter W_k. Now we can perform pruning of these solutions based on a criteria similar to the one in Figure 4. Note that in this case higher workload with small latency and temperature are desirable properties.

(3) Now we have $n-1$ tasks left. Repeat the above steps using the current set of co-optimal solutions and choosing the next task from the left-over set of $n-1$ tasks. Note that the set of leftover tasks depends upon which tasks were scheduled in the current solution being considered from the co-optimal set.

(4) Repeat step (3) till all tasks are considered.

(5) Among all solutions pick the one with smallest latency and final temperature less than T_0.

However, we found that even after our heuristic pruning, there are still many co-optional solutions left. So in the next section, we develop a heuristic method to further reduce the number of co-optional solutions.

3.3.2. *Heuristic to Improve the Computation Speed*

The heuristic works as follows.

(1) Sort the tasks in ascending order of their workload.

(2) Divide the tasks from the previous sorted list $\mathcal{G} = \{g_1, g_2, ..., g_z\}$ into z contiguous groups . Each group has n/z tasks.

(3) When choosing the next task to be executed (see step (4) of the previous algorithm), instead of considering all the tasks that has not been executed, here we only consider one task from each group. That is, we pick one task from each group that has not been scheduled yet as candidates for the next task to be executed (the task in each group is selected in ascending order of workload). This guarantees that both tasks with smaller and larger workload are in the candidate tasks each time.

With this grouping, the number of choices can be greatly reduced since the number of groups z is much smaller than the number of tasks n. Besides, we can flexibility select the number of groups z to make a tradeoff between the optimality, i.e., total task execution time and the computation complexity, i.e., total number of co-optimal solutions.

4. Simulated Test Case

We obtained the thermal model parameters R, C from.[22] The initial temperature $T_0 = 50$°C and the maximum temperature constraint $T_{max} = 80$°C. In our simulation, we use three discrete frequencies $f_1 = 1 \times 10^9 Hz, f_2 = 3 \times 10^9 Hz, f_3 = 5.5 \times 10^9 Hz$ and two sleep states whose lengths are $20sec$ and $50sec$, respectively.

4.1. *Comparison 1: With Existing Leakage-temperature Models*

We first compare the frequency schemes and corresponding temperature profile generated by the dynamic programming based algorithm when used in conjunction with 1) our leakage aware thermal model, 2) the model which does not consider the interdependence of leakage and temperature (assumes leakage as a constant value calculated at room temperature) and 3) linear leakage-temperature model.[1,33] For this comparison, we only focus on the fixed task order case.

We took a set of 20 tasks with workloads ranging from 1×10^{11} to 2×10^{12} units. The tasks are scheduled in ascending order of their workload. The optimal frequency scheme obtained by using our model is shown in Figure 5(a). The first three tasks run at the highest frequency f_3 so that the temperature heats up to as high as possible (without violating the temperature constraint), and then the frequency reduce to f_2 since continue running at f_3 at the fourth task will make the temperature exceed T_{max}. From the fourth task, the processor runs at f_2 continuously until at the last task, the frequency reduces to f_1 so that the final temperature goes back to T_0. Besides, no sleep state is used between any two tasks. **The total task execution time is 8242sec.** By applying this frequency scheme to the accurate thermal-leakage model in,[35] the temperature profile is shown in Figure 5(b). As we expect, the temperature will never violate the temperature constraint for this scheme (Figure 5(b)). The total runtime is 0.2sec.

For comparison we use the model which does not consider the interdependence between leakage and temperature (leakage is assumed to be a constant value computed at room temperature). This model under-estimate the actual temperature for a given frequency assignment. Hence it overclocks the system thereby leading to higher temperature. Figure 6(a) shows the frequency scheme and Figure 6(b) highlights the real temperature pro-

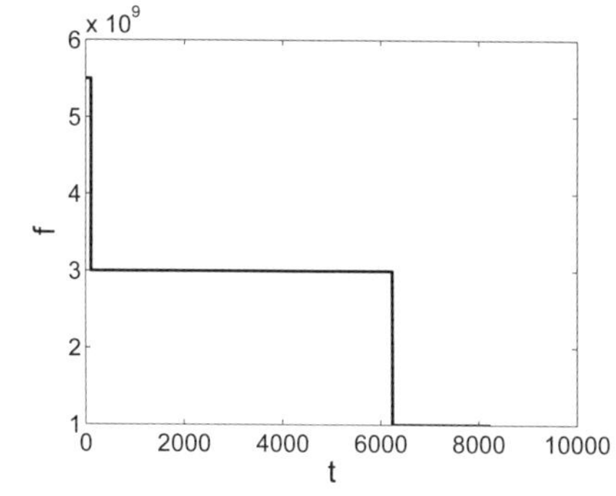

(a)

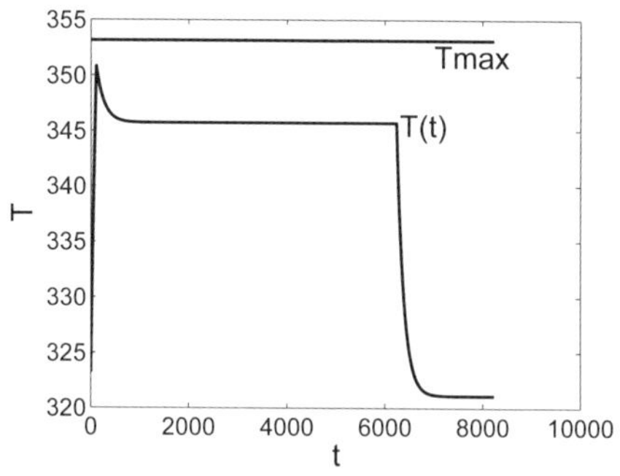

(b)

Fig. 5. (a) Optimal frequency scheme for fixed order tasks using our model, (b) Real temperature profile for this scheme

file of this speed assignment scheme when applied to the accurate leakage model in.[35] It can be seen that the solution considerably violates the thermal constraint.

Finally we also compare the results obtained by using the linear leakage-thermal interdependency model.[1,33] This linear model was generated by performing linear regression on the real thermal-leakage model.[35] The frequency policy and the associated real thermal profile is shown in Figure 7. Although this model did not violate the thermal constraint, the overall execution latency of the tasks was 9742.4 seconds. **This is 18.2% more than our model**.

Hence it can be seen that our model captures the thermal transient more accurately thereby leading to improved solutions.

Table 1. Task latency and computational complexity comparison (runtime: s, gain: %)

Bench	Fix order		Fully reschedule			Group (z=5)			Group (z=10)		
mark	Latency	Runtime	Latency	Runtime	Gain	Latency	Runtime	Gain	Latency	Runtime	Gain
1	1	0.20	0.8504	1055.22	14.96	0.8638	24.91	13.62	0.8561	121.86	14.39
2	1	0.07	0.8459	574.65	15.05	0.8713	11.15	12.87	0.8534	69.87	14.66
3	1	0.02	0.8495	150.41	15.05	0.8580	5.57	14.20	0.8504	16.42	14.96
4	1	0.01	0.8480	60.30	15.20	0.8448	1.97	15.52	0.8528	6.90	14.72
5	1	0.01	0.8480	42.77	15.20	0.8510	1.73	14.90	0.8510	6.03	14.90
6	1	<0.01	0.8480	15.86	15.20	0.8569	0.65	14.31	0.8412	2.99	15.88
7	1	<0.01	0.8480	6.75	15.20	0.8644	0.29	13.56	0.8406	1.30	13.94
8	1	<0.01	0.8480	6.63	15.20	0.8402	0.24	15.98	0.8375	1.29	15.95

4.2. *Comparison 2: Fixed Order, Full Rescheduling and Grouping*

Now we compare the results obtained with and without task rescheduling. We generated a set of benchmarks with total task workloads ranging from 2×10^{13} to 1×10^{15}. We compared the results obtained by 1) fixed task order assuming the task schedule was increasing w.r.t. workloads, 2) full task rescheduling in which no task grouping was done (see section 3.3) and 3) task rescheduling when they are grouped into 5 and 10 clusters (see section 3.3).

In table 1, we compare the total task execution time (*latency*), which is normalized w.r.t. the latency of fixed task order scheme, and the program running time for these schemes. None of these schemes violate the thermal constraint. From the table, we can see that when we fully consider task rescheduling the task latency can reduce about 15.13%, however, the runtime is about 239 sec on average which is 5000 more than the fixed order method. But the grouping works quite well. For example, when the number of group $z = 5$, computing the best schedule needs only 5.8sec on average, while the average task latency can still improve 14.37%.

5. Conclusion and Future Work

5.1. *Conclusion*

In this paper, we developed an accurate dynamic thermal management approach considering the interdependency of temperature and leakage power. We also present a discrete dynamic programming algorithm that performs thermal-aware task and speed scheduling using the model we derived. The experimental result shows that compared to a linear leakage-thermal model,

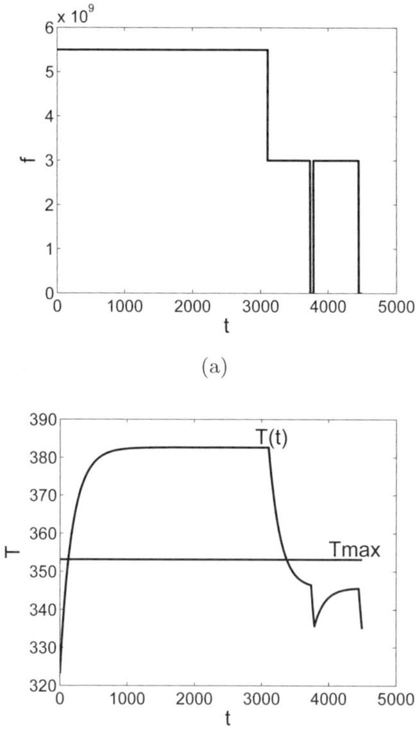

(a)

(b)

Fig. 6. (a) Optimal frequency scheme for fixed order tasks using model without leakage, (b) Real temperature profile for this scheme

owing to our more accurate model, our scheme resulted in 18.2% better performance while maintaining the temperature below constraint.

5.2. *Future Work*

Moore's law has predicted a spectacular exponential growth in chip performance. However, in recent years, such performance improvements are slowing down, leading the research community to investigate alternative technologies that can restore the expected Moores law rhythm in the functionality and cost of electronic products. The three dimensional integrated circuit (3D-IC), which contains two or more layers of active electronic components that are stacked vertically, has become a significant technology for achieving continued performance improvements. The 3D-IC allows a signif-

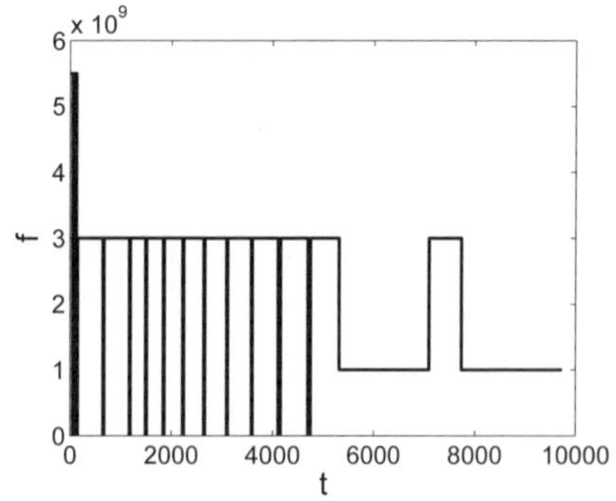

(a)

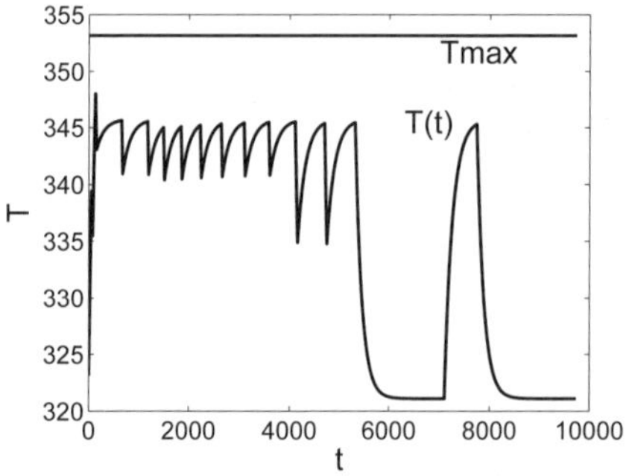

(b)

Fig. 7. (a) Frequency scheme for fixed order tasks using linear leakage model, (b) Real temperature profile for this scheme

icant increase in device densities, as well as faster on-chip communications compared with equivalent 2D circuits due to the shortening of interconnection length and increased bandwidth.[39,40] Despite these advantages, the thermal problems are exacerbated in 3D-ICs due to the stacked structure. Hence an important issue with 3D-IC is the removal of high density heat resulting from several stacks of microprocessor chips. Although current 3D-IC designs are limited to partitioning of memory and datapath across layers, future 3D-IC designs are expected to have significantly complex architectures and integration levels that would be associated with very high power dissipation and heat density.

As illustrated earlier, DTM schemes basically manage the chip temperature through controlling the heat dissipation rate/pattern. For example, in dynamic voltage and frequency scaling (DVFS), the supply voltage and operating frequency of micro-processors are dynamically controlled to reduce the chip power consumption, thereby reducing the temperature as well. However decreasing the supply voltage or operating frequency causes a potential performance reduction. Hence in these conventional DTM schemes, constraining the chip temperature usually limits the performance. Due to such strong thermal-performance interdependency, simply investigating optimal tradeoffs between performance and temperature through DTM while still relying on the conventional air cooling will hurt the overall performance and thus cannot fully exploit the advantage of new technologies (such as 3D integration technology), even with sophisticated DTM schemes. As a result, more effective active cooling schemes (in addition to DTM) are necessary.

Micro-channel cooling, which integrates micro-channel heat sinks into silicon substrates of the chip and uses liquid flow to remove heat inside the chip, is an effective active cooling scheme, especially for 3D-ICs. It has been reported to support heat dissipation higher than $700W/cm^2$.[18] Despite the excellent cooling capability, an overhead associated with micro-channel based heat removal technology is that the cooling system needs to consume extra energy for pumping the coolant through the channels. This has motivated a body of work that attempts to improve the micro-channel cooling effectiveness (thereby reducing the cooling energy consumption) by: a) controlling their dimensional parameters such as channel width, height and aspect ratio,[18,41] and b) investigating more sophisticated micro-channel infrastructures such as cross-linked micro-channels,[42] micro-pin-fins,[43,44] tree- or serpentine-shaped micro-channels.[45,46] Recently, micro-channel cooling is also adopted in dynamic thermal management to control the runtime CPU performance and chip temperature by tuning the fluid flow rate through channels.[47,48] However, most of these works fail to account for the temporal and spatial variations on-chip. The 3D-IC structure (such as through-silicon-vias) also imposes constraints on where the micro-channels can be placed. In the future work, we will try to address these problems.

Nomenclature

C	thermal capacitance
T	temperature
J_n	n-th task
P_d	dynamic power
P_l	leakage power
P_{total}	total power

R	thermal resistance
T	temperature
T_0	initial temperature
T_{amb}	ambient temperature
T_{max}	thermal constraint
d_k	latency of task k
f_m	level m-th frequency
g_z	task group z
l_l	length of l-th sleep state
w_n	workload of task n
$\mathcal{C}$	set of co-optimal solution
$\mathcal{F}$	set of available frequency level
$\mathcal{G}$	set of task group
$\mathcal{J}$	set of tasks
$\mathcal{W}$	workload of all tasks

Acknowledgments

We would like to thank NSF grant CCF 0937865 for supporting part of this research.

References

1. G. Quan and Y. Zhang. Leakage aware feasibility analysis for temperature-constrained hard real-time periodic tasks. In *21st Euromicro Conference on Real-Time Systems (ECRTS '09)*, pp. 207–216 (2009).
2. M. Pedram and S. Nazarian, Thermal modeling, analysis and management in VLSI circuits: Principles and methods, *Proceedings of the IEEE.* **94**, 1487–1501 (2006).
3. K. Laker and W. Sansen, Design of analog integrated circuits and systems, *New York: McGraw-Hill* (1994).
4. Y. Tsividis, *Operation and Modeling of the Mos Transistor.* Oxford University Press (2004).
5. M. A. Ismail, I. M. Nasir, and R. Ismail. Modeling of temperature variations in mosfet mismatch for circuit simulations. In *Quality Electronic Design, 2009. ASQED 2009. 1st Asia Symposium on*, pp. 357–362 (2009).
6. J. R. Black, Electromigrationa brief survey and some recent results, *IEEE Transactions on Electron Devices.* **16**, 338–347 (1969).
7. Z. Lu, J. Lach, M. Stan, and K. Skadron. Banking chip lifetime: Opportunities and implementation. In *Proceedings of the 1st Workshop on High Performance Computing Reliability Issues (HPCRI05)* (2005).

8. D. Brooks and M. Martonosi. Dynamic thermal management for high-performance microprocessors. In *Proc. of the 7th Intl. Symp. on High-Performance Computer Architecture (HPCA'01)* .
9. A. Cohen, L. Finkelstein, A. Mendelson, R. Ronen, and D. Rudoy, On estimating optimal performance of cpu dynamic thermal management, *IEEE Computer Architecture Letters.* **2**, 6 (2003).
10. A. K. Coskun, T. S. Rosing, and K. C. Gross. Temperature management in microprocessor socs using online learning. In *Design Automation Conference (DAC'08)* .
11. A. K. Coskun, T. S. Rosing, and K. C. Gross. Proactive temperature management in MPSoCs. In *Proceedings of the 2008 International Symposium on Low Power Electronics and Design*, pp. 165–170 (2008).
12. S. M. Martin, K. Flautner, T. Mudge, and D. Blaauw. Combined dynamic voltage scaling and adaptive body biasing for lower power microprocessors under dynamic workloads. In *IEEE/ACM Intl. Conf. on Computer Aided Design (ICCAD'02)* .
13. R. Rao, S. Vrudhula, and N. Chang. An optimal analytical solution for processor speed control with thermal constraints. In *Proceedings of the 2006 International Symposium on Low Power Electronics and Design*, pp. 292–297 (2006).
14. R. Rao and S. Vrudhula. Performance optimal processor throttling under thermal constraints. In *Proc. of Intl. Conf. on Compilers Architectures and Synthesis for Embedded Systems (CASES'07)*, pp. 257–266 (2007).
15. T. Schubert, Ł. Ciupiński, J. Morgiel, H. Weidmüller, T. Weissgärber, and B. Kieback, Advanced composite materials for heat sink applications, *Euro PM* (2007).
16. L. Stappers, Y. Yuan, and J. Fransaer, Novel composite coatings for heat sink applications, *Journal of The Electrochemical Society.* **152**, C457–C461 (2005).
17. J.-M. Koo, S. Im, L. Jiang, and K. E. Goodson, Integrated microchannel cooling for three-dimensional electronic circuit architectures, *ASME Trans. Journel of Heat Transfer.* pp. 49–58 (2005).
18. D. B. Tuckerman and R. F. W. Pease, High-performance heat sinking for VLSI, *IEEE Electron Device Letters.* pp. 126–129 (1981).
19. J. Li and H. Miyashita. Efficient thermal via planning for placement of 3d integrated circuits. In *IEEE International Symposium on Circuits and Systems (ISCAS'07)*, pp. 145–148 (2007).
20. A. Bar-Cohen. Thermal management of on-chip hot spots and 3d chip stacks. In *IEEE International Conference on Microwaves, Communications, Antennas and Electronics Systems*, pp. 1–8 (2009).
21. T. Chantem, X. S. Hu, and R. P. Dick. Online work maximization under a peak temperature constraint. In *Proceedings of the 14th ACM/IEEE international symposium on Low power electronics and design (ISLPED 09)*, pp. 105–110 (2009).
22. K. Skadron, T. Abdelzahery, and M. R. Stan. Control-theoretic techniques and thermal-rc modeling for accurate and localized dynamic thermal manage-

ment. In *Proc. of Intl. Symp. on High-Performance Computer Architecture (HPCA'02)* .

23. V. Hanumaiah, S. Vrudhula, and K. S. Chatha, Performance optimal online dvfs and task migration techniques for thermally constrained multi-core processors, *Computer-Aided Design of Integrated Circuits and Systems, IEEE Transactions on.* **30**, 1677–1690 (2011).
24. T. D. Burd, T. A. Pering, A. J. Stratakos, and R. W. Brodersen, A dynamic voltage scaled microprocessor system, *Solid-State Circuits, IEEE Journal of.* **35**, 1571–1580 (2000).
25. A. Klaiber et al., The technology behind crusoe processors, *Transmeta Technical Brief* (2000).
26. J. Yang, X. Zhou, M. Chrobak, Y. Zhang, and L. Jin. Dynamic thermal management through task scheduling. In *Performance Analysis of Systems and software, 2008. ISPASS 2008. IEEE International Symposium on*, pp. 191–201 (2008).
27. N. Kim, T. Austin, D. Blaauw, T. Mudge, K. Flautner, J. Hu, M. Irwin, M. Kandemir, and V. Narayanan, Leakage current: Moores law meets static power, *IEEE Computer Society.* **36**(12), 68–75.
28. W. Liao, L. He, and K. Lepak, Temperature and supply voltage aware performance and power modeling at microarchitecture level, *IEEE Trans. on Computer-Aided Design of Integrated Circuits and Syst.* **24**, 1042–1053 (2005).
29. H. Su, F. Liu, A. Devgan, E. Acar, and S. Nassif. Full chip leakage estimation considering power supply and temperature variations. In *Proceedings of the 2003 international symposium on Low power electronics and design*, pp. 78–83 (2003).
30. H. Huang, G. Quan, and J. Fan. Leakage temperature dependency modeling in system level analysis. In *11th International Symposium on Quality Electronic Design (ISQED)*, pp. 447–452 (2010).
31. T. Sato, J. Ichimiya, N. Ono, and M. Hashimoto. On-chip thermal gradient analysis considering interdependence between leakage power and temperature. In *IEICE Transactions on Fundamentals of Electronics, Communications and Computer Sciences*, pp. 3491–3499 (2006).
32. H. Su, F. Liu, A. Devgan, E. Acar, and S. Nassif. Full chip leakage estimation considering power supply and temperature variations. In *Proceedings of the 2003 International Symposium on Low Power Electronics and Design (ISLPED'03)*, pp. 78 – 83 (2003).
33. Y. Liu, R. Dick, L. Shang, and H. Yang. Accurate temperature-dependent integrated circuit leakage power estimation is easy. In *Proceedings of the conference on Design, automation and test in Europe (DATE'07)*, pp. 1–6 (2007).
34. R. Ayoub and T. Rosing. Predict and act: Dynamic thermal management for multi-core processors. In *Proceedings of the 2009 International Symposium on Low Power Electronics and Design (ISLPED'09)*, pp. 99–104 (2009).
35. L. He, W. Liao, and M. R. Stan. System level leakage reduction considering the interdependence of temperature and leakage. In *Design Automation Conference* .

36. S. Zhang and K. S. Chatha. Approximation algorithm for the temperature-aware scheduling problem. In *Proceedings of the 2007 IEEE/ACM international conference on Computer-aided design (ICCAD 07)*, pp. 281–288 (2007).
37. S. Murali, A. Mutapcic, D. Atienza, R. Gupta, S. Boyd, and G. D. Micheli. Temperature-aware processor frequency assignment for mpsocs using convex optimization. In *Proceedings of the 5th IEEE/ACM international conference on Hardware/software codesign and system synthesis (CODES'07)*, pp. 111–116 (2007).
38. S. Wang and R. Bettati. Delay analysis in temperature-constrained hard real-time systems with general task arrivals. In *Proceedings of the 27th IEEE International Real-Time Systems Symposium*, pp. 323–334 (2006).
39. P. Jacob, O. Erdogan, A. Zia, P. M. Belemjian, R. P. Kraft, and J. F. McDonald, Predicting the performance of a 3d processor-memory chip stack, *IEEE Design & Test of Computers*. **22**, 540–547 (2005).
40. D. H. Woo, N. H. Seong, D. L. Lewis, and H.-H. Lee. An optimized 3d-stacked memory architecture by exploiting excessive, high-density tsv bandwidth. In *IEEE 16th International Symposium on High Performance Computer Architecture (HPCA)*, pp. 1–12 (2010).
41. R. W. Knight, D. J. Hall, and et al, Heat sink optimization with application to microchannels, *IEEE Trans. on Components, Hybrids, and Manufacturing Technology*. pp. 832–842 (1992).
42. L. Jiang, J.-M. Koo, and et al. Cross-linked microchannels for vlsi hotspot cooling. In *ASME 2002 International Mechanical Engineering Congress and Exposition* (2002).
43. C. Marques and K. W. Kelly, Fabrication and performance of a pin fin micro heat exchanger, *Journal of Heat Transfer*. pp. 434–444 (2004).
44. Y. Peles, A. Kosar, C. Mishra, C.-J. Kuo, and B. Schneider, Forced convective heat transfer across a pin fin micro heaet sink, *International Journal of Heat and Mass Transfer*. pp. 3615–3627 (2005).
45. S. Senn and D. Poulikakos, Laminar mixing, heat transfer and pressure drop in tree-like microchannel nets and their application for thermal management in polymer electrolyte fuel cells, *Journal of Power Sources, Vol. 130*. pp. 178–191 (2004).
46. L. Ghodoossi, Thermal and hydrodynamic analysis of a fractal microchannel network, *Energy Conversion and Management, Elsevier*. pp. 771–788 (2005).
47. A. K. Coskun, J. L. Ayala, D. Atienzaz, and T. S. Rosing. Modeling and dynamic management of 3D multicore systems with liquid cooling. In *17th Annual IFIP/IEEE International Conference on Very Large Scale Integration*, pp. 60–65 (2009).
48. A. K. Coskun, D. Atienza, T. S. Rosing, and et al. Energy-efficient variable-flow liquid cooling in 3D stacked architectures. In *Conference on Design, Automation and Test in Europe (DATE'10)*, pp. 111–116 (2010).

CHAPTER 4

ENERGY REDUCTION AND PERFORMANCE MAXIMIZATION THROUGH IMPROVED COOLING

David Copeland

Oracle
4160 Network Circle
Santa Clara CA 95054
United States of America
david.copeland@oracle.com

Improved cooling can result in reduced power, higher frequency operation and/or higher reliability of server processors. A variety of packaging and system internal and external cooling technologies are evaluated, resulting in a wider range of total thermal resistance. Energy and economic computational efficiency are defined. A case study of a typical modern enterprise server is performed, showing that improvements in energy economic efficiency as high as 15% and in economic computational efficiency as high as 7% can be provided through improved cooling.

Introduction

The International Technology Roadmap for Semiconductors (ITRS) [1] predicts high performance processor power density to more than double by the year 2024, while during the same time allowable junction temperature will decrease from 90°C to 70°C, reducing the junction-to-ambient temperature difference to nearly half. This combined challenge to cooling will require the total thermal resistance to decrease by almost a factor of four. Even if these changes are only half as great, significant improvements in cooling will be required. The recent edition of Datacom Equipment Power Trends and Cooling Applications [2] includes a detailed discussion of both cost performance and high performance processor die size, power density, junction temperature and resulting maximum power and required thermal resistance.

Yet independent of increases in power density, advantages to improved cooling can be demonstrated. Leakage current has become an increasing fraction of processor power with each technology node. As leakage current is strongly temperature dependent, processor power dissipation can be reduced through improved cooling. Achievable processor frequency is strongly dependent on temperature and voltage. The voltage dependence is approximately proportional, while temperature dependence of performance has reduced with each technology node. In the near future, the temperature dependence of performance may almost disappear.

Through improved cooling, temperature can be reduced while voltage and frequency are increased, resulting in higher system performance. For a given cooling configuration, a combination of voltage and temperature exists which maximizes system performance per watt. At higher powers, additional performance is achieved at the expense of energy. Such increases may be limited by electromigration and other temperature- and voltage-dependent failure mechanisms. The temperature dependence of leakage current has become stronger through recent technology nodes. Significant decreases in energy consumption and/or increases in processor frequency can be achieved through improved cooling.

Leakage Current Effects

Leakage current is current which bypasses the transistor gate, which is always present regardless of whether the gate is active, and which increases with both voltage and temperature. It can be considered as wasted energy, unlike energy used to perform computation.

Leakage current effects have increased as semiconductor lithography has progressed well into the subcontinuum range [3]. In some application specific integrated circuits (ASICs), leakage current can be over half of the total power. As the industry moves from 45 nm to 32 nm to 22 nm and beyond, designs are intended to limit leakage current to about one-third of total power dissipation. Some of this will be achieved by lower junction temperatures. The dependence of leakage current on junction temperature is also becoming stronger. At 65 nm, leakage current would change by a factor of two over a temperature range of about 45°C; at 32 nm, the range may shrink to 22°C [4]. So the energy savings due to a given temperature reduction will increase with technology improvements. From the viewpoint of effectiveness of improved cooling, a relevant metric is the percent reduction in total power versus temperature. Recent observations of processors on the market show this to range from about 0.5%/°C to almost 2%/°C [5].

Improved Cooling

Ellsworth and Simons [6] discussed a variety of package cooling improvements which could be used to cool die with power densities up to several hundred W/cm^2. These began with traditional air cooling of a lidded package as shown in Figure 1. As the thermal path through the package substrate offers a high resistance, nearly all the heat flows upwards and outwards while spreading through the package lid and

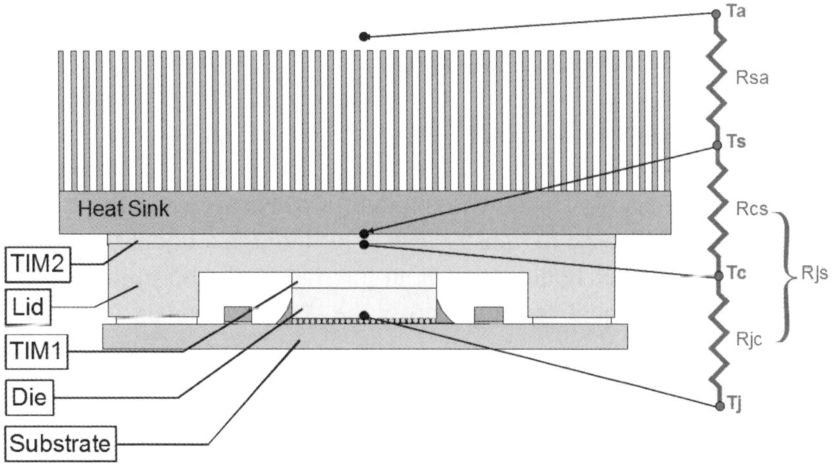

Fig. 1. Package and Heatsink.

heatsink base, then finally to the heatsink fins and into the air stream. The heat flow path is through the following components:

Silicon die
Thermal interface material 1 (TIM1)
Package lid
Thermal interface material 2 (TIM2)
Heatsink base
Heatsink fins
Coolant

Nonuniformity of power dissipation within the die increases package thermal resistance above that of a package with a uniformly powered die by a significant factor, typically two or more [7, 8]. For a given nonuniform power distribution, this factor will increase as the thermal path is improved, moving asymptotically toward the ratio of the highest local power density to the average power density.

Recent developments in packaging and cooling largely consist of replacing traditional materials with those having higher thermal conductivity. In the package, organic TIM1 with metallic or ceramic fillers can be replaced by a completely metallic material, typically indium or indium alloy. Copper package lids can be replaced by composite materials, such as aluminum-graphite, copper-graphite, aluminum-diamond or silicon carbide-diamond. External to the package, heatsinks feature embedded heatpipes or vapor chambers, and in some cases the vapor chamber shares a continuous vapor/liquid space with heatpipes extending into the fins. A cold plate can replace the heatsink or become an integral part of the package, acting as the lid. Ultimately, coolant can contact the silicon with no interfaces between. At the system level, liquid can be used to transfer heat to air through a liquid-to-air heat exchanger, to another liquid through an interface or liquid-to-liquid heat exchanger, or the liquid can flow across the system boundary to transfer heat to the datacenter and beyond. Such technologies are described in detail by Kang and Miller [9].

Performance Modeling

Consider a processor with a nominal power dissipation of 240W, operating at nominal conditions of 0.85V and 95°C.

Thermal resistance components of various combinations of packaging and cooling are shown in Figure 2. Values shown are typical for a die approximately one square inch (645 mm^2) in modern flip-chip packaging. Design variations consist of the following:

- Package thermal interface material (TIM1) - The TIM1 can be organic or metallic in a traditional lidded package.
- Coolant on silicon removes the TIM1 and lid or cold plate, simplifying package construction.
- Package types can be lidded, which requires another thermal interface material (TIM2) or cold plate lid, which eliminates TIM2.

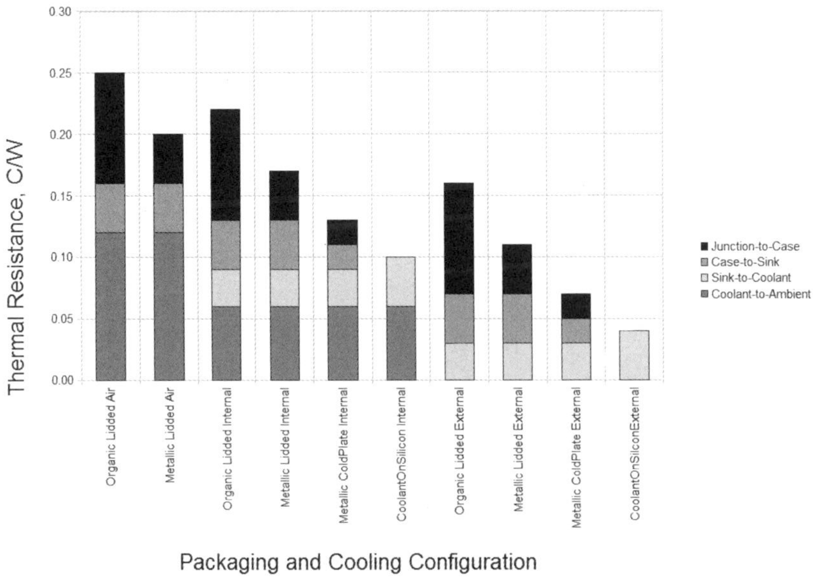

Fig. 2. Packaging and Cooling Configuration and Thermal Resistance.

The system cooling configuration can be one of three types:

- Air, in which all heat is dissipated to the air stream without liquid as an intermediary. In this case the air stream is fixed at 35°C, the allowable limit for a typical datacenter.
- Internal liquid, in which liquid is used to transfer heat to the air stream.

- External liquid, in which heat is transferred to liquid without air as an intermediary. In this case, the external liquid temperature is fixed at 25°C, slightly above the allowable maximum dewpoint of a typical datacenter.

Electrical performance is based on a generic technology representative of recent and upcoming process nodes, and is modeled by a few simple assumptions:
In the region of interest on gate-dominated paths, achievable frequency is assumed to be proportional to voltage:

$$\text{Freq} \propto V \qquad [1]$$

Dynamic power to be proportional to frequency times voltage squared:

$$\text{Pd} \propto \text{FreqV2} \qquad [2]$$

So dynamic power becomes proportional to voltage cubed:

$$\text{Pd} \propto V^3 \qquad [3]$$

Static power (leakage current) is also proportional to voltage cubed.
At nominal conditions, static power is fixed at 35% of of dynamic power:

$$P_s / P_t = P_s / (P_s + P_d) = 0.35 \qquad [4]$$

Static power is assumed to change by a factor of two every 22°C, expressed by:

$$P_s \propto \exp[0.0315(T_j - 95)] \qquad [5]$$

At nominal frequency (and voltage) the temperatures of various packaging components are shown in Figure 3. While most of the temperature reduction is due to improved cooling, some is due to the reduction in dynamic power. This positive feedback converges toward a lower operating temperature.

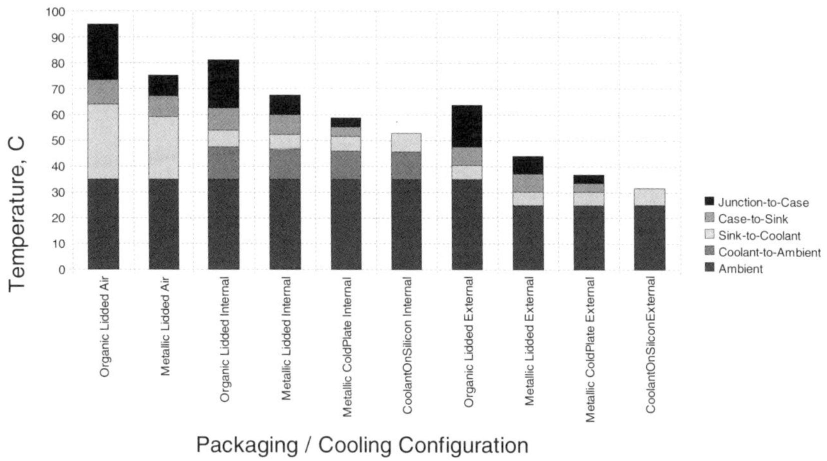

Fig. 3. Packaging / Cooling Configuration and Temperature.

Reliability Concerns

One practical limit to increasing power is reliability, which is typically dominated by gate oxide failure. This is assumed to change by a factor of two every 10°C (typical of modern technology nodes, and happens to be equal to an often misapplied rule of thumb which has been around for decades) and/or every 15mV in the present study, resulting in:

$$\text{Fail} \propto \exp\{[0.0693(T_j - 95)] + [46.4(V - 0.85)]\} \qquad [6]$$

As the cooling performance is increased and thermal resistance is reduced, the reliability limit is reached at decreasing temperatures, higher voltages and slightly higher power than nominal. Ultimately leakage current is reduced to only one-seventh of its nominal value.

Frequency Improvements

Permitting the processor to operate at the highest combination of temperature and voltage, while maintaining gate oxide reliability equal or better than that of the nominal case, results in frequency improvements as high as 19%, as shown in Figure 4. The slopes of the lines from the origin to the operating point are proportional to computational efficiency, defined as frequency divided by power. In this case, as voltage is lowered, efficiency improves monotonically, with packaging and cooling technology making very little difference at the lowest values.

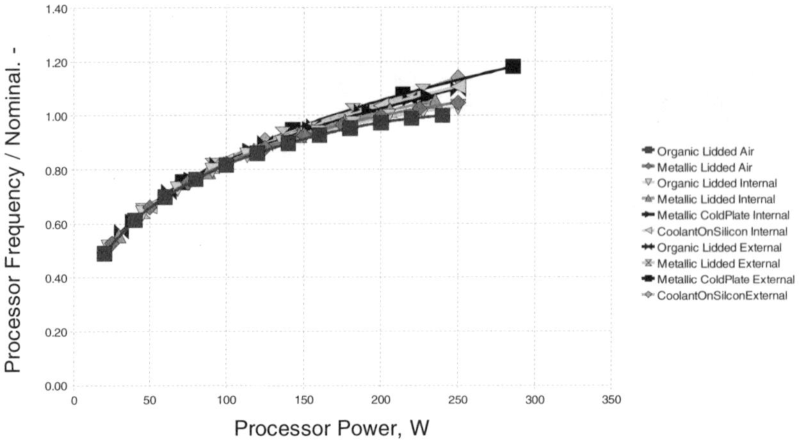

Fig. 4. Processor Power and Frequency / Nominal.

System Performance

If the processor provided all or nearly all of the power dissipation in a system, very low voltage (and frequency) operation would be most efficient. But processors consume, in all but a few cases, less power than memory. Volume systems typically feature fewer dual inline memory modules (DIMMs) per socket than high-end systems. High-end systems have been described as large boxes of memory with other hardware to drive it, and this is a reasonable description in terms of power dissipation, volume occupied and component cost.

Adding a constant memory power load changes power per socket and system performance to that of Figure 5. In this case memory power is equal to twice that of nominal processor power, typical of a modern high-end system. System performance is assumed to scale with the square root of processor frequency. This exponent relating processor frequency and system performance is at the low end of the range seen, which approaches a value of one at the high end and is strongly dependent on the program being exercised.

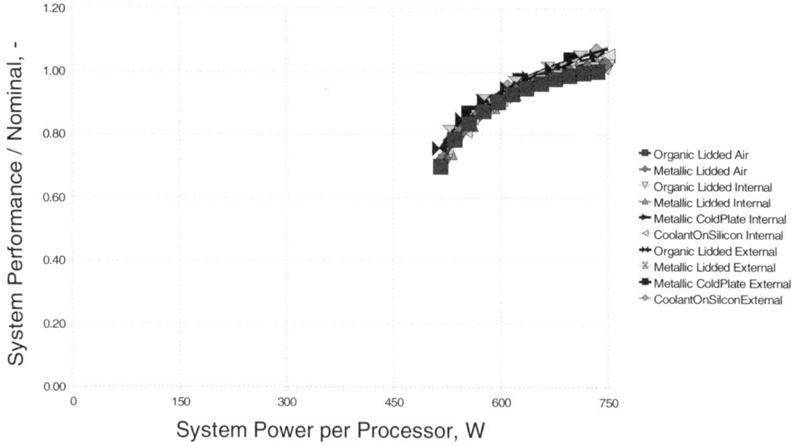

Fig. 5. System Power per Processor and Performance / Nominal.

Unlike the example of a processor only, an optimum operating point at which the slope of power versus performance is maximized falls close to the midpoint of the range evaluated. This is slightly different for each packaging and cooling configuration. Dividing performance by power provides computational efficiency, which is shown compared to the nominal configuration in Figure 6. Here the optimum operating condition shifts to lower power as cooling is improved.

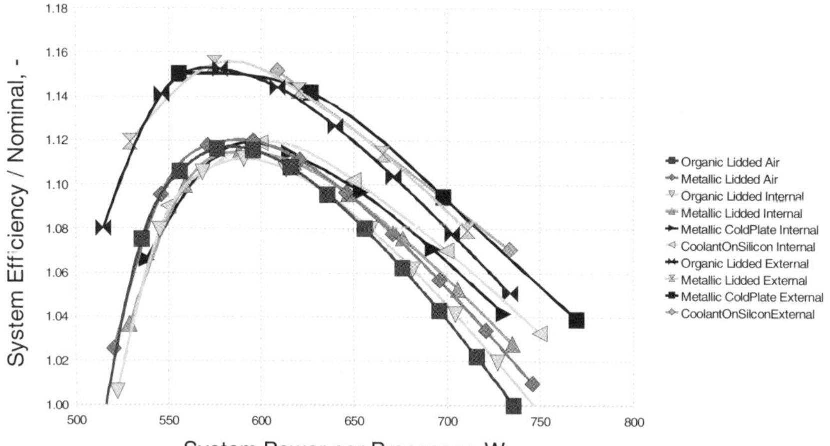

Fig. 6. System Power and Efficiency / Nominal.

Total Cost of Ownership

While maximization of computational efficiency in terms of energy has been demonstrated, the total cost of ownership should be considered. Until just a few years ago, capital expenditure on computing equipment dominated energy cost, and the two expenses were the concern of different organizations with little communication. Recently higher energy costs, much higher energy usage and environmental concerns have resulted in a more holistic view of datacenter cost and efficiency.

In the present study, computing equipment cost is fixed to be equal to that of energy cost. At 11.4 cents per kilowatt-hour, a continuous watt of electricity costs exactly one dollar per year. This is close to an average datacenter's electricity cost, so a rough approximation can be made that lifetime energy cost of a server in dollars is equal to its average power consumption in watts times its operational lifetime in years. Equal capital energy costs could thus be a server with a five year lifetime and capital cost equal to five times annual energy cost, for example.

Figure 7 plots normalized total cost of ownership (TCO) versus economic computational efficiency, a term invented for this study and defined as system performance divided by TCO. All optimum values feature TCO below the nominal value, and improvement in economic computational efficiency can be as high as 7%. As this is a simple

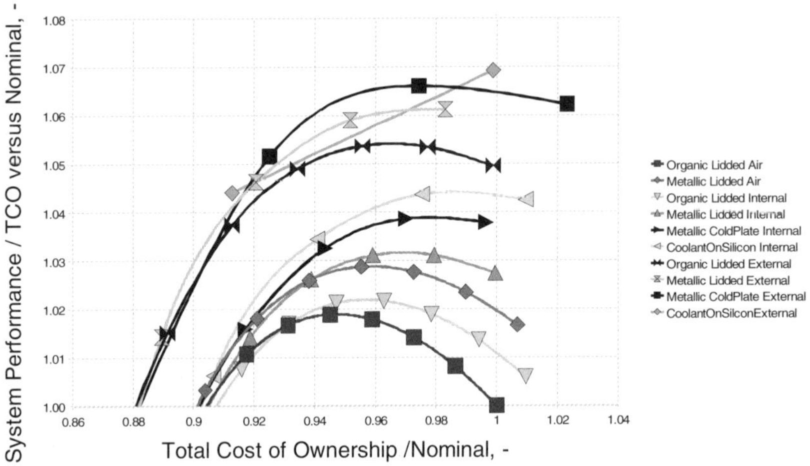

Fig. 7. Total Cost of Ownership and System Performance / TCO.

model, some factors are neglected, which will alter the results if included. The cost of floorspace may be an additional factor, tax incentives and deductions will alter TCO calculations. Additional equipment external to the computing must be added or removed.

Conclusions

A case study showing energy reduction and/or performance maximization through improved cooling of a typical modern high-end server has been performed. In both energy efficiency and economic computational efficiency or TCO minimization, an optimum frequency (and voltage) exists for each packaging and cooling configuration. As cooling is improved, the advantage becomes more favorable. Energy computational efficiency can be increased to 15% above nominal, and economic computational efficiency can be increased to 7% above nominal. More detailed models will shift the values but should follow the same general trend. Lower temperature operation of processors is an available design option with immediate rewards.

Nomenclature for Figure Legends

Air	using a lidded package and heatsink cooled by air at 35°C
Cold Plate	cold plate is attached to the die by a thermal interface material with no lid
Coolant On Silicon	coolant contacts the inactive surface of the die, with no interface material
External	cooled by liquid from outside the rack at 25°C
Internal	cooled by liquid contained within the rack, exchanging heat with air at 35°C
Lidded	die is covered by a lid which acts as a heat spreader
Metallic	thermal interface material (TIM1) consists of metal
Organic	thermal interface material (TIM1) consists of particle-filled epoxy

Nomenclature for Equations

T_j	junction temperature	°C
Freq	frequency / nominal	-
V	voltage	V
P_d	dynamic power	W
Ps	static power	W
Pt	total power	W
Fail	failure rate / nominal	-

Acknowledgments

This paper is based on the author's keynote presentation [10] of the same title at The Heat Is On: Performance and Cost Improvements through Thermal Management Design and a subsequent article [11] in Electronics Cooling. The author thanks his Oracle colleagues Vadim Gektin (now at Huawei), who provided Figure 1, and Bruce Guenin, who invited him to write the magazine article for Electronics Cooling.

References

[1] International Technology Roadmap for Semiconductors, 2010, http://www.itrs.net/Links/2010ITRS/Home2010.htm.

[2] American Society of Heating, Refrigeration and Air-Conditioning Engineering (ASHRAE), 2012, Datacom Equipment Power Trends and Cooling Applications, Second Edition, pages 113–119.

[3] N. S. Kim, T. Austin, D. Blaauw, T. Mudge, K. Flautner, J. S. Hu, M. J. Irwin, M. Kandemir and V. Narayanan, 2003, Leakage Current: Moore's Law Meets Static Power, IEEE Computer, Vol. 36, No. 12, pp. 68–75.

[4] S. Mukhopadhyay, A. Raychowdhury and K. Roy, 2003, Accurate Estimation of Total Leakage Current in Scaled CMOS Logic Circuits Based on Compact Current Modeling, Proceedings of the Design Automation Conference, pp. 169–174.

[5] J. Wei, 2007, Challenges in Package Cooling of High Performance Servers, Proceedings of the 2007 International Electronic Packaging Technical Conference and Exhibition, IPACK2007-33637, ASME.

[6] M. J. Ellsworth and R. E. Simons, 2005, High Powered Chip Cooling - Air and Beyond, Electronics Cooling, Vol. 11, No. 3, http://www.electronics-cooling.com/2005/08/high-powered-chip-cooling-air-and-beyond/

[7] J. Deeney, 2002, Thermal Modeling and Measurement of Large High Power Silicon Devices with Asymmetric Power Distribution, Proceedings of the 35th International Symposium on Microelectronics, IMAPS.

[8] D. Copeland, 2005, 64-bit Server Cooling Requirements. Proceedings of the IEEE Twenty-First Annual IEEE Semiconductor Thermal Measurement and Management Symposium (Semi-Therm), pp. 94–98.
[9] S. Kang and D. Miller, 2011, Thermal Management Architectures for Rack Based Electronics Systems, Proceedings of The Heat Is On: Performance and Cost Improvements through Thermal Management Design, MicroElectronics Packaging and Test Engineering Council (MEPTEC).
[10] D. Copcland, 2011, Energy Reduction and Performance Maximization through Improved Cooling, Proceedings of The Heat Is On: Performance and Cost Improvements through Thermal Management Design, MicroElectronics Packaging and Test Engineering Council (MEPTEC).
[11] D. Copeland, 2011, Energy Reduction and Performance Maximization through Improved Cooling, Electronics Cooling, November 2011, http://www.electronics-cooling.com/2011/11/energy-reduction-and-performance-maximization-through-improved-cooling/

CHAPTER 5

OPTIMAL CHOICE OF HEAT SINKS FROM AN INDUSTRIAL POINT OF VIEW

Clemens J.M. Lasance

Philips Research Emeritus, Consultant@SomelikeitCool, Nuenen
The Netherlands
lasance@onsnet.nu

Temperature plays a significant role in optimizing the quality of electronic products, for example the light output and reliability of LED-based systems. An important way of controlling the temperature is by a proper choice of a heat sink. Since literally thousands of heat sinks are available, many designers are confronted with the question: which one? Very often, the designer's choice is only based on cost and manufacturer's data. Unfortunately, these data cannot be used with confidence because they are almost exclusively based on measurements in a closed duct, thereby disregarding bypass effects and inflow conditions. The paper will discuss the usefulness of a variety of approaches that are commonly in use and some that are not. The ultimate goal is to enable the designer to make the right choice for the right heat sink for her application[a].

Note added in proof: In the mean time, the subject of this paper has been expanded to result in a book chapter, interested readers are encouraged to consult ref.[1]

1. Introduction

Trends over the past fifty years indicate that all electronic product and systems will be smaller, lighter, smarter, more efficient, and in particular

[a] In case you wonder why I use 'she' everywhere, you may wonder why you wouldn't wonder when I would use 'he' everywhere.

cheaper. It is interesting to have a closer look at Figure 1, showing the cost breakdown for a LED-based system. All LED based applications have the following parts in common next to the LED itself. A (Metal Core) PCB to hold and interconnect the LED, a lensing system, driver electronics to power the led, mechanics to hold the whole together and a thermal management solution to cool both the LED and driver electronics. While the cost of the LED package is continuously shrinking, the mechanical/ thermal part becomes the major part of the total price of the LED application in the nearby future.

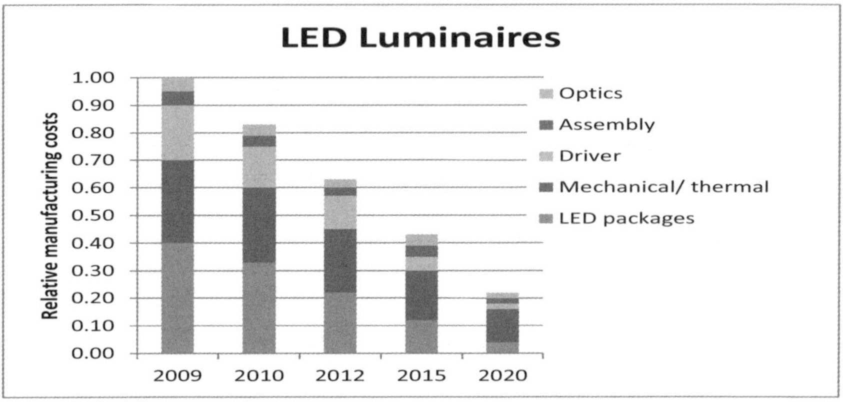

Fig. 1. Cost reduction roadmap – source: Department of Energy, SSL Manufacturing Roadmap, 2010.

The paper will focus on heat sinks at the system level, not heat sinks per se. This introduces many complications for which the vendors cannot be held responsible, such as the fact that in many cases of practical interest we cannot speak about natural convection to characterize a heat sink because of the presence of other sources, resulting in buoyancy driven forced convection. Additionally, most academic studies leading to the often-used (semi)empirical correlations focus on one heat sink only, with a well-defined approach flow, and either fully-ducted or fully non-ducted flow. Testing is only a serious alternative if a prototype is available, usually too late in the design process.

In summary, the designer is faced with the following problem areas:

- Many new developments, in manufacturing as well as in layout (extruded, corrugated, pin fin, metal foam, narrow channel, enhanced plastic, etc.),
- Many criteria to base an optimal choice on (performance, weight, volume, cost),
- Manufacturer's data are often of little use in practice,
- Standard correlations fail often in practice,
- Pressure drop data required for proper fan selection are often lacking.

How about using CFD? Detailed CFD modeling with the purpose of arriving at the best heat sink for the application, especially at the system level, is in many designer's environments no option unless expert knowledge, time, a supercomputer and a calibration lab are available. Let us make an estimation of the number of cells that are required to cover at least the basic physics causing the transitional turbulence phenomena that govern heat sink thermal behavior. Assume a heat sink base area of 10 cm^2 square, with 20 fins of 10 mm length. Between the fins we need at least 10 cells, symmetrically non-equidistant, starting at 100 micrometer. We have 10 cells over the fin height, and 30 in the streamwise direction. This boils down to 200*10*30=60,000 cells per heat sink. However, because of bleeding we are forced to add quite a number of cells around the heat sink, resulting in probably 200,000 cells. This is nowadays not a big deal when we have only one heat sink, but for sure we run into problems when we are dealing with tens of heat sinks, or a couple of big ones that are common in many LED based systems. Problems exacerbate when we have to take into account the variety of shapes that populate the catalogues. Hence, full CFD is not a serious option in many cases in practice. We need to enter the domain of compact models, such as has been the case for a long time in the IC world[2]. The additional difficulty is that we do not only need a compact model for the thermal part but also for the fluid flow part. In addition, every heat sink compact model to be used in CFD models should also contain information about the pressure losses.

Note: To prevent misunderstanding about the use of CFD in the design process: the author has emphasized many times that without CFD no proper and timely design at the system level is possible. However, we are talking here about potential paths that may guide a designer to the best heat sink for a specific application.

In summary, designers of LED systems need other approaches, in addition to the ones that are abundant in literature.

2. Heat Sink Basics

While this paper will not discuss heat sink basics in detail, some issues deserve a short paragraph. Many familiar design equations are based on often hidden simplifications, the so-called Murray-Gardner assumptions, and the author feels that a designer should at least be familiar with this approach that often fails in practice. Another topic frequently underestimated is heat spreading. Heat spreading is by no means a trivial issue, the problem being that there are no simple design rules that are accurate enough for practical purposes.

2.1. *The Murray-Gardner assumptions*

Murray (1938) and Gardner (1945) formulated limiting assumptions to allow relatively simple fin analysis, now known as the Murray-Gardner assumptions. Designers need to be aware of these simplifications and should check regularly if their heat sinks still stay within these limits, especially for heat sinks that deviate from standard parallel plate extruded heat sinks. Notably, for LED applications often heat sinks are chosen with short and widely spaced fins for which the base area plays a relatively important role.

1. The heat flow into the fin and the temperature at any point on the fin remains constant with time.
2. The fin material is homogeneous, its thermal conductivity is the same in all directions and remains constant.
3. The heat transfer coefficient between the fin and the surrounding medium is uniform and constant over the entire surface of the fin.
4. The temperature of the medium surrounding the fin is uniform.

5. The fin width is so small compared with its height that temperature gradients across the fin width may be neglected.
6. The temperature at the base of the fin is uniform.
7. There are no heat sources within the fin itself.
8. Heat transfer to or from the fin is proportional to the temperature excess between the fin and the surrounding medium.
9. There is no contact resistance between fins in the configuration or between the fin at the base of the configuration and the prime surface.
10. The heat transferred through the outermost edge of the fin (the fin tip) is negligible compared to that through the lateral surfaces (faces) of the fin.

2.2. *Heat spreading*

This section assumes the reader is familiar with the basics of conduction and convection.

Heat spreading is essentially area enlarging: the larger the area that is in contact with the heat exchanger (often the ambient air), the more heat can be removed at the same temperature difference. Unfortunately, even the simple case of heat spreading of a single source on a plane with uniform boundary conditions does not have an explicit mathematical solution. Hence, we have to rely on clever approximations or suitable computer codes.

A recent review paper[3] and a subsequent update[4] discuss many issues that are related to heat spreading, generated by one or more heat sources on top of a single or multilayer assembly. It is argued that problems arise with the interpretation of heat spreading effects due to a misconception about the meaning of often-quoted flux limits and especially the physical meaning of thermal resistance. The papers compare also the usefulness of a number of approaches that are generally in use to analyze heat spreading effects. It is shown that the popular series resistance approach has severe limitations in many cases when heat spreading plays a significant role. The final conclusion is that only in very special cases it is recommended to use analytical approaches, by far the safest way for a designer is to use specialized software such as a CFD code in conduction-only mode. When we focus on heat spreading in heat sinks, the effects are less than when considering heat spreading in PCBs, but are not absent. Everywhere where we have a heat sink area that is larger than the source area, we face potential heat spreading problems. As a rule-of-thumb, we may state that when we have non-negligible temperature gradients in the base of the heat sink we should take into account heat spreading. Fortunately, in many cases of practical interest the temperature gradients can be neglected. *Note*: We are not talking here about the gradients in the fins! Exceptions are: large-area heat sinks serving many LEDs at the same time, heat sinks with a very thin base due to space restrictions, heat sinks that are much larger than the LED plus (MC)PCB. However, designers tend to forget another important role of heat spreading. One should realize that a heat sink performs two different functions:

1. To enlarge the surface area
2. To spread heat

It is the second function that needs attention when we face a temperature gradient over the surface of a LED module that requires a heat sink. Suppose that we split the surface of our LED module in two areas: the central area, and the periphery. The heat spreading reduces the maximum temperature of the central area and increases the minimum temperature at the edges. Effectively, in terms of heat transfer coefficients, this means that the heat transfer coefficient of the central area is significantly increased. This effect is different from the effect caused by the first function of the heat sink, namely enlarging, effectively inducing a higher heat transfer coefficient over the whole area. For more details please consult Lasance[5]. When heat spreading should be taken into account is mainly dependent on the ratio between the source and baseplate areas. The reader who is interested in calculating this effect is recommended to consult Lee[6] and Lasance[7].

2.3. *Radiation*

Usually the designer should not bother too much about the emissivity of the heat sink material because its contribution is generally less than for flat plates, unless the distance between fins is relatively large while the fin height is relatively small, such as often occurs in natural-convection driven LED systems. It is difficult to give some rules-of-thumb. We recommend a simple test when a global choice has been made: measure the temperature drop after painting the heat sink. Alternatively, when CFD codes are available: use the radiation option. However, in most forced convection applications radiation can be neglected. One should realize that even for heat sink materials with a low emissivity the effective emissivity would approach 1 because the radiation entering the space between the fins cannot escape. Please realize also that the effective emissivity for short-wide channels could well be larger than one because the area available for radiation is larger than the projected area.

3. Heat Sink Modeling Approaches from a Designer's Perspective

Let us make a distinction between approaches that are useful for acquiring insight in the physics including the fluid dynamics, approaches that can be used when it comes to optimization of individual heat sinks, and finally approaches that are required for the final design stages.

This brings us the following list of potential approaches that will be discussed in more detail in subsequent sections.

- Approaches not recommended
    - Vendor data
    - Convective resistance models
    - Two-resistance model
    - Empirical correlations
- Approaches to acquire insight in the basic physics
    - Heat exchanger model
- Approaches to acquire insight in optimization parameters
    - Literature study
    - Complex shapes, finned arrays
    - Pin fin heat sinks
    - Inclined plates
    - Semi-analytical correlations
    - Web-based tools
    - Constructal theory
    - Dedicated heat sink software

- Approaches to acquire insight in fluid dynamics
    - CFD analysis of single heat sinks, boundary conditions assumed
    - Pros and cons of CFD
- Approaches suited for final design of heat sinks in LED applications
    - Dedicated tests to rank a series of heat sinks
    - CFD with calibrated heat sink compact models, including optimization
    - Prototype testing

3.1. *Approaches not recommended*

Vendor data

Around the mid-nineties, it was realized that there was a serious problem with the various methods of heat sink characterization by the vendors. The ASME K16 committee took the initiative to try to solve this problem, and Belady[8] reported first results.

Unfortunately, as far as the author knows, no final report has been published, despite the impressive list of highly qualified members of the committee. It is enlightening to copy a few highlights of Belady's paper. The main reasons for the founding of the committee are:

- The data presented are application specific.
- There has never been a standard method for characterizing the behavior of heat sinks.

Belady goes on with discussing the limitations of the traditional approach.

- The performance of the heat sink is also a function of the size of the wind tunnel relative to the size of the heat sink.
- The effect on fan performance is unknown since the airflow resistance of the heat sink is not characterized.
- Data are not in a form that can be used in system modeling.

- The tests are not normalized and, thus, comparing the performance between heat sinks is difficult.
- Wind tunnel measurements have limited reusability. Published data by one manufacturer cannot be used by another manufacturer unless the applications are identical.

One of the solutions proposed is to use a ducted wind tunnel test. While this solves many of the problems cited, one big disadvantage remains: when bypass is present the data are useless. Another serious shortcoming is that wind tunnel approach flow is significantly different from fan-driven approach flow. Not much has changed since then. The conclusion still is that vendor data at best can be used for comparison purposes but should be treated with great care when these data are going to be used for design purposes.

Convective resistance models

The following sections are mainly based on an article by Moffat[9]. Convective resistance approaches make use of a single thermal resistance invoking an effective surface area and/or an effective heat transfer coefficient:

$$R_{th}=1/h_{eff}/A_{eff} \quad (1)$$

This is a legacy from the days of simple, low-density, finned heat sinks, but it does not accurately describe the thermal resistance of high fin-density heat sinks. In an age of searching for the most optimal design, this approach is not recommended except for very early design phases in order to get some idea about the heat transfer coefficient when the area is more or less fixed or vice versa, hence only for back-of-the-envelope calculations. Another application field could be LED-based systems in case free convection and wide/short fins are applied, but this should always be checked.

A marginally better approach is to include the fin efficiency in the equation and add the surface of the prime area (the base area exposed to air):

$$R_{th} = \frac{(T_{base} - T_{air,in})}{q} = \frac{1}{hA_{eff}} = \frac{1}{h(A_{prime} + \eta_{fin} A_{fin})} = \frac{1}{h\eta_0 A_{total}} \quad (2)$$

A better 'compact' model is the two-resistance model where the air temperature rise is taken into account.

Two-resistance model

Azar and Tavassoli[10] proposed adding a 'fluid resistance' term to the convective resistance to get the 'total resistance.'

$$R_{tot} = \frac{1}{h\eta A_{fin}} + \frac{1}{2\dot{m}c_p} \tag{3}$$

Simons[11] described this as a "caloric, or apparent fluid resistance, R_{fluid}" and showed its derivation in terms of the change in air temperature within the heat sink. Moffat shows clearly that the use of this equation leads to wrong results in some limiting cases. However, there is a more fundamental objection against using equation (3). Calling $1/(2\dot{m}c_p)$ a 'resistance' simply violates physics. Analog elements such as resistors and capacitors must be based on the rate equations that describe the flow of a conserved property (one that is neither created nor destroyed within the component) in response to a potential drop that causes that flow. The physical laws of fluid mechanics, heat transfer, and mechanics identify the potentials and flows. Regarding equation (2), the term (T_{base}-$T_{air,in}$) is clearly the potential that drives the convective heat transfer rate and the convective resistance can be taken to be 1/hA. On the other hand, the term ($T_{air,out}$-$T_{air,in}$) is not a potential in any recognized rate equation; it is simply the consequence of an energy balance on the fluid. Maybe even more convincing, a resistance is per definition symmetric: it does not matter in which direction the current flows. This is clearly not the case for the 'fluid' resistor: the heat flows in one direction only, and even more serious, in the wrong one, because it flows from cold to hot. The only way to circumvent this problem from a circuit analysis point of view is to make the fluid resistance voltage-dependent, but then we do not talk about a linear circuit anymore.

Empirical correlations

First of all, we should make a distinction between (semi)analytical correlations and empirical correlations. While *analytical* correlations are more or less based on physics, albeit often of the too-simplified kind, most *empirical* correlations are of the type: There is a linear dependence between the number of churches and pubs. A thorough analysis of the problems associated with the use of correlations in real-life is provided by Lasance[12]. Focusing on heat sinks only, it is instructive to read the myriad of papers that are devoted to increasingly sophisticated correlations, claiming an increasingly accurate correspondence with sophisticated tests. While this conclusion is certainly valid for the conditions tested, one may wonder what the usefulness is in practice when one considers the fact that many phenomena that occur in practice are not taken into account, all of them leading to a decrease in accuracy. Why try to improve the accuracy of some correlation from 90 to 95%, when obviously the inaccuracy in practice is below say 70%? My advice for designers: stay away from empirical correlations when accuracy and/or insight is required, and focus on some other approaches from the list to follow.

3.2. *Approaches to acquire insight in the basic physics*

Heat exchanger model

Moffat again: Heat exchanger theory was developed to deal with two-fluid systems: a hot fluid transferring heat to a cold fluid. The nomenclature reflects this history. The three most important terms used in applying heat exchanger theory to heat sinks and cold plates are: Effectiveness (ξ), Capacity Rate ($\dot{m}c_p$), and NTU ($hA_{eff}/\dot{m}c_p$). The term NTU (Number of Transfer Units) is sometimes referred to as the "dimensionless size" of the heat exchanger; the ratio of the convective conductance to the flow capacity. The higher the NTU, the more effective the heat exchanger. Space is lacking to go into the details, only the end result is given:

The heat sink thermal resistance can be represented as:

$$R_{th} = \frac{1}{\dot{m}c_p\left\{1-e^{-\frac{hA_{eff}}{\dot{m}c_p}}\right\}} \tag{4}$$

Equation (4) shows the complex way in which the flow affects heat sink performance.

There is nothing new in the idea of using heat exchanger theory in electronics cooling. Moffat (op.cit.) provides a literature overview. Finally, Moffat shows a convincing table in which he compares the 1R and 2R models, and he concludes that in the region between 25% and 90% effectiveness, where modern high density heat sinks are designed, the 2R convective resistance model greatly over-predicts the heat transfer.

3.3. *Approaches to acquire insight in optimization parameters*

Literature study

To get insight in heat sink behavior, or to start research to acquire this insight, the first step is of course to check the literature. All heat sink handbooks and review papers start with the treatment of parallel-plate channels, the basic thermal building block of the most common heat sinks in practice. The problem we face here is that, while this knowledge is mandatory for understanding what the parameters are that influence the performance of heat sinks, the designer tends to forget that optimization in practice might well lead to entirely different heat sinks concepts.

Natural convection

The interested reader may consult the following papers, for a more detailed discussion please sce Lasance[1], Starner and McManus[13], Welling and Wooldridge[14], Van de Pol and Tierney[15], Bilitzky[16], and very recently, Kim *et al.*[17] All cited papers suffer from one or more of the following shortcomings: heat sinks far too big for mainstream IC and LED systems, unwarranted assumption of fully-developed flow, experimental data set too limited, boundary conditions too far from reality, no 3D effects, sometimes very complex (but still limited) correlations and correction factors. However, it is interesting to note that most experimental data showed a significant increase in h compared to numerical channel flow data.

Forced convection

Usually the treatment of forced convection heat transfer related to heat sinks follows a certain order: laminar fully developed flow, turbulent flow, the temperature field, the Nu correlations, developing flow, the analogy between heat transfer and fluid friction, etc. A good introduction can be found in the book of Kraus and Bar-Cohen (op.cit., page 275). Because of the dominance by the computer industry in electronics cooling, especially the cooling of CPUs and graphic cards, the emphasis is often on relatively high-speed flow and big heat sinks with small interfin spacing and relatively long fins. This approach leads to heat sink design equations based on fully-developed turbulent channel flow. It should be realized that in case of LED applications where forced convection is used we face often another situation: relatively short fins, and small fans. Additionally, formulating proper design equations is more difficult than for natural convection because of the following reason. h is a function of the air velocity, which is dependent on the available pumping power, which is dependent on the system air resistance. Hence, contrary to natural convection, it is not possible to define a single optimum fin array geometry. Optimization must be performed for each combination of parameters. Available numerical codes can help considerably in finding an application-dependent optimum. To illustrate the complexity for the designer further, it is sufficient to study one of the attempts that have been described in literature to address the problem by some clever but complicated means, namely the work by Holahan *et al.*[18]

This approach is unusual among the analytic approaches in dealing both with the impact of developing flow and axial (typically upstream) conduction in the fins (conjugate heat transfer), typical of thin fins combined with relatively low flow rates. They discretized the plate fin into a large number of "patches," and the local heat transfer coefficient was determined from the straight duct flow correlation for developing thermal and hydrodynamic laminar flow between parallel plates with uniform wall temperature, provided in Kakac *et al.*[19] The authors used a matrix approach to solve the many patch-related equations. However, one should realize that this rather complex procedure is formally only valid for 'fully shrouded, laminar, parallel plate heat sinks'. When we are dealing with a practical situation, in most cases the fan is rather close to the fins, especially when we consider CPU cooling. This simply means that the approach flow is highly turbulent, and thus violating any laminar-based correlation, except maybe in cases where the interfin spacing is very small compared to the heat sink length.

Biber and Belady[20] realized a long time ago that one of the most important parameters is the pressure drop through the heat sink, because it sets the velocity and hence the heat transfer. They compared correlations for fully-developed flow and developing flow with CFD and test results, for various hat sink types, see Figure 2.

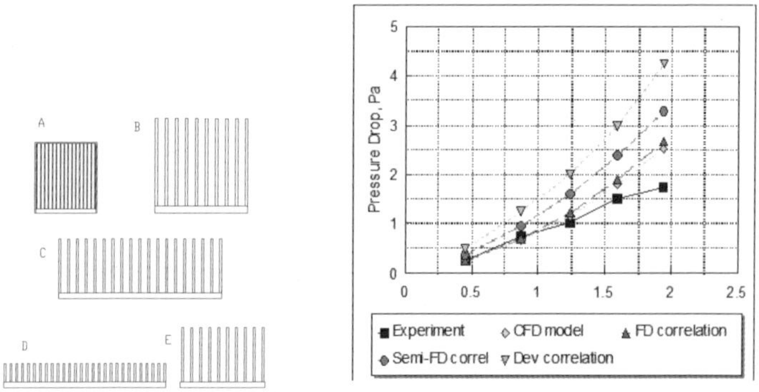

Fig. 2. A Heat sinks tested 2b Pressure drop results for typical heat sink.

For purpose of illustration, the pressure drop results for a typical heat sink are shown at the right side. Obviously, the results are not very encouraging. The authors concluded that not only the correlations were pretty useless when it comes to accuracy but also that the prevailing test methods should be improved since no standardized test method did exist. However, for a first guess fully-developed correlations are recommended because of their ease of use. Introductory texts using design equations for forced convection heat transfer can be found in Biber[21,22,23], and in an overview of Calculation Corner topics that appeared in Electronics Cooling over the past ten years or so[24].

Complex shapes, finned arrays

When it comes to the analysis of heat sinks with complex geometries such as the one shown in Figure 3 we face a serious problem when we are after a first-order guess.

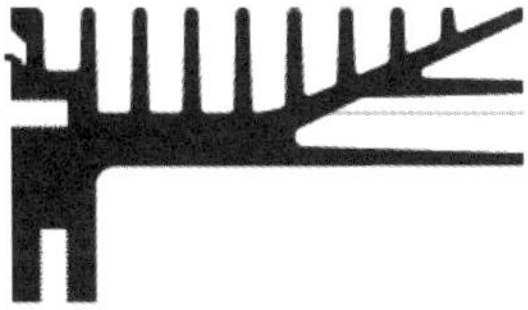

Fig. 3. Example of complex heat sink shape.

Kraus and co-workers (see e.g. chapters in the book by Kraus and Bar-Cohen, op.cit.) offered a solution based on matrix theory and published various algorithms to calculate finned arrays of the complexity shown in the Figure.

Pin fin heat sinks

Pin fin heat sinks are mostly used when the orientation of the system can vary in practice, for example indoor floor lamps. The focus of this paper is on parallel-plate heat sinks and space is lacking to treat pin fin heat sinks with the same attention. One of the problems is that they are considerably more difficult to treat analytically, complicated by the fact that literature data are sometimes contradictory when it comes to performance comparison. The interested reader may consult the following literature: Kraus and Bar-Cohen (op.cit., p 320, natural convection), Bahadur[25] (natural and forced convection), Bar-Cohen *et al.*[26] (natural convection), and Khan *et al.*[27,28] (forced convection).

Inclined plates

Especially in application fields where it is not clear upfront how the end-user will position the product, such as in consumer lighting, care should be taken in designing the heat sink in natural-convection driven cooling. In addition, it is not always obvious what the most optimal position is. When we are dealing with only one heat source, the direction of the air flow will be opposite to gravity. However, when more than one source is present, it might well be the case that some critical components will experience so-called buoyancy induced forced convection, and the mean flow direction can be any direction. Positioning a plate fin heat sink in such a way that the air is parallel to the base plate becomes then a challenge. In those cases, preference could be given to pin fin heat sinks, see above. Below the reader will find some relevant literature on inclined heat sinks.

Papers by Starner/Marcus and Bilitzky have already been mentioned in the section called Literature Study. Both papers discuss different orientations and inclinations and are very useful because they show accurate test results for a variety of heat sinks under various circumstances and can be used to calibrate CFD models. Further literature recommended: Ledezema and Bejan[29], Naidu *et al.*[30]

Malhammar[31] did observe an interesting phenomenon.

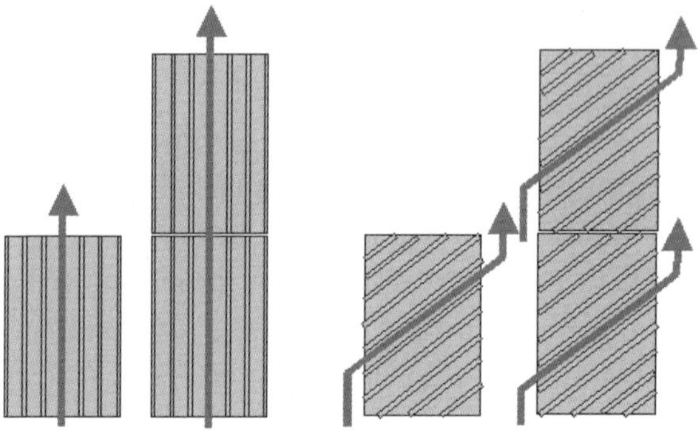

Fig. 4. For stacked heat sinks, obviously the inclined stack experiences the smallest air temperature rise. *Permission obtained from author.*

In his own words: "Figure 4 shows a heat sink with vertical fins and another one with inclined fins. It is somewhat difficult to compare their performance on single unit bases but if they are stacked things get much clearer. For the vertical arrangement, it is apparent that the top heat sink is less effective than the bottom heat sink, because it receives heated air from below. For the inclined arrangement, this is not the case. Each stacked unit will in fact approximately contribute as much as the units below. My first idea was therefore that there must be a critical height above which inclined fins perform better than the vertical fins. That turned out to be wrong! What I found was that inclined fins *always* perform better than vertical fins." End quote. For ratios of Height/Width > 4 he shows the gain to be 50%, but some disadvantages are also discussed.

Semi-analytical correlations

This section discusses approaches that may give some insight in the influence of common design parameters, such as optimal distance between fins, optimal fin cross section, least-material fin, for both natural and forced convection. While the literature abounds with papers on how to optimize the shape of fins to ensure maximum heat transfer, in practice more mundane matters such as ease of manufacturing and cost usually result in the choice for a parallel plate-fin heat sink. In this section, we will limit the discussion to this type.

An important part of parallel plate fin heat sinks are the parallel plates. Especially when the fin height is large, heat sink fluid dynamics can be characterized by the flow between parallel plates only, hence without the base. The well-known book by Kraus and Bar-Cohen[32] provides a good introduction into these matters. Much theoretical effort has been devoted to finding the optimal parameter settings for the so-called least material fin. However, this topic is only of academic interest because the shape of the most optimal fin will significantly add to the costs of manufacturing. Fortunately, the next best optimal shape performance is not too far away from the mathematical optimum.

Kraus and Bar-Cohen [*op. cit., p326*] showed how to optimize the rate of heat transfer from a shrouded array of longitudinal rectangular fins, making a distinction between the fin average h and the base average h_w. They found iteratively:

$$(mb)_{opt} = 1.42 + 1.13\frac{h_w b}{k} \quad (5)$$

with $= \sqrt{\frac{2h}{k\delta}}$, δ the fin thickness and b the fin height. Further analysis shows that the second term on the right hand side that describes the influence of the base only marginally influences the result for the optimum value for a single fin, $(mb)_{opt} = 1.42$.

Other useful equations to get insight in the parameters that determine the heat transfer performance of singe fins are the following.

For the longitudinal fin of rectangular profile, the optimum fin thickness and fin height are given by:

$$\delta_{opt} = \left(\frac{hA_p^2}{k}\right)^{1/3} \tag{6}$$

and

$$b_{opt} = \left(\frac{kA_p}{h}\right)^{1/3} \tag{7}$$

with the fin cross-sectional area $A_p=b.\delta$. These equations are valid for specified A_p. Rewriting when the dissipation Q and the base temperature rise θ_b are given:

$$\delta_{opt} = \frac{0.63}{hk}\left(\frac{Q}{\theta_b}\right)^2 \tag{8}$$

and

$$b_{opt} = 0.8\frac{Q}{h\theta_b} \tag{9}$$

The optimum profile area is given by:

$$A_p = \frac{0.5Q^3}{h^2 k\theta_b^3} \tag{10}$$

It is interesting to compare the optimum profile areas for various cross sections: rectangular, triangular and parabolic. The conclusion is that the triangular fin requires only 60% of the material of the rectangular fin for the same performance, while the concave profile is only marginally better. From the equation, it becomes also obvious that the weight is proportional to the cube of Q. Hence, when the dissipation should be doubled, it is better to strive for two fins rather than extend a single fin. Because A_p is proportional to the volume V, the mass becomes proportional to ρ/k, meaning that the material with the lowest ρ/k gives the lightest fin.

Other optimum parameters discussed by Kraus and Bar-Cohen are the optimum aspect ratio and the optimum single-fin heat transfer rate, respectively:

$$\left(\frac{\delta}{b}\right)_{opt} = \frac{hb}{k} \tag{11}$$

and

$$\left(\frac{q}{L}\right)_{opt} = 1.26\left(h^2 A_p k\right)^{1/3} \tag{12}$$

The equations presented are valid for both natural and forced convection. It is worthwhile to mention that the book by Kraus and Bar-Cohen also treats pin and radial fins and fin arrays. The next sections focus on finding the approximate values for h.

<u>*Natural convection*</u>

Let us first discuss parallel-plate channel flow only. Elenbaas[b], in a famous paper from 1942[33], was the first to discuss natural convection between parallel plates. Elenbaas used the incompressible natural convection form of the Navier-Stokes equations and formulated semi-analytically a Nu-Ra correlation based on bounding relations for the isolated-plate limit in the entrance region and for fully-developed flow after the boundary layers merge. In recognition of his pioneering work, a modified gap-based Ra number multiplied with the ratio of the gap width and the plate length, is referred to as the Elenbaas number El, for an isothermal channel according to:

$$El = \frac{\rho^2 g \beta c_p z^4 (T_w - T_\infty)}{\mu k L} \tag{13}$$

with z the gap width, L the plate length, T_w the wall temperature and T_∞ the inlet air temperature. The Elenbaas correlation is:

$$Nu_0 = \frac{hz}{k} = \frac{El}{24}\left(1 - e^{-35/El}\right)^{3/4} \tag{14}$$

---

[b] Dr. Elenbaas was with Philips Research when he wrote this article but he became much more famous as the father of the high pressure mercury vapor lamp.

with Nu_0 the gap-based Nu number.

When limiting expressions are known, often use is made of a formulation proposed by Churchill and Usagi resulting in a function varying smoothly between the two bounds. Bar-Cohen and Rohsenow[34] proposed the following composite relation:

$$Nu_0 = \left[\frac{576}{(El)^2} + \frac{2.87}{(El)^{1/2}}\right]^{-1/2} \tag{15}$$

To give the reader an idea of what can be expected, let us take a temperature difference of 50°C and a plate length of 15 cm (e.g. resembling a board). For a spacing of 20 mm we find h to be 6.1 W/m^2K, for 5 mm this becomes 3.3 W/m^2K, and for 2 mm h reduces to 0.2 W/m^2K, showing the strong dependence of h on the gap width z. When we reduce the plate length to 1 cm, resembling a fin, these values become resp. 12.1, 11.8, and 3.6 W/m^2K. Notice the difference with L=15 cm. The first two values clearly indicate developing flow, almost independent of the plate distance over a large range. The last value points again at fully-developed flow.

Kraus and Bar-Cohen also discuss the optimum fin spacing and the optimum Nusselt number, to result in:

$$h_{opt} = 1.25\frac{k_g}{z_{opt}} \tag{16}$$

with

$$z_{opt} = \left(\frac{Lv^2}{\eta g\beta\theta_b Pr}\right)^{1/4} = 3P \tag{17}$$

where η is the fin efficiency of the optimum single fin equal to 0.63, and P an array parameter:

$$P = \left(\frac{Lv^2}{g\beta\theta_b Pr}\right)^{1/4} \tag{18}$$

Another useful optimum equation presented is the one that maximizes heat dissipation from a least-material rectangular fins with a prime area of given length L, width W, and specified temperature rise θ_b:

$$\left(\frac{Q}{LW\theta_b}\right) = 0.24\left(k\frac{k_{air}}{P^2}\right)^{1/2} \tag{19}$$

It is to be understood that results based on channel flow analysis can only be at best a reasonable approximation of heat sink heat transfer. This is especially true for LED based applications because of their frequent use of short-wide fins. When we introduce a base, the problem becomes inherently asymmetric and 3D, leading to lateral inflow and/or outflow effects. These effects introduce often a significant reduction in the average interfin air temperature. Furthermore, near the base the local h will be reduced, violating the Murray-Gardner assumptions. Finally, because the fin temperature changes along its height, (semi)analytical calculations become impossible because of the coupling between temperature and flow. Usually the approach taken is to assume an average fin temperature using the fin efficiency as a correction factor. All these added complexities have resulted in many empirically-based and CFD-based papers. An overview can be found in Kraus and Bar-Cohen [*op.cit., p.304*]. The main conclusion to remember is that the experimental Nusselt numbers are often substantially higher than those obtained from the Elenbaas correlation due to the strong impact of 3D flow, at least for the vertical-plate vertical-fin orientation. This observation has led several research groups to propose correction factors to the Elenbaas correlation, and other groups to propose correction factors to these correction factors, etc., resulting in a set of increasingly complex equations. One research group examined 12 different heat sinks in various orientations, fin spacings, etc., to result in different Nu correlations for all 12 heat sinks! This is not the path to walk for the average designer. To get some insight in the main parameters governing vertical parallel plate heat sinks it is better to use the Bar-Cohen-Rohsenow composite relation. Further reading regarding the use of optimal individual plate and pin fins to compose multiple fin arrays: Iyengar and Bar-Cohen[35], Bar-Cohen *et al.*[36] and Bahadur and Bar-Cohen[37].

Forced convection

Kraus and Bar-Cohen [*op.cit.*] treat forced convection-driven heat sinks fin in a certain order starting with parallel-plate channels, the transition from laminar to turbulent flow, the transition from developing to fully-developed flow, and the analogy between heat transfer and fluid flow. The interested reader is encouraged to consult this book for further details.

Open a heat transfer handbook and one will find a plethora of Nu-Re correlations. The author is of the opinion that the differences between them are minor in relation to the differences that occur in real-life. Hence, only some typical and relatively simple equations to guide the designer in understanding the relative importance of various parameters will be presented.

For laminar developing flow along plates we recommend to use:

$$h = 0.66\left(\frac{k_{air}}{L}\right) Re^{0.5} Pr^{0.33} \tag{20}$$

For turbulent flow along plates:

$$h = 0.036\left(\frac{k_{air}}{L}\right) Re^{0.8} Pr^{0.33} \tag{21}$$

For fully-developed channel flow we get:

$$h = \frac{4k_{air}}{d_h} \tag{22}$$

with d_h the hydraulic diameter: $d_h = 4A/P_w$ with A the flow area and P_w the wetted perimeter of the flow channel For parallel plates the hydraulic diameter becomes equal to twice the plate spacing z and we find h = 2k/z. For fully-developed flow, the pressure drop per unit length is:

$$\frac{\Delta P}{L} = \frac{48\mu v}{d_h^2} \tag{23}$$

with v the average velocity. The friction factor f is found to be:

$$f = \frac{24}{Re} \tag{24}$$

For turbulent flow, the friction factor can be written as:

$$f = 0.046Re^{-0.2} \tag{25}$$

To get an idea about the hydrodynamic entrance length x, for laminar flow this may be obtained from:

$$\frac{x}{d_h} = 0.05Re \tag{26}$$

with Re based on d_h. For turbulent flow, because there is no satisfactory general expression for the entry length, fully developed flow is usually assumed when:

$$\frac{x}{d_h} > 10 \tag{27}$$

As argued before, it is not possible to define a single optimum forced convection array. In other words, only in case the fan characteristics are given we are able to find the optimum fin array geometry. Within the framework of this paper is not possible to present even a limited overview of all correlations and optimization strategies that can be found in literature for various kinds of operating conditions. It is highly recommended to use a spreadsheet approach to be able to vary the multitude of parameters that are governing the world of forced convection driven heat sinks. This way of working has been described by Biber and Fijol[38]. They conclude that practical design considerations often place significant limitations on heat sink optimization, and that using a fan curve-based design method sometimes results in performance effects that run counter to conventional wisdom.

To get a flavor of what has been published, please consult papers by Knight *et al.*[39], Teerstra *et al.*[40], Iyengar and Bar-Cohen[41] and Copeland[42]. Optimization strategies based on the entropy generation minimization as proposed by Bejan[43] is treated by Shih and Liu[44].

Another popular trend is to take into consideration the impact heat sink design has on sustainability. Not only maximum heat transfer, but also the energy required to operate the heat sink and to manufacture it are part of the objective function to be optimized, see e.g. the paper by

Bar-Cohen *et al.*[45] Worth mentioning is the attention paid to enhanced-conductivity polymers to replace aluminum as a possible material for natural convection-driven systems.

Constructal theory

Bejan[c] developed a way of identifying and minimizing losses in the field of thermodynamics through the method of entropy generation minimization (EGM), which has its origins in his doctoral thesis. The following introduction is a mix of what can be found on his very interesting web portal[46]. Around the mid-nineties he took this philosophy one step further, coining it Constructal Theory. The Constructal Law is a first principle of physics that accounts for all design and evolution in nature. The constructal law asks the question: Why does this design arise at all? Why can't the water just seep through the ground? The law provides this answer: Because the water flows better with design. It covers the tendency of nature to generate designs to facilitate flow.

Existing flow configurations are replaced by better flowing configurations, smoothly or stepwise, in animal design, river basin design, automobile design, and geopolitical design. While constructal theory gives rise a fractal-like deterministic structure is space, constructal theory should not be confused with fractal theory. Fractals are mere descriptions but not explanations of natural phenomena.

The following principle is the essence of the Constructal Law:

For a finite-size flow system to persist in time (to survive) its configuration must evolve such that it provides greater and greater access to the currents that flow through it.

Recent Constructal Theory predictions reported are the scaling laws of all animal locomotion (running, flying, swimming), the distribution of the sizes of human settlements, the scaling laws of river basins, the shape

[c] Bejan is also the author of the book titled Convection Heat Transfer [Wiley, 2004]. A.o, he derives common analytical correlations starting from simple scaling laws, and criticizes the common interpretation of numbers such as Re on the grounds that it is total nonsense. Highly recommended!

of all Egyptian pyramids, vegetation, and the evolution of science and civilization. Pretty impressive.

Now, what has all this to do with the subject at hand? One of the many recently published papers[47] is devoted to maximizing the heat transfer density, by applying constructal theory, for natural convection between vertical plates. Instead of optimizing only the distance between the plates, the authors sought for an optimized multi-scale flow structure. The key to higher performance is the increased design freedom, because more length scales (to be optimized) represent more degrees of freedom in the design. More length scales mean that the flow structures can morph in more design directions. In the end, the optimized multiple scales are distributed non-uniformly through the available volume. The basic idea is sketched in Figure 5.

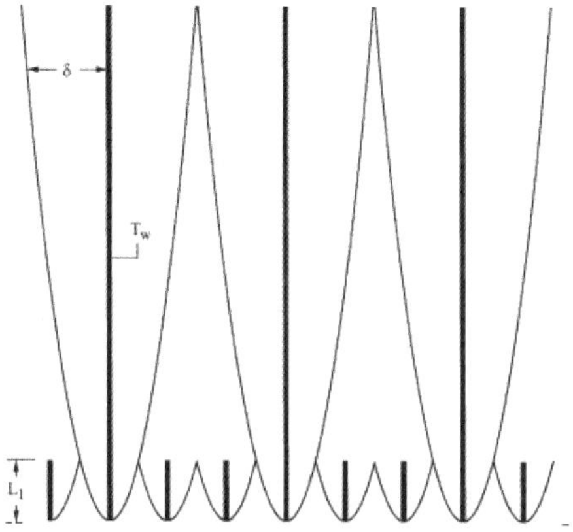

Fig. 5. Space between fins provided with much smaller fins.

For developing flow, the area in between the boundary layers is more or less lost for heat transfer. Filling this area with fins of a certain optimized height, and repeating this process, leads to better performance. The authors reported improvements up to 20%.

While I don't expect that constructal theory will become a mainstream method used by designers to optimize their heat sinks, the method is considered enough thought-provoking to be included in our list.

Web-based tools

While there are many heat sink calculators to be found on the web, the majority does not show background information. Providing this service is the major advantage of the tools generated by the University of Waterloo. An article in ElectronicsCooling[48] discusses a web based tool for the calculation of natural-convection driven heat sinks that is very easy to use and saves the designer from the burden to enter a lot of equations into a spreadsheet. Interactive, real time solution tools, plus an extended set of complementary tools are available at the Microelectronics Heat Transfer Laboratory web site[49]. Unfortunately, the website does not offer an equivalent tool for forced-convection driven heat sinks. A first-order guess can be found at another website[50] that is based on research by Kays and London.

Dedicated heat sink software

Apart from the free web-based tool discussed above, there exists also commercial heat sink software[51]. This code can address natural and forced convection, including bypass. One step further, something between this tool and CFD, is software that can handle also fans, interface resistances, radiation etc.[52]

The main objection in using these tools is that they exclude all heat sinks that deviate from the shape imposed by the extrusion process, neglecting many alternative design options. They also fail in taking into account the significant 3D fluid flow phenomena that occur often in practice. Nevertheless, they can be fruitfully used in what-if studies.

3.4. *Approaches to acquire insight in fluid dynamics*

In the past 20 years, we have observed a significant increase in the use of Computational Fluid Dynamics (CFD) codes to calculate the thermal behavior of electronic systems. The benefits are undisputed when it comes to performing parametric studies in early design phases. However, when the objective is accuracy (at the 5-10% level), the discussion about what we can expect in practice becomes far from trivial. The heat transfer behavior of electronic systems is very complex when compared to the standard canonical cases usually treated in fluid dynamics classes. The next section discusses the CFD analysis of single heat sinks, and the second section formulates answers to the question about the accuracy that can be expected.

CFD analysis of single heat sinks

When the purpose is to design the most optimal heat sink given a number of design constraints the best approach is to use a full-blown CFD code. The standard approach is to model a wind tunnel with uniform approach flow, set a certain bypass, and model a heat sink with any complexity that comes to mind, be it oblique fins or interrupted fins, fin arrays, pin fins, etc. Next, the best available turbulence model is chosen and a very fine mesh. Chances are high that one not only acquires fundamental insight in many aspects of the fluid dynamics that rule heat sink heat transfer but is also able to find the best heat sink given the constraints, but clearly this the approach for the heat sink manufacturer, not the end-user. Another very useful outcome of such a CFD analysis could be insight in the factors that determine outflow of air as a function of fin distance and fin dimensions. Finally, useful data are generated concerning pressure losses. Some codes allow full-factorial optimization. Please note that the discussion is about a **single** heat sink. A better approach is discussed in the Section called *Approaches suited for final design of heat sinks in LED applications.*

Pros and Cons of CFD for Heat Sink Analysis

A natural question is how accurate numerical simulations are when compared to well-designed experiments. Two papers[53,54] focus on the following question: Can we predict the junction temperature of critical components with sufficient accuracy? The conclusion is: not without fitting! The major reason is the lack of sufficiently accurate input parameters and boundary conditions combined with complex geometries. One example is given by Lasance and Rind[55]. This paper deals with instabilities and 'mini-jets' generated by a grille the details of which of course are never modeled. However, these effects persist long enough to cause a measurable influence on downstream local heat transfer. For LED based systems heat sinks are often located near luminaires with openings, causing significant interactions with the created turbulence near the walls.

Transitional and turbulent convection pose another problem. Sophisticated turbulence models are not expected to be implemented in system-level analysis design tools for quite some time. To capture low-Re transitional flow requires a very fine grid and time-dependent analysis. Rodgers and Eveloy[56] discuss this topic in more detail and conclude that significant errors are unavoidable. A last complex issue that deserves attention in LED based systems that often combine heat sinks and fans or synthetic jets is the modeling of these fans and jets. For sure not every detail of a fan, both dimensionally and a velocity-wise, can be taken into account in a system level analysis. Synthetic jets pose even a more difficult problem, see a series of papers by Lasance *et al.*[57,58,59]. Hence, accurate prediction of local heat transfer phenomena is considered problematic and requires often a lot of time in a research lab to formulate adequate conclusions, time that is normally lacking in a design environment. Focusing on heat sinks only, in most cases, numerical modeling is not an option because reality is too complex. One of the conclusions of a panel on heat sink modeling held at Interpack 2003[60] was that the errors using commercial codes could easily exceed 30%. Because this problem is not related to CPU power we believe that not much has changed since 2003. Please note again that the subject of the discussion is here: how to arrive at the best heat sink for the application, not about the usefulness of CFD for system level design.

How to Improve?

We may expect that on the numerical side some errors will decrease to more acceptable levels. The errors related to the treatment of complex fluid phenomena will not be reduced to the same extent unless Direct Numerical Simulation (DNS) or Large Eddy Simulation (LES) techniques become feasible for 3D complex systems. While it is true that some improvement could be gained by implementing more sophisticated turbulence models it is by no means a panacea for the problems mentioned. Suppose we want to model spray cooling. It makes a lot more sense to design appropriate experiments to estimate the local heat transfer instead of relying on very complex 3D highly turbulent two-phase CFD simulations. One way to enhance the predictability of CFD analyses is to use a pragmatic approach by employing correction factors. A recommended way to measure correction factors is by developing 'ideal' experimental benchmarks for complex geometries where all boundary conditions and material properties are under control and are well known to within a few percent.

The most important conclusion is: *A numerical analysis of an electronic system may or may not be correct, and no one can tell.* Another way to put it: *An accurate match (± 10%) is not possible without some kind of fitting.*

Despite the conclusions, CFD tools are vital in design environments! However, do not expect or claim high accuracy in a direct comparison with experiments.

3.5. *Approaches suited for final design of heat sinks in LED applications*

When it comes to business objectives such as cost reduction, time-to-market, optimization of weight, height and size from a sustainability point of view, and temperature reduction, it is obvious that the designer needs tools that allow her to predict with sufficient accuracy what will happen in real-life. None of the approaches discussed above are suited to meet this goal. This section treats three approaches that may provide the designer with all relevant information that is required to make an optimal choice out of the thousands of heat sinks that are offered on the market. The first approach discusses a test dedicated to ranking real heat sinks under real-life boundary conditions. The second approach is purely numerical, applying user-friendly CFD codes at the system level using suitable heat sink compact models.

The final approach treats prototype testing, the proof of the pudding as it were. The combination with CFD modeling will bring expert experience to a higher level, useful for optimizing the next generation.

Dedicated tests to rank a series of heat sinks

This Section briefly discusses a method to rank heat sinks given a certain application and is drawn from a paper by Lasance and Eggink[61]. The measurement is based on the extraction of the average heat transfer coefficient from time-dependent temperature curves as a function of velocity and bypass. Scaling the measured effective heat transfer coefficient by mass, volume, weight or height provides several performance metrics allowing designers a novel way of ranking heat sinks in conditions that resemble the application. The basic idea is to measure the cooling curve of a heat sink and to extract an average heat transfer coefficient from this curve. The setup is shown in Figure 6. It consists of a channel with two doors that allow access to the test section. Inside the channel, a foil heater has been installed upon a holder that can be moved vertically. The heater is covered with copper-based grease and provided with a 0.1 mm K-type thermocouple to which a heat sink can be attached. If the heat sink is larger than the heater heat spreading effects are taken into account. To measure the reference temperature of the inlet air, a copper cube provided with a thermocouple is used to damp the always-present 'high'-frequency temperature fluctuations[d].

[d] In case the inlet temperature is used as the reference temperature and is subject to variations over time the only physically correct way is to use a dummy set up with zero dissipation. In our case the 'low-frequency' fluctuations are minimized by controlling the temperature within the lab.

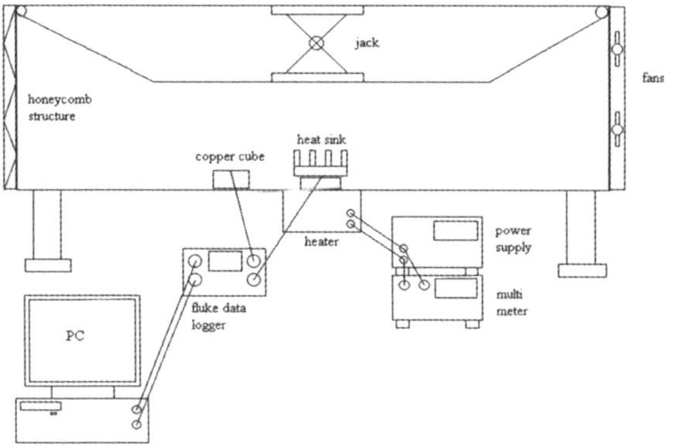

Fig. 6. Side view. The copper cube dimensions are not to scale.

Six fans were installed operating in suction mode to vary the air velocity. (In contrast to blowing, suction does not create approach-flow turbulence.) To more or less straighten the flow, a honeycomb structure is positioned at the entry of the channel. The bypass can be adjusted by changing the height and width of the channel. The height is changed by adjusting the jack. The width can be adapted by adding plates of various heights[e]. In practice, upstream objects are often present influencing the inflow conditions. The designer could easily address their influence by placing dummies of these objects at the desired position in the measurement section. No attempt has been made to improve the flow conditions of this low-quality wind tunnel, on the contrary. The reason is simple: these effects are not addressed in real-life either. The objective of the proposed method is *ranking,* and we believe that the effects mentioned above do not influence the ranking. Further details of the test setup are provided in the cited paper. The measurement procedure is as follows. First, the heat sink is raised above the substrate and heated to say 60 °C. Next, the heater is switched off, the heat sink lowered onto the substrate and the fans are turned on. The resulting cooling curve is used to calculate a number of performance metrics. The curve can be described by:

---

[e] While more complicated methods enabling continuous adjustment of the bypass dimensions have been considered, we decided to keep it as simple as possible.

$$\frac{T - T_\infty}{T_0 - T_\infty} = \exp\left[-\frac{hA}{\rho c_p V}(t - t_0) \right] \tag{28}$$

Let us define a time constant β to be equal to $\rho c_p V/hA$. From the slope of a logarithmic plot of the cooling curve β can be extracted. In practice, the performance metric hA is calculated that can be considered as an effective heat transfer coefficient. This parameter shows the effective performance of a heat sink, excluding for example weight and volume influences.

The experiments were performed using seven heat sinks taken from the collection that was present in 2005, based on more or less equal base areas. The heat sinks are different in shape, area, volume, material and cost.

The fundamental performance metric used to compare the heat sinks is hA, the heat transfer coefficient times the effective (wetted) area of the heat sink. To calculate hA, $\rho c_p V$ has to be known. ρV (the mass) can be measured by weighing. For the specific heat c_p, we need to know the material of the heat sink. Heat sink 4 is made of magnesium, the others of aluminum. Here is a suggestion when the c_p is not known: make a copy of the heat sink base) using a known material, the ratio of the cooling curves provides the required value. While hA can be considered as the most 'fair' performance metric, in some cases other arguments to choose the heat sink play a role. For example, weight and height are selection criteria in portable applications, and volume in many others. By dividing hA by the mass, volume or height one finds three additional performance metrics[f]. By way of illustration, Figure 7 shows the results for the four performance metrics for the seven heat sinks for a velocity of 2.2 m/s and a relatively large bypass. As is obvious from Figure 7, the results for the other velocities tested show the same trend. *Note*: It is very well possible that a much smaller bypass results in a different ranking!

[f] Iyengar and Bar-Cohen use the term 'array heat transfer coefficient' for our 'effective heat transfer coefficient', and they use thermal metrics such as the space-claim heat transfer coefficient for our volume –weighted h (Least-Material Optimization of Vertical Pin-Fin, Plate-Fin, and Triangular-Fin Heat Sinks in Natural Convective Heat Transfer, Proceedings of the ITHERM Conference, May'1998, Seattle, USA, pp. 295–302)

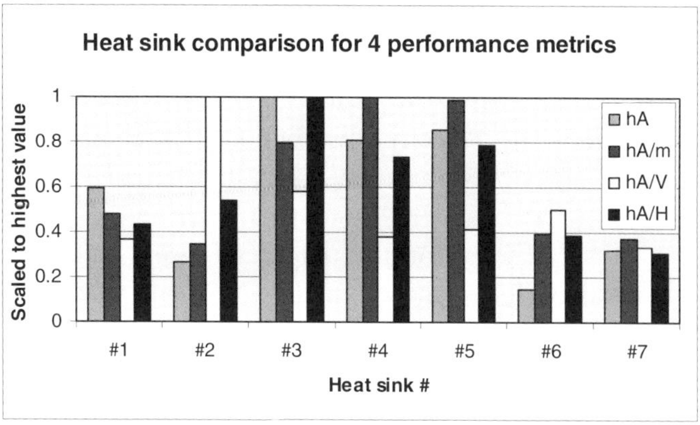

Fig. 7. Heat sink comparison for v=2.2 m/s

Of course, it is up to the designer to invent other metrics, for example based on performance and cost, or by combining several metrics using weighting factors. *Note*: When the heat source is (much) smaller than the base area of the heat sink, the effect of heat spreading will be included, but this measuring method cannot deal with the individual contribution of the spreading effect. This effect introduces an additional thermal resistance that can be easily addressed using the tricks outlined in [6].

It is interesting to observe that the best scores for all metrics are associated with three different heat sinks, and also that some heat sinks score much better on one metric than on the others.

A rigorous explanation of the observed differences is rather difficult because of the complex physics involved.

CFD with calibrated heat sink compact models

There is not much literature on compact heat sink models that are in use at the big CFD companies that market software for the electronics cooling market, such as Flotherm, Icepack and 6SigmaET.

Coetzer and Visser[62] introduced a compact model to predict the interfin velocity and the resulting pressure drop across a longitudinal fin heat sink with tip bypass. The compact model is based on results obtained from a comprehensive study into the behavior of both laminar and turbulent flow in longitudinal fin heat sinks with tip bypass using

CFD analysis. The new compact flow prediction model is critically compared to existing compact models as well as to the results obtained from the CFD simulations. The results indicated a good match. The improved accuracy in velocity distribution between the fins also increased the accuracy of the calculated heat transfer coefficients applied to the heat sinks.

It is beyond the scope of this paper to discuss the details and differences between the compact models that are generated by the software companies mentioned above.

In addition to what has been argued about the pros and cons using CFD for heat sink optimization, it should be stressed that using CFD at the system level is the recommended approach to support the designer in selecting the most suitable fan. For this choice, the designer needs a number for the total system air resistance in order to conclude about the most suitable operating point, and CFD is second to none in finding this number.

Prototype testing

When a prototype becomes available, in many industrial cases a lot of verification tests are performed. Unfortunately, oftentimes there is no time for a thorough thermal analysis except for the measurement of critical temperatures. Once the system has been approved, often a sigh of relief is heard and one moves on to the next job at hand. One tends to forget that useful insight can be obtained by comparing pressure losses and local air velocities and temperatures with the results of a CFD model, insight that can be fruitfully used to save considerable design time in developing the next generation. Whenever possible, measure a certain heat sink temperature, switch off (or reduce) the power of the component to which the heat sink is attached, measure the unpowered temperature, and deduce the adiabatic effective heat transfer coefficient for the component in question. Such acquired values can be used in conduction-only analyses of part of the system, focusing on details such as vias and multilayer boards that cannot be addressed at the system level.

3.6. *Pitfalls to avoid*

The author had some interesting discussions with Prof. Bar-Cohen from which some caveats popped up. In addition, he found more non-trivial issues related to heat sink analysis that deserve highlighting.

Caveat 1. <u>Know the limits of the Elenbaas correlations</u>
The Elenbaas correlation was developed for isothermal plates and works very well for that configuration, as long as the Elenbaas Number is not too small and the fins are much thinner than the gaps. When the Elenbaas number falls below 10 or so, the inflow from the open edge of the inter-fin gap keeps the channel from realizing the fully-developed limit and Nu is much higher than predicted. Generally, however, this is not a desirable operating point, so this does not have much practical impact and does not affect the optimum spacing that is around Elenbaas of 50 and well above the values at which inflow is important. However, your design may force low Elenbaas numbers and then the use of these correlations becomes doubtful.

Caveat 2. <u>Complex composite correlations versus simple Elenbaas correlations</u>
The composite equations were develop to deal with the effect of the plate/fin thickness and isoflux plates, as well as asymmetrically heated channels. For standard natural convection applications, the general rule is to use the Elenbaas correlations when fin thickness can be neglected with respect to fin spacing. However, when you may decide what can be neglected? The recommendation is to acquire insight to try both to calculate the optimum spacing and heat transfer relations to quantify the effect of thickness.

Caveat 3. <u>Be very careful with choosing the right temperature difference in calculating h</u>
The heat transfer coefficient h is always based on some temperature difference, and using published equations one should be very careful in selecting the right one. For example, heat exchanger theory is a simple and recommended way to get insight in many parameters that govern heat sink thermal behavior. However, the designer needs to get a value

for h. Maximum dissipation is often calculated using the T_{wall}-T_{inlet} difference, but then using h derived from a fully-developed constant value that is based on $T_{wall} - T_{bulk}$ is incorrect. However, because h extracted from the Elenbaas equation does include the air temperature rise, the consequence is that when the heat exchanger concept is used this effect is taken into account twice. Our advice is to take an educated guess for h: 5 W/m^2K.

Caveat 4. Nu=C can also be used for natural convection
It is not generally known that the fully developed constant Nu value is also valid for natural convection. The only reference we know of can be found on p286 in the Kraus/Bar-Cohen book (op.cit.). However, be aware of the temperature difference that is associated with this value! All other definitions state something like Nu=C.Ra.(d/L^3) or equivalent, thereby obscuring its constant value, the reason being that this correlation is linked to another temperature difference. According to Bar-Cohen, this was one of Eckert's pet peeves, in later life, that the use of enclosure and channel aspect ratios in the various correlations obscured the underlying physics. For example, in some of the enclosure correlations, using the width as the length scale in the Nusselt number requires that you add an aspect ratio term in the correlation but, in fact, for a width that is greater than 2x the boundary layer thickness, there is no dependence on width. On the other hand, one may question the physics behind a constant Nu number, hence a constant h, hence independent of the velocity. This topic is discussed by Lasance[63].

Caveat 5. Be aware of the useful range of your correlations
An example is given in Caveat 4. If you use a correlation based on the fin spacing, this correlation is not valid beyond a certain width.

3.7. *Why more expensive heat sinks can save money*

The question arises why to pay more for a higher-efficient heat sink that is also smaller and lighter in weight, while a simple larger but more heavy cast or extruded heat sink can also do the job? In most cases, the single-piece part cost price is the main driver why engineers and purchasers stay with lesser effective solutions because they are of the opinion that they save money, but is this really true? Let us compare the price of two single heat sinks, a highly efficient versus a standard extruded type with the same thermal performance, but not taken into account the flow performance, effect on neighbor and downstream components, weight, volume usage etc. Indeed, the single heat sink will win, the standard type will always be cheaper, if the other issues are accounted as less important.

To arrive at a fair comparison, one has to look at other related effects that will result in a system price increase due to a non-optimal cooling solution:

1. Effect on the flow, especially pressure drop
2. Heat sink weight
3. Heat sink attachment
4. Mechanical adjustments required to handle the weight on board and system level, and to fulfill mechanical requirements for shock and vibration
5. Raw material usage to manufacture the cooling solution
6. Required fan performance
7. Product reliability
8. Transportation cost
9. End of life cost

Including these effects, it might well turn out that choosing the cheapest fan will increase the system costs.

4. Conclusions

Whenever cooling a certain electronic component or system asks for the addition of a heat sink, the designer is confronted with the question: which one? Apart from the plethora of choices in shapes, sizes, materials, volumes etc., most handbooks and papers focus on parallel plate heat sinks and wind tunnel flow. The objective of this paper is to provide designers with the right mindset to enable an optimal choice for their application. The body of the paper is devoted to heat sink modeling approaches from a designer's perspective. A distinction is made between approaches that are not recommended, approaches that are useful to acquire either insight in the physics, or in optimization parameters, or in fluid dynamics, and finally approaches that are suited for the final design stage in LED applications. Some pitfalls to avoid are discussed. The paper closes with showing that often in industry heat sinks that are more expensive can save money at the system level. The final conclusion boils down to the following: the only way that leads to optimization of the whole system is to use CFD modeling supported by dedicated tests. The use of correlations is only recommended in the very first stages of a design.

References

1. Lasance C., Chapter 9 in Thermal Management for LED applications, C. J. M. Lasance, A. Poppe (eds.), Solid State Lighting Technology and Application Series 2, Springer New York, expected for Q4 2013.
2. Lasance C., Ten Years of Boundary-Condition-Independent Compact Thermal Modeling of Electronic Parts: a Review, Heat Transfer Engineering Vol. 27, pp. 149–169, 2008.
3. Lasance C., How to Estimate Heat Spreading Effects in Practice, J. Electron. Packag., 132, 031004, (2010).
4. Lasance C., Two-layer heat spreading revisited, Proc. SEMITHERM 28, March 2012, San Jose.
5. Lasance C., "The Influence of Various Common Assumptions on the Boundary-Condition-Independence of Compact Thermal Models", IEEE CPT 27, pp. 523–529), 2004.
6. Lee S., "Calculating spreading resistance in heat sinks", Electronics Cooling, Vol. 1, January Issue, 1995.
7. Lasance C., Heat spreading: not a trivial problem, ElectronicsCooling, Vol. 14, May issue, pp. 24–30 (2008).
8. http://www.electronics-cooling.com/1997/09/standardizing-heat-sink-characterization-for-forced-convection/
9. Moffat, R, "Modeling Air-Cooled Heat Sinks at Heat Exchangers," ElectronicsCooling, February 2008.
10. Azar K. and Tavassoli B., How Much Heat can be Extracted from a Heat Sink?, ElectronicsCooling, May 2003.
11. Simons R., "A Simple Thermal Resistance Model — Isoflux Versus Isothermal", Electronics Cooling, February 2006.
12. Lasance C., 2005, "Sense and Nonsense of Heat Transfer Correlations Applied to Electronics Cooling", Proc. EUROSIME 2005, pp. 8–16.

13. Starner K. and MacManus H., An experimental investigation of free convection heat transfer from rectangular fin arrays, J Heat Transfer Vol. 85, pp. 273–278 (1963).
14. Welling W. and Wooldrigde C.,Free convection heat transfer coefficients from rectangular vertical fins, J Heat Transfer Vol. 87, pp. 439–444 (1965).
15. Van de Pol, D.W., and Tierney, J.K., "Free Convection Heat Transfer from Vertical FinArrays", IEEE Trans. Part, Hybrids, and Packaging, Vol. PHP-10, no. 4, Dec. 1974, pp. 267–271.
16. Bilitzky, A., The Effect Of Geometry On Heat Transfer By Free Convection From A Fin Array, MS Thesis, Dept. Of Mechanical Engineering, Ben-Gurion University Of The Negev, Beer Sheva, Israel, 1986.
17. Kim T., Do K., Kim D, Closed form correlations for thermal optimization of plate-fin heat sinks under natural convection, IJHMT Vol. 54, pp. 1210–1216 (2011).
18. Holahan, M. F., Kang S. S., Bar-Cohen, A., 1996, A Flowstream Based Analytical Model for Design of Parallel Plate Heat Sinks, ASME Proceedings of the 31st National Heat Transfer Conference, HTD-Vol. 329, Volume 7, pp. 63–71.
19. Kakac, S., Shah, R. K., Aung, W., 1987, Handbook of Single-Phase Convective Heat Transfer, 2nd Ed., Wiley, NY, pp. 3–35, 3–42.
20. Biber C. and Belady C., Pressure drop prediction for heat sinks: what is the best method?, Proceedings of Interpack '97 Conference, ASME, Mauna Lani, Hawaii, 1997.
21. http://www.coolingzone.com/library.php?read=489
22. http://www.coolingzone.com/library.php?read=490
23. http://www.coolingzone.com/library.php?read=491
24. Simons B., Calculation Corner: a useful catalog of calculation corner articles, ElectronicsCooling September 2011.
25. Bahadur R., Characterization, modeling, and optimization of polymer composite pin fins, Thesis University Of Maryland, 2005. (http://drum.lib.umd.edu/handle/1903/2821)
26. Bar-Cohen, A., R. Bahadur and M. Iyengar, 2006, "Least-Energy Optimization of Air-Cooled Heat Sinks for Sustainability-Theory, Geometry and Material Selection," Energy, Vol. 31, pp. 579–619.
27. Khan, W. A., Culham, J. R., and Yovanovich, M. M., "Optimization of Pin-Fin Heat Sinks Using Entropy Generation Minimization," IEEE Transactions on Components and Packaging Technologies, Vol. 28, No. 2, 2005, pp. 247–254.
28. Khan, W.A., Culham, J.R., and Yovanovich, M.M., Modeling of Cylindrical Pin-Fin Heat Sinks for Electronic Packaging, IEEE Transactions on Components and Packaging Technologies, Vol. 31, No. 3, September 2008.
29. Ledezma, G. and Bejan, A., Heat sinks with sloped plate fins in natural and forced convection, International Journal of Heat and Mass Transfer, Vol. 39 no. 9 (1996), pp. 1773–1783.
30. S. V. Naidu, V. Dharma Rao, B. Govinda Rao, A. Sombabu And B. Sreenivasulu, Natural Convection Heat Transfer From Fin Arrays: Experimental And Theoretical Study On Effect Of Inclination Of Base On Heat Transfer, ARPN Journal Of Engineering And Applied Sciences, Vol. 5, No. 9, 2010.
31. http://frigprim.x10.mx/articels2/parallel_pl_Inc.html

32. Kraus A. and Bar-Cohen A., Design and Analysis of Heat Sinks, John Wiley, 1995.
33. Elenbaas W. 1942. Heat dissipation of parallel plates by free convection. Physica. 9: 1–28.
34. Bar-Cohen A. And Rohsenow W., Thermally-optimum Arrays of Cards and Fins in Natural Convection, IEEE CHMT, p.154–158, 1983.
35. Iyengar M., and Bar-Cohen, A., Least-Material Optimization of Vertical Pin-Fin, Plate-Fin, and Triangular-Fin Heat Sinks in Natural Convective Heat Transfer, Proceedings of the ITHERM Conference, May'1998, Seattle, USA, pp. 295–302.
36. Bar-Cohen A., Iyengar M., and Kraus A.D., 2003, "Design of Optimum Plate Fin Natural Convection Heat Sinks," ASME Transactions – Journal of Electronic Packaging, Vol. 125, Number 2, pp. 208–216.
37. Bahadur R. and Bar-Cohen A., "Thermal Design and Optimization of Polymer-Based Pin Fin Natural Convection Heat Sinks," IEEE Transactions on Components and Packaging Technologies, Volume 28, Issue 2, June 2005 pp. 238–246.
38. Biber C. and Fijol S., Fan-plus-Heatsink "Optimization" – Mechanical and Thermal Design with Reality, Proceedings of the International Systems Packaging Symposium, pp. 285–289, 1999.
39. Knight R., Goodling J., Hall D., Optimal thermal design of forced convection heat sinks-analytical, JEP Vol. 113, pp. 313–321 (1991).
40. P. Teertstra, M. M. Yovanovich, J. R. Culham, and T. Lemczyk, "Analytical forced convection modeling of plate fin heat sinks," in Proc. 15th IEEE SEMI-THERM, 1999.
41. Iyengar, M., and Bar-Cohen, A., 2001, Design for Manufacturability of SISE Parallel Plate Forced Convection Heat Sinks, Trans. of the IEEE-CPT, June, Vol. 24, No. 2, pp. 150–158.
42. Copeland D., Fundamental performance limits of heatsinks, JEP Vol. 125, p. 221 (2003).
43. Bejan A., Entropy Generation Minimization. Orlando, FL: CRC, 1996.
44. Shih C. and C. Liu G., Optimal Design Methodology of Plate-Fin Heat Sinks for Electronic Cooling Using Entropy Generation Strategy, IEEE Transactions on components and packaging technologies, Vol. 27, no. 3, pp. 551–559 (2004).
45. Bar-Cohen A., Bahadur R.and Iyengar M., 2006, "Least-Energy Optimization of Air-Cooled Heat Sinks for Sustainability-Theory, Geometry and Material Selection," Energy, Vol. 31, pp. 579–619.
46. http://www.constructal.org/
47. Silva A.K. da, and Bejan A., Constructal multi-scale structure for maximal heat transfer density in natural convection, , International Journal of Heat and Fluid Flow 26 (2005) 34–44.
48. Teerstra P., Yovanovich M, Culham J., Natural convection modeling of heat sinks using web-based tools, ElectronicsCooling, Sept 2000.
49. http://www.mhtl.uwaterloo.ca
50. http://www.novelconceptsinc.com/calculators-forced-convection-heat-sink-thermal-resistance.cgi
51. http://www.heatsinkdesigner.com/
52. http://www.qfin.net/drupal/qfin47

53. Lasance C., "The Conceivable Accuracy of Experimental and Numerical Thermal Analyzes of Electronic Systems", IEEE CPT 25, pp. 366–382, 2002.
54. Lasance C., 2005, "CFD Simulations in Electronic Systems: A Lot of Pitfalls and a Few Remedies", Electronics Cooling, Vol. 11, no. 2, pp. 20–27.
55. Lasance C., Rindt C., Accuracy Comparison of a Standard CFD Code for the Thermal Analysis of Non-Simple Geometries with Baseline Experiments, Proc. 23rd SEMITHERM, San Jose, CA (2007).
56. Rodgers P., Eveloy V., 2004, 'CFD Prediction of Electronic Component Operational Temperature on PCBs', Electronics Cooling Vol. 10, No. 2, 22–28.
57. Lasance C., Aarts R., Synthetic Jet Cooling Part I: Overview of Heat Transfer and Acoustics, Proc. SEMITHERM 24, March 2008, San Jose.
58. Lasance C., Aarts R., Ouweltjes O., Synthetic Jet Cooling Part II: Experimental Results of an Acoustic Dipole Cooler, Proc. SEMITHERM 24, March 2008.
59. Lasance C., Nicole C., Aarts R., Ouweltjes O., Kooijman G., Nieuwendijk J., Synthetic Jet Cooling Using Asymmetric Dipoles, Proc. Semitherm 25, San Jose, 2009.
60. Copeland D. (moderator), "Modeling, Optimization and Performance of Heat Sinks with Non-Traditional Flow and Fin Geometries", Panel Session at Interpack 2003, Maui.
61. Lasance C. and Eggink H., 2005, "A Method to Rank Heat Sinks in Practice: The Heat Sink Performance Tester", Proc. 21st SEMITHERM, pp. 141–145, San Jose, CA.
62. Coetzer C. and Visser J., Compact Modeling of Forced Flow in Longitudinal Fin Heat Sinks With Tip Bypass J. Electron. Packag.,Vol. 125, pp. 319–324, 2003.
63. Lasance C., Thermal facts and fairy tales: fully-developed channel flow: why is Nu constant?, ElectronicsCooling September 2010.

CHAPTER 6

SYNTHETIC JETS FOR HEAT TRANSFER AUGMENTATION IN MICROELECTRONICS SYSTEMS

Mehmet Arik* and Enes Tamdogan

Ozyegin University
College of Engineering
Istanbul, Turkey
**marik06@gmail.com*

Rapid progress in science and information technology, growing manufacturing activities and increase in globalization have boosted the demand for advanced electronics devices. Moreover, increase in microprocessor power dissipation coupled with the reduction in feature sizes due to manufacturing process improvements have resulted in continuously increasing heat fluxes. Thus, ever increasing heat fluxes have required the development of novel, reliable and affordable thermal management technologies. Although some of those proposed solutions for high flux cooling problems based on liquid cooling methods such as spray and evaporative cooling; air cooling is still commonly preferred due to its availability, reliability, easiness and low cost. Therefore, over the last decade microfluidics devices such as synthetic jets have been investigated as an alternative to conventional air moving devices, and have been shown as highly effective for cooling of clectronics in compact thermal real estates.

Synthetic jets are meso scale fluidics devices, which operate on the "zero-net-mass-flux" principle. However, they create a positive net momentum flux to the external environment, and are able to produce effective cooling similar to a fan without posing ducting, reliability, and large dimension issues. These devices, which alternate suction and ejection of fluid through an orifice bounding a small cavity, by the time periodic motion of a diaphragm built into one of the walls of the cavity.

A unique feature of these jets is that they are formed entirely from the working fluid in which they are deployed. Moreover, unlike conventional jets, synthetic jets produce fluid flow with no mass addition to the system and without the need for complex plumbing. Since, the rate of heat removal from the thermal source is expected to depend on the location, orientation, strength, and shape of the jet, in the current paper, an overview of the state of the art for synthetic jet technology, research, design approaches and also possible applications is presented. Heat transfer enhancements in natural and forced convection, recent progresses in modeling and heat transfer predictions are presented. Finally, recent practical applications of synthetic jets are discussed.

Keywords: *Synthetic jets, electronics cooling, thermal management, impingement, jet design.*

Introduction

Recent advances in science and technology have enabled compact, fast and high functionality electronics devices. With the growing use of these devices by businesses and especially by research and teaching institutions, the thrust of computing power on faster computations and solving complex problems is rapidly rising. However, most of the electronic devices generate excessive heat during the operation that poses significant challenges for thermal architects. This excess heat may cause general or localized thermal degradation or even catastrophic failures. Thus, finding new solution techniques for removing heat to enable the safety and reliability of electronics systems is vital for reliable operations.

There is an urgent need for low cost and high performance heat transfer solutions in many electronics systems. This is especially true in today's world where feature sizes of packages are getting smaller due to manufacturing process improvements coupled with the increase in microprocessor power dissipation resulted in continuously increasing local heat fluxes. Although, in some of these devices the generated heat is dissipated somehow sufficiently by enclosure, package, header, or by leads, but the remaining and most of the heat still requires additional apparatus and techniques due to the limited capability.

While some of the solutions for high flux thermal problems based on liquid such as spray, pool boiling, micro-channel, and evaporative cooling; air cooling is still commonly used due to its availability and low cost. Although forced convective air-cooling can address some of those problems either alone or in conjunction with heat sink devices, fan-based cooling systems were found to be sometimes undesirable due to electromagnetic interference, noise, large moving parts, and reliability concerns. Moreover, since the thermal management of tight space, low power electronics applications are one of the current challenging engineering problems; new thermal management technologies have been developed. One of those technologies is microfluidics synthetic jet, which is more energy efficient than comparable fan-based systems, and also requires no piping or plumbing without any electromagnetic issues. Thus, over the last several years synthetic jets have been investigated as an alternative to fans. As novel and compact air moving devices, they have been shown to be highly effective for cooling of electronics in a very small form factor.

In this study, we are presenting a recent research overview of synthetic jet technology including fundamentals, modeling, natural and forced convective cooling followed by potential applications. Although there is currently limited practical applications utilizing those high performance jets, scientists have been performing significant amount of research that was driven by industrial interest. Therefore, it is soon expected to see those jets in applications such as LEDs, tight space electronics and mobile devices in a close future.

Microfluidics jets and cooling fundamentals - Synthetic jets are zero net mass flow devices that synthesize unsteady jet flow from an ambient fluid. They are useful for flow control, triggering turbulence in boundary layers, but especially for heat transfer applications [1,2,3,4,5]. In heat transfer applications, the cooling process is facilitated by direct impingement of vortex dipoles, created by a synthetic jet, on heated surfaces, or by employing a synthetic jet to enhance the performance of existing cooling circuits [6,7].

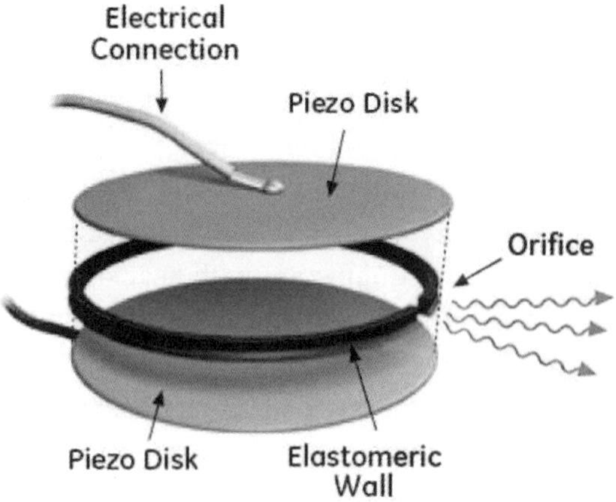

Fig. 1. The synthetic jet has a pair of piezoelectric disks that deform to change the volume of the cavity of air between them and force air in and out of the orifice [8].

A synthetic jet is created when a fluid is periodically forced back and forth through a tiny orifice in such a way that the net mass discharged through the orifice is zero, while generating a large positive momentum in the jet fluid stream. A typical synthetic jet configuration shown in Figure 1 has two piezoelectric actuators with a cavity of air in between. The disks are energized to reduce or expand the cavity volume at a low driving frequency in the 50 to 3000 Hz range to ingest and expel the surrounding air through an orifice. This frequency range is of particular interest due to the human sensitivity to noise that may result in due to jet operation. Effective heat transfer performance at lower perceived noise levels (<17 dBA) is critical to implement in practical applications.

Moreover, the resulting pulsating jet has positive momentum due to the asymmetry of the flow in and out of the orifice. The jet is directed on a surface to enhance the heat transfer. Though most of the research efforts spent at impingement heat transfer, synthetic jets can also augment heat transfer in a parallel flow configuration such as a tight channel flow. In addition, the unique feature of these jets is that they are formed entirely from the working fluid of the flow system. Unlike conventional jets, synthetic jets are "zero-net-mass-flux" in nature and produce fluid flow with no mass addition to the system and without the need for complex plumbing, an external fluid source as well as no sliding or rotating parts.

In summary, the application of synthetic jets for the active cooling of microelectronics is particularly compelling due to trends towards high heat densities as a result of increased functionality in small packages. Furthermore, the small size of these devices accompanied by the high exit air velocity provides an exciting opportunity to significantly reduce the size of thermal management hardware, which is the most desirable features of future electronics.

Background

Microfluidics technology has been instrumental for many science and technology fields. These small-scale dense turbulent jets have been routinely investigated for flow control applications [1-5], thrust vectoring of jets, triggering turbulence in boundary layers [6,7], and heat transfer applications [9,10,11]. Two closely related science branches, fluid flow and heat transfer, have been taking advantage of synthetic jets. Especially, the jets result in tremendous heat transfer performance in comparison with the natural convection cooling. Since they do not require an external fluid source as well as being passive with non-moving parts, these make the technology very attractive for a number of heat transfer applications. Typically, these devices are built in tightly sealed housings where space is at a premium. To find new heat transfer paths as a possible replacement for the natural convection heat transfer, many researchers in both academia and industry have presented different types of heat transfer studies using micro/meso scale devices like synthetic jets.

The first appearance of synthetic jets in heat transfer applications dates back to early 1980s. Gutmark *et al.* [12] presented the first heat transfer data on using synthetic jet for both natural convection and forced convection and revealed that the acoustically excited flow increases the overall heat-transfer coefficient. As a continuation of their earlier work, Yassour *et al.* [13] then published a set of synthetic jet heat transfer results. Their experimental study reported that the acoustic 'puffs' emanating from the orifice produced about 4x enhancement in the heat transfer. Coe *et al.* [14] generalized the above cooling concept by making use of actuation technologies (piezo-electric actuation) other than loudspeaker to create the synthetic jet action[15]. Since then active cooling with synthetic jet has been an active topic of research [7,9,11,15]. Garg *et al.* [16] performed an experimental study to characterize the cooling power of the jet, by contrasting its heat transfer capability under a given temperature difference, with that of natural convection, with the aid of an enhancement factor. Recently, Utturkar *et al.* [17] investigated the sensitivity of synthetic cooling performance to its operating conditions and alignment by experimental and computational strategies.

Synthetic jet technology is a strong candidate for the next generation tight space electronics thermal management applications. An interesting study of synthetic jets, for thermal management of electronic devices such as cell phones was presented by Minichielo *et al.* [18]. They used synthetic jet technology to promote active cooling in a cell phone mock-up. The authors observed heat transfer enhancement of 2-4 times over natural convection. A similar enhancement over natural convection in air was also observed by Gillespie [19].

Synthetic jets can also affect the liquid cooling performance. Synthetic jet enhancement of natural convection and pool boiling heat transfer was presented by Garg *et al.* in an enclosure filled with a dielectric liquid (FC-72) and the augmentation of convective thermal transport from the heated surface was found to peak around 300 Hz and 120 V_{RMS}, yielding nearly four-fold improvement over the natural convection. The highest enhancement of pool boiling was observed near boiling incipience, with the synthetic jet producing an earlier transition to nucleate boiling than encountered in a quiescent pool.

Synthetic Jets Design and Measurement Approaches

Numerical Modeling of a Synthetic Jet

Synthetic jet heat transfer strongly depends on fluid dynamics, specifically vortex generation. To provide insight into the flow structure, numerical codes are invaluable. Thus, a number of studies have been presented by using various commercial softwares. A commercial CFD software (Fluent) is employed to perform a two-dimensional, incompressible, time-dependent, Navier-Stokes computations over an idealized geometry [20]. The second-order upwind scheme is employed for the convective terms and central differencing is used for the viscous terms. The energy equation is solved in addition to the mass (continuity) and momentum equations. Time stepping is performed via the first-order implicit Euler discretization. The commonly used, non-iterative, operator-splitting technique PISO is selected for the pressure-velocity coupling to expedite the computational time without sacrificing the formal accuracy.

The schematic of this 2D synthetic jet model is depicted in Figure 2 . The jet is modeled by a 23 mm tall cavity. The motion of the side diaphragms is suitably modeled by a spatially varying velocity profile, as shown below in Eqs. (1) and (2).

$$u_{\text{left}} = A/2[1-\cos(2\pi y/D)]\sin(2\pi ft) \tag{1}$$

$$u_{\text{right}} = A/2[1-\cos(2\pi y/D)]\sin(2\pi ft+\pi) \tag{2}$$

Moreover, a CFD study with direct numerical simulations demonstrated turbulent flow transitions for synthetic jets operating around Re = 1150 and S = 17 [21]. It is also indicated the tendency of the flow to stabilize to laminar regime in the jet orifice. Thus, the flow regimes for synthetic jet actuators could range from being laminar in the orifice, to turbulent in the external environment.

Though use of RANS-based turbulence models appears as an intuitive option, the resultant high effective viscosity may damp the small vorticity scales in the jet orifices and blur the flow physics. Considering this theory, laminar simulations were investigated by two grids: 16300 nodes (40 across the orifice) and 26800 nodes (60 across the orifice) to provide a grid-independent solution.

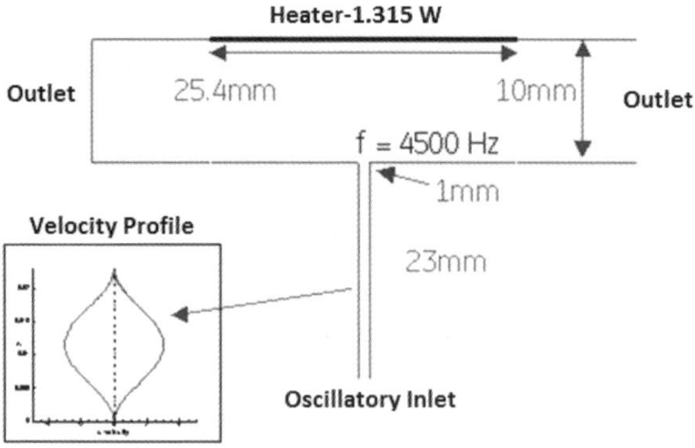

Fig. 2. Schematic of a 2D synthetic jet model.

The time-dependent results shown in Figure 3 not only present that they are fairly grid-independent, but also illustrate a reasonable match with the experimental measurements, where the average temperatures and average heat transfer coefficients, which are likewise calculated, agree within 5% between the experiments and simulations for both mesh densities. This agreement instills confidence in the computational procedure.

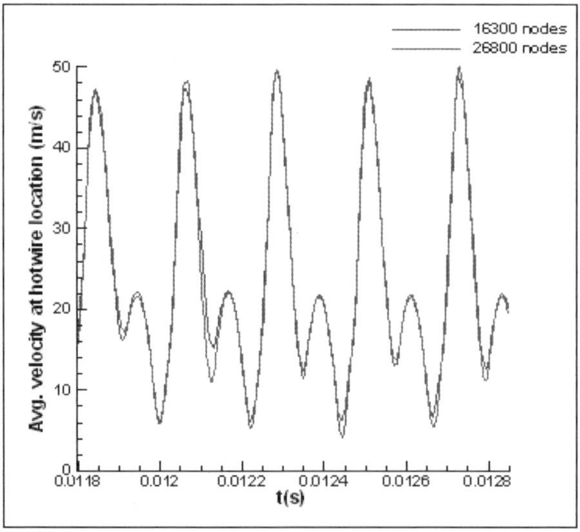

Fig. 3. Time-dependent computational velocity magnitude at the hotwire location.

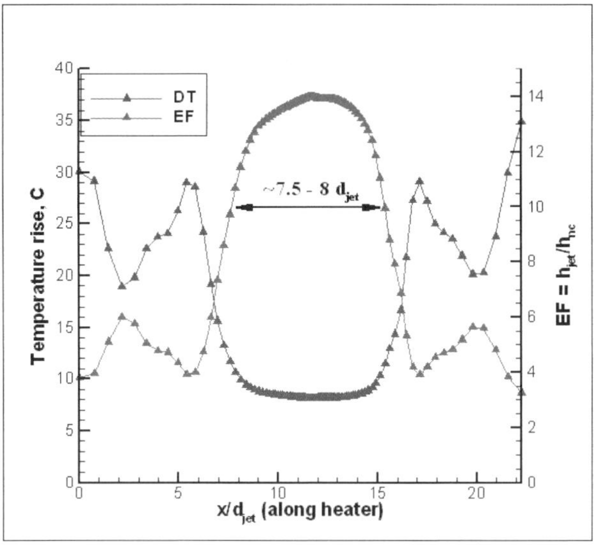

Fig. 4. Variation of temperature rise and heat transfer enhancement along the heater surface.

The results for overall natural convection heat loss experiment yielded a natural convection heat transfer coefficient equal to 18 W/m^2K at 40 C heater temperature. By using this value it is depicted as seen in Figure 4 that, though the overall enhancement on the heater is about 5X, the middle portion of the heater is subjected to a 14X enhanced heat transfer. Furthermore, almost eight jet diameter wide area on the heater experiences at least 10X heat transfer. This information is highly useful in sizing the complete thermal system.

Another special contribution has been made to the related topic by Arik *et al.* [22]. In this study, the computational domain is presented in Figure 5, synthetic jets had a fundamental frequency of around 600 Hz, where the disk displacement and the exit velocity reach their peak values. They were capable of producing periodic jet streams with peak velocities in excess of 20 times the air velocities caused by comparable size conventional fans [23]. Any further increase in the frequency has expected to result in reduced cooling efficiency and increased power consumption.

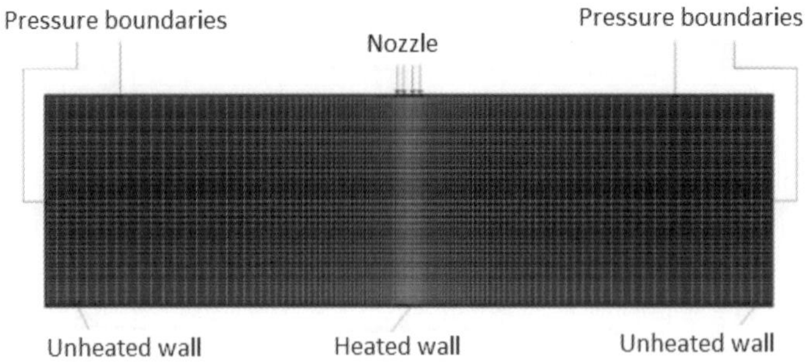

Fig. 5. Computational domain and grid.

Thus, as an output of that application, it was found out that synthetic jets can provide more than 20% better cooling than a steady impinging jet over vertical surfaces. It was also shown that vortex generated synthetic jet cooling was capable of providing much higher performance than a comparable steady jet for the same jet exit velocity and orifice size. Thus, based on the published literature, these jets enhance the heat transfer in both natural and forced convection significantly. For recognizing the heat transfer physics over surfaces require a fundamental understanding of the flow physics caused by pulsating coolant flow, this computational study provided a further understanding for the flow physics of a small scale synthetic jet. A second-order temporal implicit scheme was used for the unsteady terms to avoid stability issues.

Consequently, although some of the negligible discrepancy between numerical and experimental studies, overall findings from a number of papers supports that 2-4 times higher heat dissipated in case of synthetic jet augmentation. Moreover, increasing heat transfer coefficient with the jet driving voltage and the peak cooling performances of synthetic jets at their resonance frequencies were also in agreement given different research studies.

Heat Transfer Experiments

While a number of experiments on synthetic jets have been proposed in the literature, first experimental studies have been on simple-thin foil heaters to measure localized surface temperatures and the heat transfer to the ambient via synthetic jet cooling. To minimize ambient air drifts to the experimental rig a large plexiglas enclosure was employed. Thin foil heaters sandwiched between thin aluminum backing plates were used in the experimental study. While the jet was impinging on one of the heater surface, the temperature measurements were performed from the back surface using an infrared (IR) camera.

The emissivity of the surface was measured by using the same methodology given in Garg *et al.* [16]. Steady state temperature profiles were captured by using an IR camera and also appropriate lenses were used to achieve a minimum pixel size. However, while IR measurements provided local temperature gradients and heat transfer information, a more robust and automated experimental setup has been necessary to accelerate the technology development. Thus, a fully automated experimental test rig was designed and built to measure the synthetic jet heat transfer performance on which the operating conditions (voltage, frequency, heater surface temperatures, etc.) were all programmed into a Data Acquisition System (DAQ) and series of tests were performed for various jet operating conditions and geometrical variables. The setup, as shown in see Figure 6, was composed of enclosure, transverse mechanism, function generator, signal amplifier, laser vibrometer, and heat source, DC power supply to energize the heater and DAQ system.

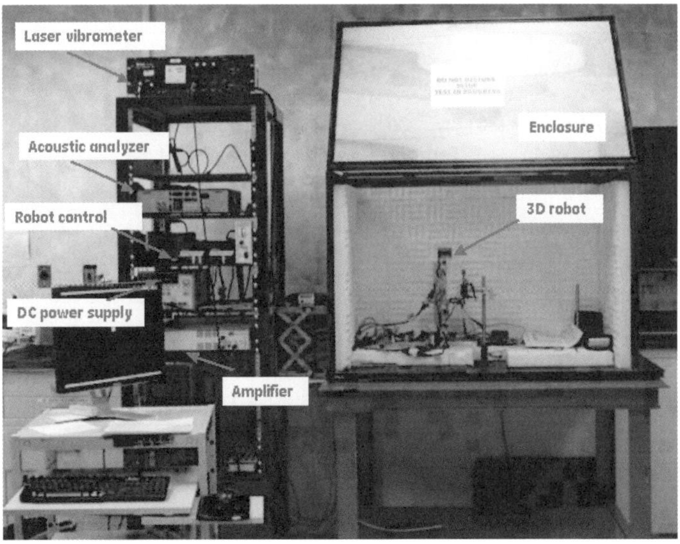

Fig. 6. Synthetic jet test rig [23].

A laser vibrometer was used to measure the piezo actuator displacement, which is an important parameter for the understanding of fluid-solid interactions. A lab windows program was used to control the equipments and the test conditions. Also T-type thermocouples was used inside the housing of the heater, and three additional T-type thermocouples were also installed on this plastic substrate to quantify the heat losses. Thereafter, a series of tests were conducted on the heater assembly to measure the heat losses and develop a heat loss equation for the system. The heat loss test results, as shown in Figure 7, were inputted to the post processing algorithm so that the net cooling effect of the synthetic jets can be accurately computed along with other parameters such as coefficient of performance (COP).

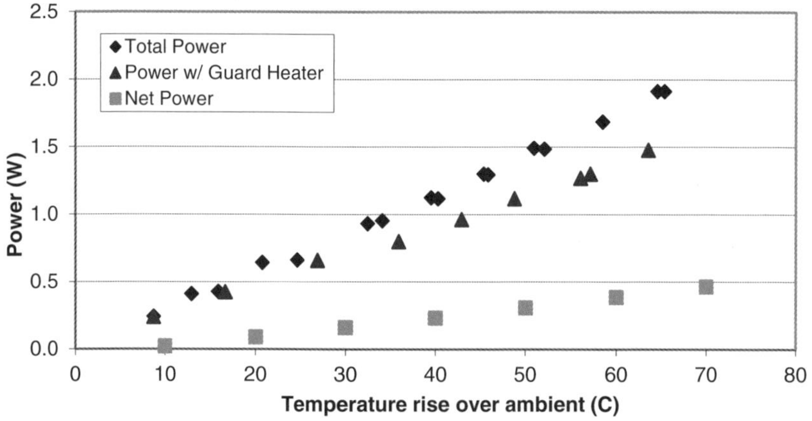

Fig. 7. Heater temperature rise with heat rejection.

Moreover, the impact of the driving frequency, voltage, and jet-to-heater spacing on the heat transfer and jet power consumption were also studied. In the first set of heat transfer augmentation experiments conducted, the power to the heater was set to such a level that the heater temperature was in the vicinity of 80 C and the total power input to the heater for this setting was 1.56 W. After the system reached steady state under natural convection conditions, the infrared camera (IR) temperature distributions were recorded. The axial distance between the jet and the heater varied between 5 and 50 mm, was set to the desired location.

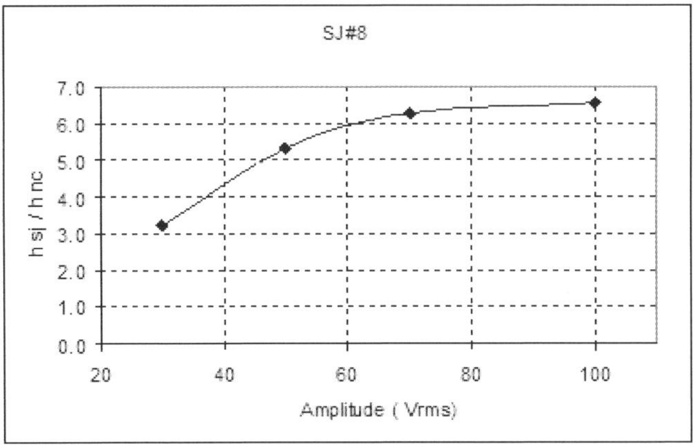

Fig. 8. Effect of driving voltage on heat transfer enhancement.

The consequences shown in Figure 8 and Figure 9 have presented the effect of driving voltage and the axial distance over the heat transfer. As seen from the figure, the increased driving voltage does not increase the heat transfer at the same rate as the jet power consumption. This is particularly important because it might cause degradation of coefficient of performance, and coarsen the advantage of the higher enhancement factors obtained at higher voltages.

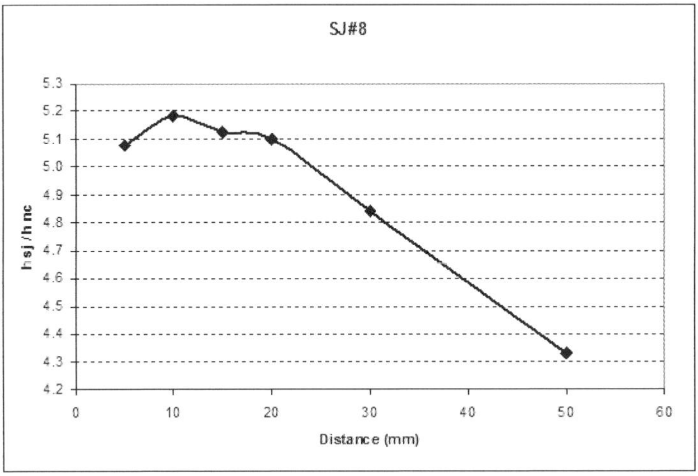

Fig. 9. Effect of axial distance on heat transfer enhancement.

Moreover, the placement of the jet from the heater was also seen as another important criteria, for which an optimum placement should be set. In an ideal practical system, the jet should not consume a large footprint area and volume. Therefore, it should be placed as close as possible as given in Figure 9. There are several reasons for an optimum spacing to provide a better heat transfer performance. First of all, at low spacings, the area of the heater being cooled by impingement is much smaller. This implies that the region of high heat transfer coefficient is smaller at smaller spacing, leading to a lower overall heat transfer coefficient. This in turn leads to a higher average heater temperature. Another important aspect is the coolant temperature. The air exiting the jet during the exhaust cycle is hotter than the ambient, mainly because of the heat generated within the jet itself. When the jet is placed very close to the heat source, the temperature of the impinging air is close to the temperature of the air exiting from the jet. When the distance between the jet and the heater is increased, ambient air gets entrained, causing the temperature of the air impinging upon the heater to decrease substantially [16].

Velocity Measurements

Similar to other convective heat transfer applications, the amount of heat transferred strongly depends on the local air velocity. Therefore, it is essential to characterize the exiting air jet velocities from the synthetic jets at various operational and geometrical conditions. Thus, in order to capture both the suction and the ejection velocities, a hotwire probe was placed close to the jet orifice [16]. Then, the jet velocity was measured by means of a TSI IFA 100 hot wire anemometry system.

A computer code was used to separately average the ejection and suction peaks as shown in Figure 10. The average exhaust and average suction velocities were calculated. But, due to the single-wire probe type, no directional output can be obtained; thus only the local velocity magnitude was possible to tap. On the other hand, the exhaust velocity was found to be increasing non-linearly with the driving voltage. Specifically, the peak exhaust jet velocity tends to an increase as the driving voltage is increased.

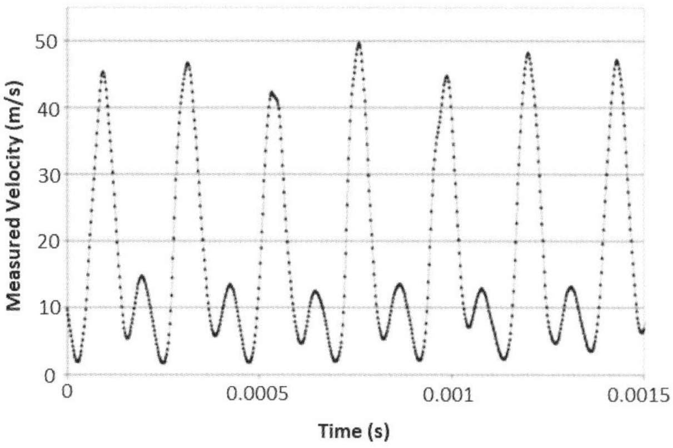

Fig. 10. Typical velocity response tapped by the hotwire (50VRMS and 4500 Hz) [16].

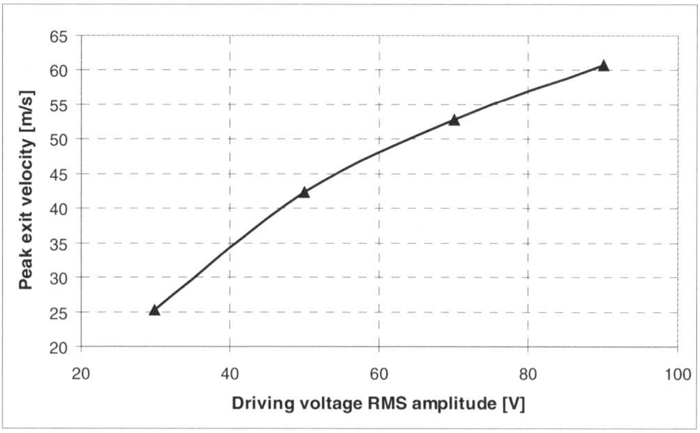

Fig. 11. Variation of peak jet exit velocity with driving voltage amplitude.

Figure 11 presents the trend between the jet exit velocity and the heat transfer enhancement. The driving voltage was kept at 50 V, while the harmonic signal was swept between 2-5.5 kHz. The definition of enhancement factor is given in Eq. (3) [16].

$$\mathrm{EF} = \frac{h_{\mathrm{jet}}}{h_{\mathrm{nc}}} = \left.\frac{q_{\mathrm{jet}}}{q_{\mathrm{nc}}}\right|_{(T_{\mathrm{s}} - T_{\mathrm{air}})} \tag{3}$$

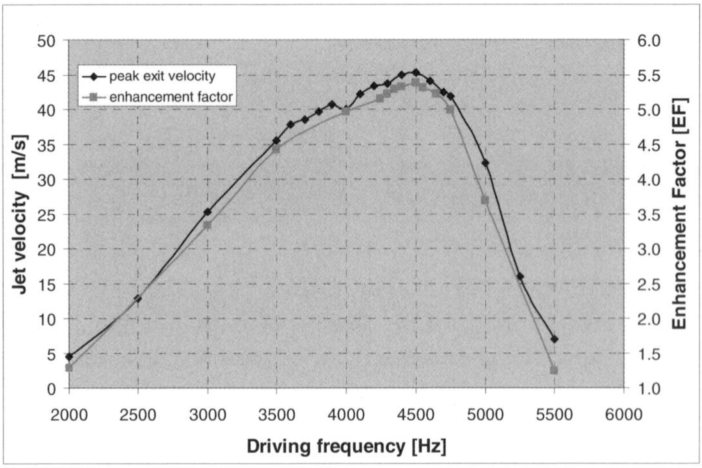

Fig. 12. Effect of driving frequency on heat transfer enhancement and peak jet exit velocity.

As seen from the Figure 12, an optimal air velocity of about 45 m/s is observed at 4.5 kHz, which is the natural frequency (or Helmholtz frequency) of the jet. The best heat transfer performance (5.5x based on Eq. (3) is also obtained at that resonant frequency.

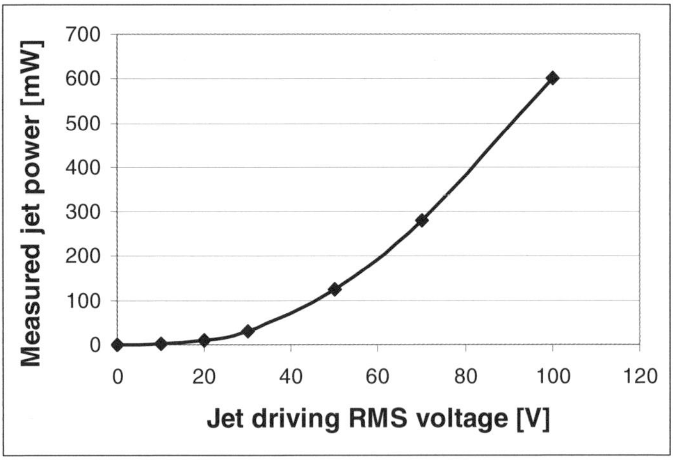

Fig. 13. Synthetic power consumption with driving voltage.

However the power consumption in an electronics cooling system is also very important since it directly affects the coefficient of performance (COP) of the cooling solution. Therefore, a careful study was carried out to obtain the true power consumption of the synthetic jets. The synthetic jet power measurements were performed under different driving voltages. In addition to the basic synthetic jet heat transfer measurement setup, a 1000 Ohm serial resistor was added to increase the accuracy of the current measurements. Voltage measurements were done before and after the resistance using a high-frequency differential probe (Tektronix 5205, 100 Hz). By using a known resistance and voltage values, electrical current for the synthetic jet was calculated. Another high-frequency probe measured the exact driving voltage of the synthetic jets.

Effect of the driving voltage on the jet power consumption is depicted in Figure 13. It is interesting to note the steady increase of power consumption over 30 V of driving voltage. While the power consumption follows a steady trend, the jet exit velocity does not show the same steady trend over 30 V. This indicates that the jet may be less efficient at higher voltages (greater than 30 V).

The effect of the driving frequency over the jet power consumption is seen in Figure 14. The jet was driven at 50 V constant voltage supply, and the frequency was swept between 2000 and 6000 Hz. This range encompasses the structural and the Helmholtz natural frequencies of the jet. The jet showed maximum power consumption at the resonance frequency of 4500 Hz. Before and after the resonance frequency, it showed lower power consumption. This is particularly significant because of the capacitive effect and the fluid-structural coupling of the jet.

As a result, it can be interpreted from Figure 12 through Figure 14 that one can drive synthetic jets between 40-100 mW and get about 4-6 times heat transfer enhancement. Furthermore, based on the consumed jet power and the measured heater input power, a coefficient of performance of about 10 is estimated for the synthetic jet actuators in the present study.

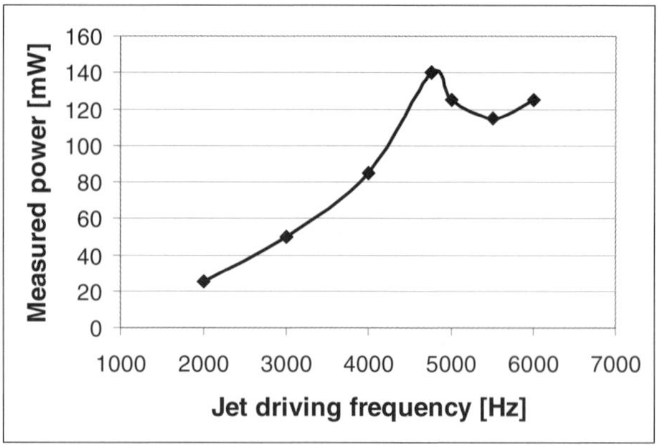

Fig. 14. Synthetic jet power consumption with the frequency.

Pressure Measurements

There are strong interactions between the air and the jet structure that effect the operation and performance of a synthetic jet. As the pressure starts to build during the intake stroke of the jet, the stress on the piezo disks starts to rise and structural forces start to influence the internal pressure and hence the exit air velocity [8]. During the exhaust stroke, the disks start to push the air out from a relatively small orifice and realizing the pressure built up inside the jet cavity. Since, the synthetic jet exit velocity strongly depends on the internal pressure created by the actuators; an experimental setup was designed for measuring jet internal pressure with a special synthetic jet having only a single piezo actuator disk jet and half the elastomeric wall material thickness as seen in Figure 15.

Then, to create the internal cavity the half jet was mounted on a substrate with a pressure sensor flush mounted at the center of the substrate. Thus, the internal pressure was measured at the center of the disc cavity. Due to its high sensitivity at very low pressures (0-1psig) Endevco piezo resistive pressure transducer 8510B-1 was selected. The output voltage signal was also acquired with the same DAQ system as used for the velocity and the heat transfer measurements. The same DAQ unit also collected data from laser vibrometer so that piezo-actuator displacement and internal pressure measurements are taken simultaneously.

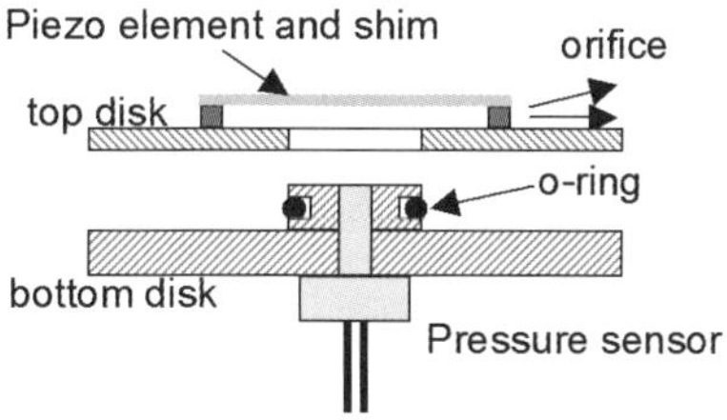

Fig. 15. Illustration of pressure measurement setup [8].

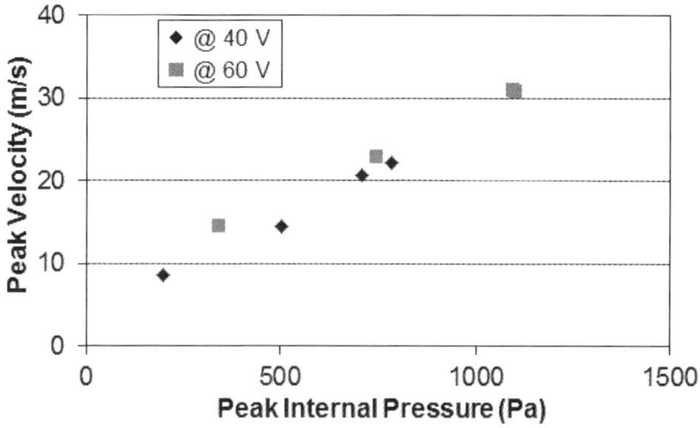

Fig. 16. Variation of synthetic jet peak exit velocities versus peak internal pressure for an 8 mm orifice.

As a result overall obtained data confirmed that since the force generated by the piezo actuator drives the jet internal pressure, there are internal losses in the jet cavity as well as orifice, while ejecting the high velocity of air. Also Figure 16 presented a variation of synthetic jet peak exit velocities with peak internal pressure for 8 mm orifice and a linear dependence of jet exit velocity with the frequency is observed.

Enhancing Heat Transfer with Synthetic Jets

Natural Convection Heat Transfer with Synthetic Jets

Natural convection air cooling is the method of choice for many low-power electronics applications due to cost, availability, and reliability considerations. This method is not only limited to low-power applications, but is also constrained by the buoyancy dependence of the flow. With the recent advancements in electronics technology leading to thinner, lighter, faster, and higher functionality products [23,24], significant heat dissipation in tight thermal real estate becomes unavoidable. A number of innovative cooling techniques [25,26] for low-cost, compact, and reliable systems have been considered and the demand for more powerful cooling systems, however, has led engineers to look for unconventional methods that further expand the limits of natural air cooling. One of these methods is the use of synthetic jets in passively or actively cooled electronic systems.

In the case of synthetic jets heat transfer applications, the cooling process can either be facilitated by direct impingement of vortex-dipoles on heated surfaces [15], or via employing the jets to enhance the performance of existing cooling systems [22]. These active cooling devices are particularly well suited not only for small-scale electronics, such as printed circuit board components, but also for others, where natural convection or pure use of fans is not practical and not sufficient due to limitations. Thus a heat transfer investigation as an enhancing synthetic jets embedded in a heat sink will be shown below with the findings, and the pertaining discussions.

Enhancing Heat Sink Thermal Performance with Embedded Synthetic Jets

Flow through a small opening (several millimeters to centimeters) led people to think that synthetic jets can be used only for local cooling and for a majority of time they are thought for extending limits of natural convection. However, ability to have many openings and a number of jets can be synchronized enabled advancing heat sink cooling. Thus, a number of studies have been published in the open literature. The effects of number of the jets, locations, arrangements, and operating ranges on heat transfer from a cylindrical heat sink were investigated both numerically and experimentally. CFD simulations were utilized to provide fundamental understanding to the fluid flow and airflow interaction with the heat sink fins, whereas the experiments were conducted for validation of CFD models and to further thermal investigation of jets.

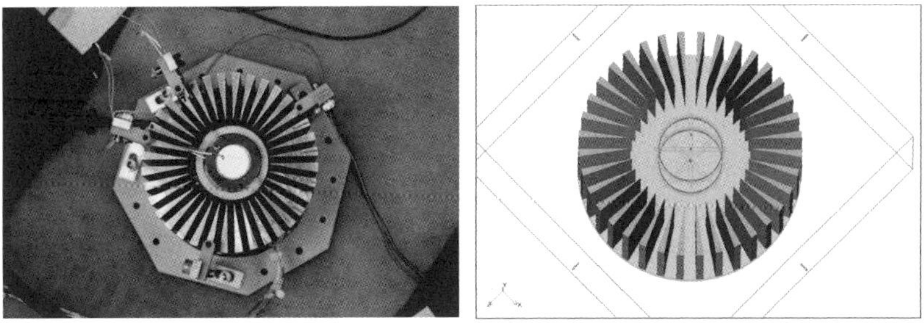

Fig. 17. Idealized CFD model and actual heat sink with s jets [27].

A synthetic embedded heat sink design was investigated in detail. Two synthetic jets mounted at the center of the heat sink equally spaced along the height of the sink. Each jet has 4 orifices, and the jets are arranged such that they have a 45-degree angular offset. The heat sink has 36 triangular fins, which have been optimized for maximum heat removal. The base of heat sink is maintained at a constant temperature and the heat dissipated from the sink is calculated as a part of the solution. As a baseline to quantifying the enhancement of the jet embedded heat sink, a conventional heat sink with plate fins, having the same overall dimensions, is considered. Naturally, the fin area for the baseline heat sink is 57% more than the jet embedded heat sink due to the availability of additional space that is occupied by the jets.

Computational (CFD) and experimental studies were independently performed for both the baseline and the jet embedded heat sink to obtain the cooling enhancements. In the numerical study, commercially available software packages (Icepak and Fluent) were used for the numerical computations. The jet movement was created using time-dependent velocity boundary conditions on the surfaces representing the jet disks. Second-order upwind scheme was used for the convection terms, while central differencing was employed for the diffusion terms. In general, 20 mesh points were employed to resolve the orifice flow and 10 mesh points were used for the fin gaps. The Pressure Implicit Splitting of Operators (PISO) methodology was employed for the pressure-velocity coupling and non-iterative time advancement.

As a result of numerical study, dissipated heat by the baseline heat sink under natural convection at two different base temperatures was illustrated in Figure 18. Note that the thermal resistance of the heat sink is respectively equal to 2.0 K/W and 1.7 C/W, at 60 C and 80 C base temperature. The average heat transfer coefficient (natural convection and radiation) over the fin surfaces, when the base is held at 80C, is estimated from the computations as 6.8 W/m^2-K.

Following the natural convection calculations, computations were performed under the activated jet conditions over a frequency range of 500 – 800 Hz at 60 V. The surface heat transfer coefficient distribution over the fin surfaces was depicted in for the 700 Hz and 60 V condition. Note that the fins that lie exactly in the wake of the orifices showed local enhancements of 4X the heat transfer coefficient. The overall enhancement of surface convection was noted around 2.6X for the chosen driving condition of the jet. This overall enhancement in heat transfer, while compensating for the loss of fin area due to the jets, additionally leads to an augmentation in cooling. The result was a thermal resistance of 0.8 C/W at 80 C base temperature, almost 2x better than the natural convection case.

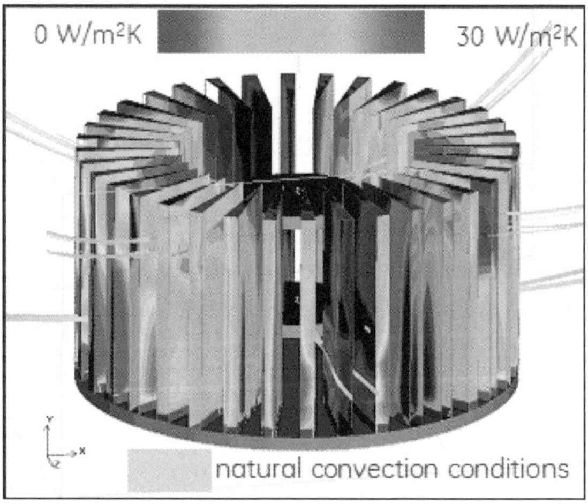

Fig. 18. Heat Transfer coefficient over synthetic jet embedded heat sink fins [27].

However, an experimental investigation was also carried out to validate single jet computational results and to evaluate thermal performance of multiple jets in the embedded heat sink design. For this purpose an experimental setup was designed and fabricated. Various combinations of number of jets, number of orifices and operating conditions were evaluated. The heat was applied to the heat sink via a Kapton film heater, which was mounted on the base of the heat sink. Synthetic jets were mounted onto the central cavity via plastic holders, which were attached to on the insulating substrate. Special consideration was given to the design of these holders such that they would not affect the air jet exiting the jet orifices. Required driving conditions for the synthetic jets were controlled using a signal generator and an amplifier. Heat input and temperature data were recorded via Agilent data acquisition system. The characterization tests revealed that natural frequencies of 4-port and 8-port jets were 600 Hz and 800 Hz, respectively. As previously stated, thermal performances of multiple jets embedded in the heat sink were evaluated in the experimental part of this study. The jets were evenly distributed and 10 mm apart from each other in the vertical direction for experimental investigation of multiple jets in the embedded heat sink concept a series of experiments are performed on 4-port and 8-port synthetic jets.

First, thermal performances of 4 and 8-port jets with increased number of jets were compared at their natural frequencies as shown in Figure 19. It was found that 8-port jets outperform 4-port jets by as high as 50%. This greater performance was attributed to the fact that net effected fin surface area by the air jet created by the synthetic jets was larger for 8-port jets than that of 4-port jets. There was more “wetted” fin area with 8-ports than 4-port jets. It was further observed that as the number of synthetic jets increased from 1 to 4, the resulting thermal resistance approaches asymptotic values. There was no additional performance gain by operating more jets.

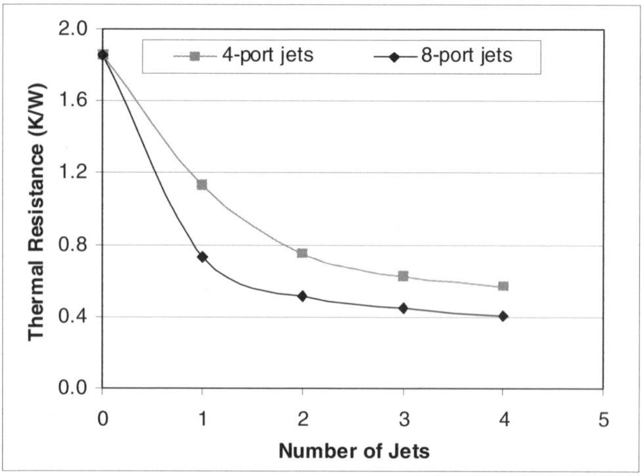

Fig. 19. Thermal resistance comparison of 4 and 8-port jets.

On the other hand, the enhancements in the heat removal from the heat sink using synthetic jets are summarized in Figure 20 and Figure 21 for 4-port and 8-port jets, respectively. The total cooling capability can be improved by a factor of 4.5 with respect to baseline, free convection heat sink. Furthermore, the effect of orientation of jet orifices on enhancement factor is investigated. For this purpose, orifices of even numbered jets are rotated by 45 degree for 4-port jets and 22.5 degree for 8-port jets.

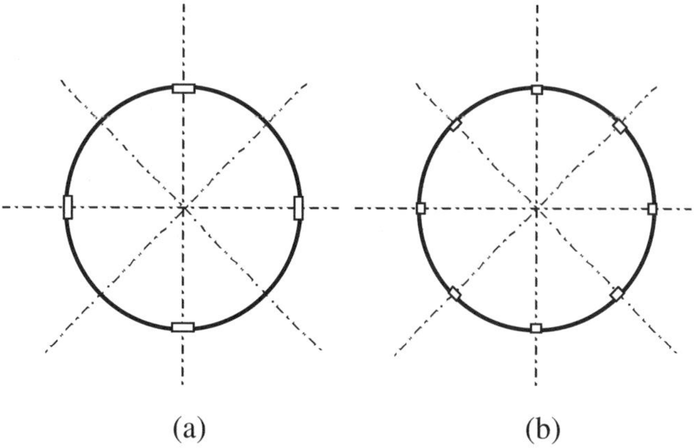

Fig. 20. (a) 4-port jet with 3 mm orifices and (b) 8-port jet with 1.5 mm orifices.

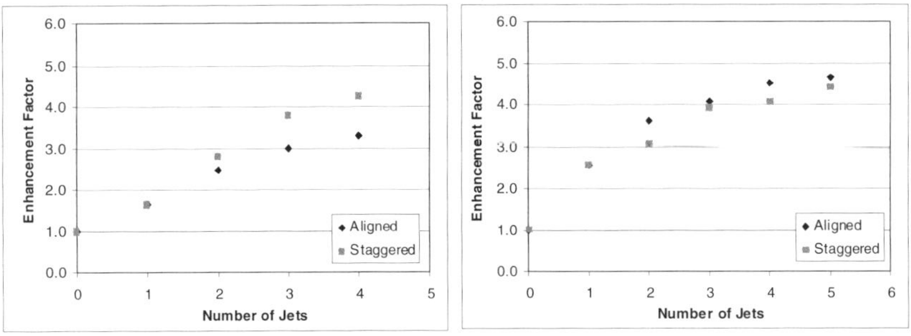

(a) (b)

Fig. 21. Enhancement factors measured with (a) 4-port jets (b) 8-port jets.

The motivation was simply maximizing the total wetted fin surface area. The results revealed that the staggered orientation of orifices increased the thermal performance by up to 30% on 4-port synthetic jets. However, the same affect could not be seen on 8-port jets. This suggests that the air stream from the eight orifices effectively distributes the flow across most of the fins such that staggered orifices make no further improvement in the thermal performance.

Thus, an experimental and computational study has been performed to justify the use of synthetic embedded heat sinks to meet the ever-rising cooling demands in the electronic industry. A baseline heat with plated fins has been compared to a heat sink embedded with two synthetic jets. The base of the heat sinks has been held at a constant temperature, and the power dissipated by the heat sink under varying operating conditions of the jet has been reported. The thermal resistance of the integrated heat sink is 0.8 C/W at 80 C base temperature, almost 2X better than the natural convection case. Furthermore, it is found that 8-port jets outperform 4-port jets by as high as 50% within the embedded heat sink.

Forced Convection Heat Transfer with Synthetic Jets

Synthetic jets have been considered as local cooling devices and mostly proposed as an amendment for natural convection. However, some researchers have considered for enhancing forced convection heat transfer. Utturkar *et al.* [28] presented an experimental and computational study to show the forced connection enhancement with synthetic jets. Figure 22 portrays the computational domain. The synthetic jet – denoted by green boundary – was located 2mm above and 1.5mm away from the heater leading edge. The disk movement was simulated by moving boundary and spring-based mesh smoothing techniques. Note that since the airflow surrounding the synthetic jet was solved as a part of the solution, the heat transfer condition on the synthetic jet disk did not need to be artificially stipulated, and was coupled with the fluid flow solution.

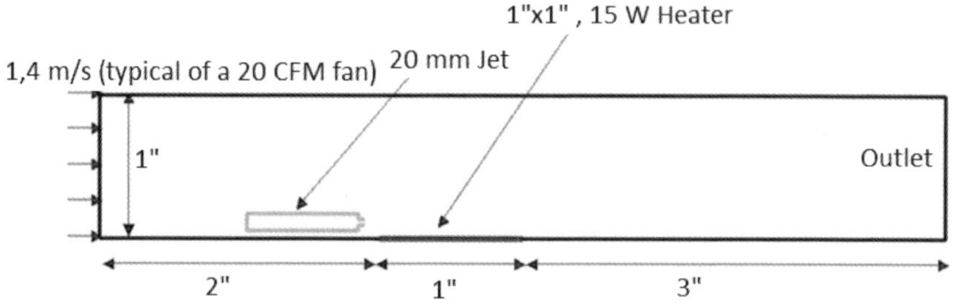

Fig. 22. Schematic of the computational model [28].

However, the upper and lower walls in the above model were adiabatic, and the heater was modeled as a heat flux patch. The displacement amplitude, *A*, was determined by converting the disk displacement and measured by laser vibrometer in the experimental setup, into an equivalent, volume-conserving, 2D amplitude. This corresponded to an experimental driving condition of 400 Hz and 90 V_{RMS} for a 20mm jet. In the stated study, the synthetic jet blows air with velocity of about 10-15m/s in the direction of the bulk flow. Note that the jet velocity was about 10x the fan velocity and this jet of air was able to penetrate through the heater's thermal boundary layer, thus yielding a greater heat transfer rate. Additionally, the dissipated heat was convected away by the fan's bulk flow, and was prevented from accumulating in the heater vicinity. Thus, the combination of a fan and the local effect of a synthetic jet were able to yield significant benefits over forced convection and up to 4x increase in the surface heat transfer coefficient.

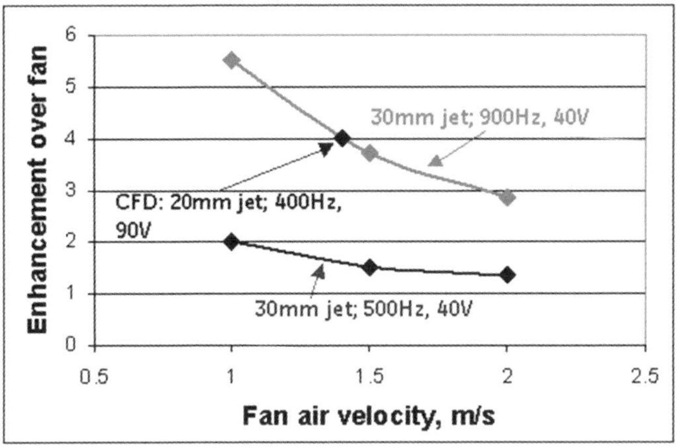

Fig. 23. Enhancement of synthetic jet vs. fan flow speed.

However, an experimental test rig was also designed to directly assess the CFD findings. Thus, a small wind tunnel was designed with a fan and honeycomb flow straightener at one end. Channel walls were made up of 12 mm wall thickness. Channel width and height were both 100 mm creating a square geometry. The fan was about 30 mm away from the heater surface. A square heater, 25.4 mm in length and 1 mm thickness, was aligned along one of the channel walls, and a synthetic jet was placed parallel to the wall such that its jet was in the direction of the fan flow.

As a result, the jet was operated at different operating conditions, while the fan velocity was gradually increased. Naturally, the enhancement values reduced as the fan air speed was increased. This was because the difference between the fan velocities and the local jet velocities reduces with increasing fan speeds. Computational results were able to help to clarify the findings pointed by the experiments. The jet's local flow was able to disrupt the thermal boundary layer over a heated body thus dramatically increasing the heat transfer rates.

Synthetic Jets in Practical Applications

Air cooling methods are still consumer-friendly systems because of their simplicity and ease of implementation (Bar-Cohen [29]). Another challenge in electronics cooling is the packaging of compact systems such as portables, where in many cases there is insufficient place to even use fans, so that synthetic jets have been studied as an alternative to fans as air moving devices and have been shown to be highly effective for cooling of electronics in a very small form factor. Although, many of these applications have been shortly mentioned before, below some of the related applications will be illustrated briefly.

Firstly, the modeling results for prediction of thermal performance using synthetic jets ejectors studied by Raghav Mahalingam [30] at Nuventix, Inc. will be illustrated. In the related research, a 2-D synthetic jet synthesized directly from the fluid and containing a jet ejector consists of a primary high momentum jet inducing a secondary flow within a channel, has been used as the primary jet causing secondary flow to be induced in a channel. Within the channel, the flow entrainment prediction for the model was based on the solution of mass and momentum equations. A resistance network model was also used to predict the thermal performance. Thereafter the modeling results were compared with the previous measurements taken from the past tests of synthetic jet cooling within channels.

The experiments were performed in a channel that was fabricated of two heated walls besides having a variable distance in order to change the channel width and aspect ratio as shown in Figure 24. The synthetic jets were placed alongside the walls and each jet spans the entire height of the channel.

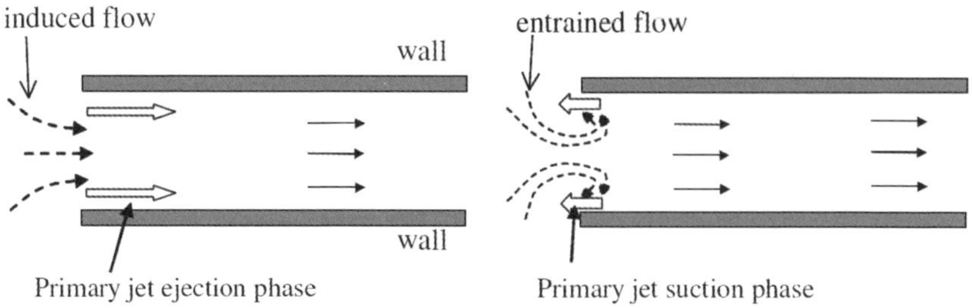

Fig. 24. Basic principle of operation of a synthetic jet ejector.

While, the channel walls were made up of an aluminum plate, they were heated with a flat foil heater that provided a constant heat flux along the wall. The wall length and height were fixed resulting in varying channel width and aspect ratio. Backside of the heated surface, the narrow top and bottom surfaces were coated with high temperature machinable glass ceramic with embedded air pockets. As used before in different applications, T-type thermocouples were monitored the temperature at several points.

The experimental study has resulted that the volumetric flow rate was accurately calculated based on the centerline velocity and exit area of the channel with the experimental points being well within 10% of the predicted data points. The heat transfer coefficient was slightly under-predicted with an average deviation from the measurements of about 15%. While the measured and computed Nusselt numbers have been compared very well, it was clear that the turbulent, pulsating flow created by the Synthetic jet ejector results in significantly higher Nusselt numbers than a hydrodynamically similar steady channel flow condition. The synthetic jet driven flow resulted inNusselt numbers that are on an average of 2.5 times of the steady flow, with higher improvement for lower Reynolds numbers. The heat dissipation was calculated by measuring the airflow rate at the end of the channels as well as the air temperature rise from the inlet [30].

In addition to the experiments, for the prediction of the overall thermal performance of the jet flow within the channels required calculation of the induced flow rate, and the calculation of the heat transfer coefficient followed by addition of the component resistances to obtain the overall thermal resistance. With the help of mass and momentum conservation equations the jet ejector model predicted the flow rate within the channel. The jet velocity used for the calculations was based on measurements of the synthetic jet prior to mounting in the channel. The data for the jet velocity calculations based on measurements of the synthetic jet in the channel has showed that the jets decay rapidly from about 10 to 3 m/s within 40 mm from the jet exit.

As a result of this investigation, where the operation of a synthetic jet ejector is described, a model developed to predict the cooling performance of a synthetic jet ejector was validated using a simple configuration of a 2-D synthetic jet ejector in a rectangular channel. The effect of channel spacing on the induced flow rate, heat transfer coefficient, thermal resistance and overall power dissipated was modeled. All the primary quantities required to describe the performance of a synthetic jet ejector in a cooling environment were predicted within $\pm 15\%$ of the experimental values. Measurements taken from experiments and predictions from computational studies, Nusselt numbers for a synthetic jet driven flow were observed on average about 2.5 times that of a steady, fully-developed turbulent flow. Moreover, by using the validated computational model, the performance of a synthetic jet ejector heat sink was predicted.

Another application has been developed by Arik *et al.* on the development of a high lumen solid state down light with LED technology [31]. The investigation is mainly focused on the development of an LED based high lumen luminaire technology enabled by novel thermal management, LED packaging, and driver electronics under the consideration of the critical subcomponents of the luminaire.

It is well known that high LED junction temperatures can result in LED degradation, because the LED chips and driver electronics performances are highly temperature dependent. Therefore, thermal design is critical for optimal performance and reliability of the LED based luminaires. Thus, the development and widespread use of high brightness LEDs and the application to the lighting industry require the development of advanced heat management systems to ensure the integrity of the LEDs and the electronics that drive them. As a response to this requirement, passive cooling with conventional aluminum heat sinks, thermoelectric or synthetic jets have been proposed by several groups [32].

Passive cooling systems account for the majority of the LED luminaire cooling solutions. However, in high lumen applications their use may be limited by size and weight constraints. Liu *et al.* proposed and tested a closed micro jet array to maintain a low chip junction temperature [33]. During this study, conventional heat management methods were evaluated such as natural convection, a heat sink and a heat pipe and the results were compared to the micro jet array cooling. It was found that the jets provided superior cooling and the junction temperature was cooler due to the use of micro jet active cooling. However, although the micro jet system provides better cooling, issues pertaining to cost and reliability need to be addressed.

Thus, a 1500 lm, 21 W, 6" downlight was analyzed using computational models, as a baseline design, to understand the thermal resistance breakdown from the substrate to the ambient/attic air, and subsequently highlight trends for varying lamp power, can size, and LED count. It is clearly shown that, the only way of obtaining a reduction in the LED count, without any advanced thermal management is by increasing the can size, which is an unattractive and unacceptable trend for the lamp design.

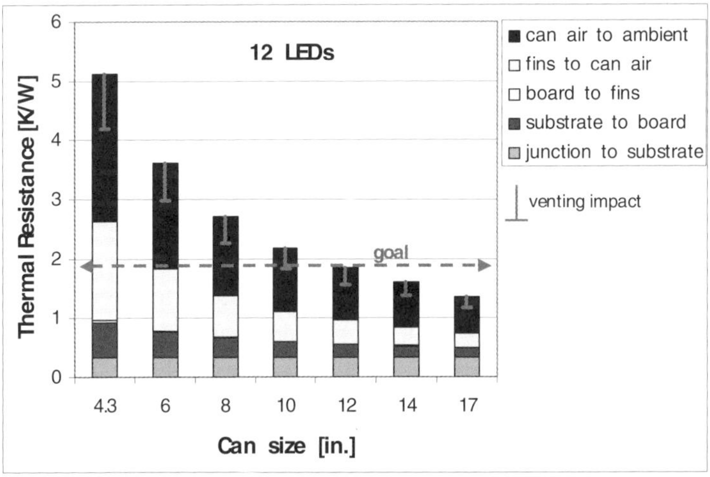

Fig. 25. Thermal resistance trend with increasing can size for a 1500 lm lamp using 12 LEDs.

Moreover, an experimental testing was performed on a surrogate 6" downlight in a simulated ceiling environment to validate the predictions made before. The test rig comprises a 6" downlight can, wrapped around with 1.5" thick insulation, and elevated 5' above the ground level by using a tripod. A heat sink integrated with an air mover is used for thermal management. Thermocouples are instrumented at various key locations to measure the temperatures at relevant points along the thermal chain. The base of the heat sink is artificially heated using a Kapton based heater. The heat sink is attached to an insulated circular plate (k < 0.2 W/m-K) having the same area at the 6" can cross-section. Experiments are performed in a complete air blockage condition (no air exchange between the can and ambient) and by drilling circular vents on the plastic board near the circumferential region.

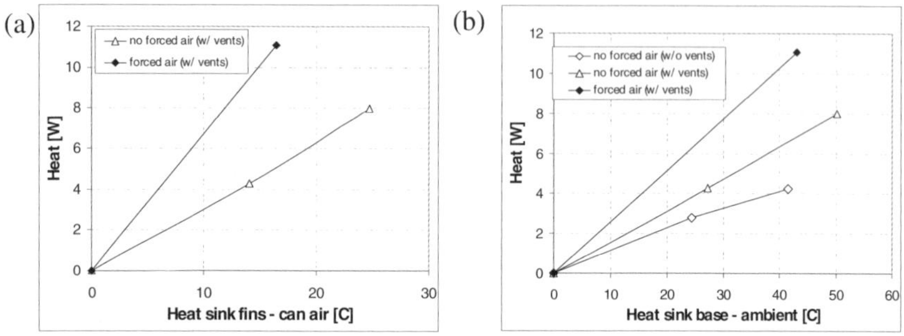

Fig. 26. (a) Heat sink base to ambient temperature drop for various scenarios (b) Heat sink base to can air temperature drop for different scenarios.

Experiments were performed at different heating loads under varying conditions of venting and forced air circulation (Figure 26). The heat sink base undergoes a rise of 40 C above ambient even with forced cooling, at 11 W of heat. Although venting and forced convection cause an increase in the cooling level, it is not adequate enough to meet the 21 W heat removal requirement. The temperature rise of the heat sink base above ambient at 11 W is 41 C, while that between the fins and can air is 16 C. The lack of air replenishment adds a substantial limitation to the thermal resistance (Figure 26).

The use of an advanced air mover integrated within a heat sink can , however, lead to radical improvements to the lamp design. Specifically, a 2X enhancement for such a heat sink will allow reaching a can size of 6" and a 50% reduction in the chip count. Using that condition as an intermediate goal, further technological improvements to the thermal management could ultimately reduce the chip count by 66% and the can volume by 40-50% of the baseline condition (high lumen]) Moreover, synthetic jets designed with piezoelectric disks have demonstrated cooling enhancements (EF) of at least 10 with low cost components and a simple design. Therefore, the lifetime of the luminaire depends on the optics (LED light engine and other optical components), the active cooling solution (synthetic jets) and the driver electronics for the LEDs as well as the synthetic jets. Continuing with the proposed hierarchical scheme, the junction temperature is controlled by the actively cooling mechanism (i.e., the synthetic jets). Apart from the synthetic jet degradation over time, the degradation of the electronics can adversely impact the driving conditions for the synthetic jets, thereby resulting in an increase in the junction temperature.

As a result, a novel luminaire design approach with thermal, light engine, driver electronics technologies has been presented for a 100W incandescent replacement lamp. The number of LEDs in the luminaire is certainly a major driver for the cost of the luminaire. It is critical to have the lowest possible number of LEDs so that the product can be affordable. In addition, different sub components must interact with each other seamlessly for the lifetime of the luminaire (>50,000 hrs) and is critical for the SSL product. The reliability of the luminaire had not been proven, but system level modeling models are progressing. We believe that active cooling is necessary to achieve high lumen, compact, reliable, and low cost SSL luminaries.

Synthetic Jet Heat Transfer Correlation

Heat transfer in synthetic jet impingement flows is different than steady jets. Thus, Garg *et al.* [34] attempted to predict synthetic jet heat transfer by using a conventional single slot jet impingement correlation given as following [35],

$$\mathrm{Nu} = \mathrm{Pr}^{0.42} G\left(\frac{r}{D_h}, \frac{z}{D_h}\right) F_1(\mathrm{Re}) \tag{4}$$

Where $F_1 = 2\,\mathrm{Re}^{1/2}\left(1 + 0.005\,\mathrm{Re}^{0.55}\right)^{1/2}$ and

$$G = \frac{D_h}{r} \frac{1 - \frac{1.1\, D_h}{r}}{1 + 0.1\left(\frac{z}{D_h} - 6\right)\left(\frac{D_h}{r}\right)} \tag{5}$$

$$\mathrm{h} = \frac{\mathrm{Nu\, k}}{\mathrm{D_h}} \tag{6}$$

They have observed that heat transfer trend was similar with the jet axial distance from the heater. However, steady jet prediction was not able to show better than 40% agreement with the experimental findings. Therefore, a correlation is developed to capture for synthetic jets cooling over a vertical surface. In this study, only air was used as the working fluid, therefore the effect of *Pr* number in the correlation was not captured. Closed form *Nu* number equation can be given as;

$$\mathrm{Nu} = \alpha\, \mathrm{Re}^{\beta} \left(\frac{\mathrm{f}}{\mathrm{f_n}}\right)^{\gamma} \left(\frac{\mathrm{A_{heater}}}{\mathrm{A_{orifice}}}\right)^{\kappa} \left(\frac{\mathrm{z}}{\mathrm{D_h}}\right)^{\xi} \tag{7}$$

where α, β, γ, κ, ξ are the constants that were computed based on the experimental data [36]. 128 distinct data points generated from four synthetic jets with different orifice sizes were utilized in this study. Reynolds number in the equation is based on the peak air velocity measured at the jet orifice and it captures the steady convective heat transfer. The second term is the normalized jet driving frequency with respect to the jet's resonance frequency. From our earlier flow visualization studies we had observed that the jet frequency affects the vortex dipole and its interaction with the surface. The next term is the ratio of the heater surface area to the jet orifice, and the last term is the dimensionless axial distance between the jet and the heated surface. The constants were computed to such that the following correlation was formed [36]:

$$\mathrm{Nu} = 0.364\, \mathrm{Re}^{0.802} \left(\frac{\mathrm{f}}{\mathrm{f_n}}\right)^{0.003} \left(\frac{\mathrm{A_{heater}}}{\mathrm{A_{orifice}}}\right)^{-0.74} \left(\frac{\mathrm{z}}{\mathrm{D_h}}\right)^{0.055} \tag{8}$$

where $Re < 2900$ $0.16 < f/fn < 1$ $5 < z/Dh < 20$

Eq. (8) shows that Reynolds number has the most profound effect on the Nu number, followed by the heater area to the jet orifice ratio. Interestingly, it was observed that dimensionless axial distance and driving frequency showed a weaker dependence compared to other two terms.

In other words, heat transfer performance is similar at those particular points. It, in fact, provides design flexibilities so that system performance will not be limited to tight tolerances. Nusselt number calculated by Eq. (8) does capture both synthetic jet and radiation heat transfer. Both heat transfer modes were not separated out as this correlation will help in many practical correlations since radiation always exist and part of overall heat transfer. Radiation contribution on the overall heat transfer coefficient is calculated between 5-15% depending on the jet operating condition (i.e. jet exit velocity). For high velocities (close to the resonance number), radiation is small, while low frequencies and low voltages it may be as much as 15% of the overall heat transfer coefficient.

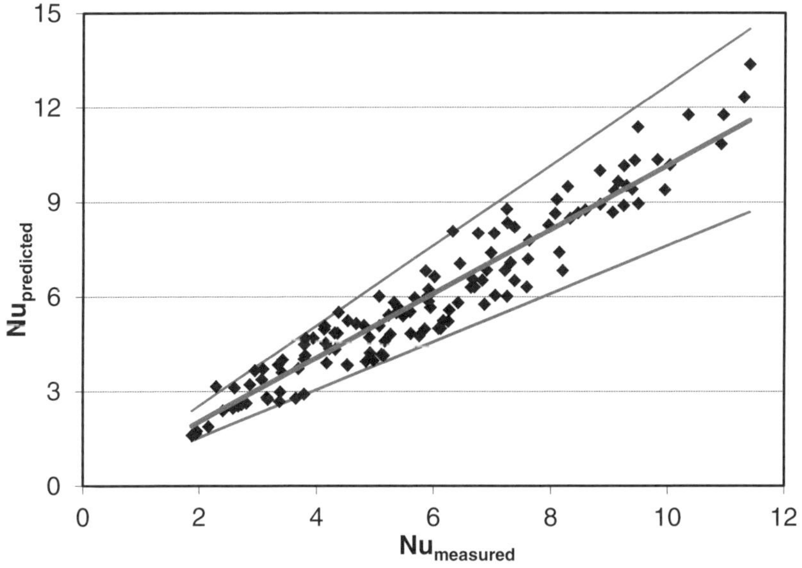

Fig. 27. Variation of predicted and measured average Nu numbers with + 25% boundaries [36].

Figure 27 presents the comparison of the measured and calculated average Nu numbers for the complete data set with ± 25% boundaries. We have observed that proposed correlation can predict synthetic jet heat transfer. Heat transfer below 200 Hz becomes so weak that experimental uncertainty dominates the measurements and it has very low value for the heat transfer enhancement research. It was also observed that proposed correlation cannot predict heat transfer below 200 Hz, and jet-to-heater axial distances less than 6 mm with less than 25% accuracy. Therefore, those data were not included in the current study. Jet to heater axial distance effect needs further investigation with flow visualization techniques to better understand the air replenishment around heater surface.

A critical review of synthetic jets technology for the thermal management of electronics have been presented in this study. With the demanding needs of electronics, novel thermal technologies have to be invented and developed that will enable high functionalities at a affordable costs. In this review, fundamentals, synthetic jet approaches, experimental techniques, and rapidly progressing applications have been presented.

Summary and Conclusions

Growing manufacturing activities, rapid progresses in science and information technology have increased the usage of electrical and electronic devices in which continuously increasing heat fluxes posing significant challanges. Thus, the ever increasing heat flux has necessitated the development of new thermal management technologies. Since, thermal management is currently one of the key limitations in the design of electronics systems, parallel to the advancements in the electronics manufacturing (nano scale feature sizes etc) and increase in the power dissipation, the development of effective, low-cost, compact heat removal solutions become extremely critical to ensure a failsafe and reliable operation. While liquid cooling is poised to provide the cooling capability for next generation electronics, its use in the present-day products is less prevalent due to risks associated with condensation, leakage, and pumping power. Consequently, air-cooling strategies still continue to meet for near-term cooling needs in the electronic industry. The need for new cooling technologies such as piezo fans, ionic winds and synthetic jets is obvious. This has led to many researchers devoting their attention to new technologies. Synthetic jets, meso scale cooling devices, have acquired noticeable attention and both academia and industry have devoted substantial efforts to these technologies. While synthetic jet technology is still progressing, a summary of the state of the art for heat transfer applications is discussed. The following conclusions are drawn:

- The heat transfer application of synthetic jets was first appeared in early 1980s
- Various synthetic jet sizes, from 10s of mm to 100s of mm, are investigated and results are presented (There is no size limitation so, one can make as large as possible).
- A number of actuation techniques are studied and each actuation method has its own unique advantages and features.
- Synthetic jets can enhance both natural and forced convection heat transfer.
- Driving voltage and frequency impact the performance of the jet, power consumption and noise. A different design optimal point may exist for each application
- A number of 2D computational models have been presented. There is a need for 3D models to better understand the flow scheme.
- While there is some data about the reliability of the synthetic jets, further research is necessary before implementing in the actual applications.
- Implementation of synthetic jets just started in a number of practical applications. Further development is necessary to make those jets low-cost and system packaging will accelerate the adoption process.

Nomenclature

A	amplitude of velocity profile
d	orifice diameter
D	length/height of the diaphragm
DAQ	data acquisition system
EF	enhancement factor
f	operating frequency
h	heat transfer coefficient
P	power consumption
R	inner radius of the jet enclosed volume
Q	volumetric flow rate
Re	Reynolds number
S	Stokes number
T	period of oscillation or temperature
u	x-velocity component
U	degrees of freedom
V	velocity scale
q	heat flux
t	time
x	coordinate axis
y	coordinate axis

Greek symbols

ν	kinematic viscosity
μ	dynamic viscosity
ρ	density
ω	vorticity

Subscripts

NC	*natural convection*
FC	*forced convection*
s	*surface*
f	*fluid property*
Jet	*side of the heater facing the jet*

References

1. Mittal, R. and Rampunggoon, P., 2002, "On virtual aero-shaping effect of synthetic jets," Phys. Fluids (Brief Communication), 14(4), pp. 1533–1536.
2. Smith, D., Amitay, M., Kibens, V., Parekh, D., and Glezer, A., 1998, "Modification of lifting body aerodynamics using synthetic jet actuators, AIAA Paper 1998–0209.
3. Crook, A., Sadri, A.M., and Wood, N.J., 1999, "The development and implementation of synthetic jets for the control of separated flow," AIAA Paper 1999–3176.
4. Mittal, R., Rampunggoon, P., and Udaykumar, H.S., 2001, "Interaction of a synthetic jet with a flat plate boundary layer," AIAA Paper 2001–2773.
5. Rathnasingham, R., and Breur, K.S., 1997, "System Identifcation and control of turbulent boundary layer, Phys. Fluids, 9(7), pp. 1867–1869.
6. Lee, C.Y., and Glodstein, D.B., 2001, "DNS of microjets for turbulent boundary layer control," AIAA Paper 2001–1013.
7. Mahalingam, R., Rumigny, N., and Glezer, A., 2004, "Thermal management using synthetic jet ejectors," IEEE Trans. Comp. Packaging Tech., 27(3), pp. 439–444.
8. Mehmet Arik and Tunc Icoz, 2010, "Predicting heat transfer from unsteady synthetic jets", ASME Journal of Heat and Mass Transfer. Smith, B.L. and Glezer, A., 2002, "Jet vectoring using synthetic jets," J. Fluid Mech., 458, pp. 1–34.
9. Mahalingam, R., and Glezer, A., 2005, "Design and thermal characteristics of a synthetic jet ejector heat sink," Trans. ASME, 127, pp. 172–177.
10. Holman, R., Utturkar, Y., Mittal, R., Smith, B., and Cattafesta, L., 2005, "Formation criterion for synthetic jets," AIAA J., 43(10), pp. 2110–2116.
11. Li, S., 2005, "A numerical study of micro synthetic jet and its applications in thermal management," Ph.D. thesis, Georgia Institute of Technology.
12. Gutmark, E., Yassour, Y., and Wolfshtein, M., Acoustic enhancement of heat transfer in plane channels, Proceedings of the Seventh International Heat Transfer Conference, Munich, Germany, pp. 441–445, 6–10 Sept. 1982.
13. Yassour, Y., Stricker, J., and Wolfshtein, M., 1986, "Heat transfer from a small pulsating jet," Proc. of the 8th International Conference, San Francisco, CA, Aug 17–22, 3, pp. 1183–1186.
14. Coe, D.J., Allen, M.G., Trautman, M.A., and Glezer, A., 1994, "Micromachined jet for manipulation of macro flow," Solid-state sensor and actuation workshop, Hilton Head Island, SC, Jun 13–16.
15. Erbas, N., Koklu, M., and Baysal, O., 2005, "Synthetic jets for thermal management of microelectronic chips," Proceeding of IMECE, Paper IMECE 2005–81419.
16. Garg, J., Arik, M., Weaver, S., Wetzel, T., and Saddoughi, S., 2005, "Meso pulsating jet for electronics cooling," J. Elec. Packaging, 127(4), pp. 503–551.
17. Utturkar, Y., Arik, M., and Gursoy, M., 2006, "An experimental and computational sensitivity analysis of synthetic jet cooling performance," ASME International Mechanical Engineering Congress and Exposition, Chicago, IL, Nov 5–10.

18. Minichiello, A. L., Hartley, J. G., Glezer, A., and Black, W. Z., Thermal management of sealed electronic enclosures using synthetic jet technology, Advances in Electronic Packaging 1997, EEP (19–2), pp. 1809–1812, 1997.
19. Gillespie, M. B., Black, W. Z., Rinehart, C., and Glezer,A., 2006, "Local Convective Heat Transfer From a Constant Heat Flux Flat Plate Cooled by Synthetic Air Jets" Journal of Heat Transfer, Vol. 128, pp. 990–1000.
20. Issa, R., 1985, "Solution of the implicitly discretized fluid flow equations by operator-splitting," J. Comp. Phys., 62, pp. 40–65.
21. Kotapati, R.B. and Mittal, R., 2005, "Time-accurate three-dimensional simulations of synthetic jets in quiescent air," AIAA Paper 2005–0103.
22. Arik, M., Petroski, J. Bar-Cohen, A. and Demiroglu, M., "Energy Efficiency of Low Form Factor Cooling Devices", ASME International Mechanical Engineering Congress and Exposition, IMECE2007-41275, November 11–15, 2007, Seattle, WA.
23. M. Arik, R. Sharma, J. Lustbader, X. He,"Comparison of synthetic and steady air jets for impingment heat transfer over vertical surfaces", ITHERM 2012 Conference, May 30, 2012, San Diego, CA.
24. Chen, Jason. 2007. Citing Website. http://gizmodo.com/286400/japanese-thermal-expert-examines-xbox-360-figures-out-whats-wrong-hint-everything (accessed August 25, 2012).
25. J. Petroski, M. Arik, and M. Gursoy, "Piezoelectric Fans: Heat Transfer Enhancements or Electronics Cooling," ASME-JSME Thermal Engineering and Summer Heat Transfer Conference 2008, HT-2008-56405, Jacksonville, FL, August 10–14, 2008.
26. T. Açikalin, I. Sauciuc, and S. V. Garimella, "Piezoelectric Actuators for Low-Form-Factor Electronics Cooling," The ASME/Pacific Rim Technical Conference and Exhibition on Integration and Packaging of Micro, Nano, and Electronic Systems (InterPACK '05), IPACK2005-73288, San Francisco, July 17–22, 2005.
27. Utturkar, Y. Arik, M. and Gursoy, M, 2009, "Thermal characteristics of a synthetic jet integrated heat sink design for air-cooled electronics, Interpack 2009, IPACK2009-89322.
28. Utturkar, Y. Arik, M. and Gursoy, M, 2007, "Assessment of Cooling Enhancement of Synthetic Jet in Conjunction with Forced Convection", IMECE 2007, IMECE2007-41446.
29. Bar-Cohen, A., "Computer-Related Thermal Packaging at the Millennial Divide", Electronics Cooling, v. 6, n. 1, January 2000, pp. 32–40.
30. Raghav Mahalingam, "Modeling of Synthetic Jet Ejectors for Electronics Cooling", IEEE Transactions on Components and Packaging Technologies, v. 27, n. 3, September 2004, pp. 439–444.
31. Mehmet Arik, M., Jennifer Jackson, J., Satish Prabhakaran, S., Seeley, C., Sharma' R., Utturkar, Y., Weaver, S., Kuenzler, G., and Han, B. "Development of a high lumen solid state down light application".

32. M. Arik and A. Setlur, "Environmental and Economical Impact of LED Lighting Systems and Effect of Thermal Management", *Int. Journal Energy Research*, in press, October 2009.
33. Arik, M., Weaver, S. "Chip scale thermal management of high brightness LED packages", *Fourth International Conference on Solid State Lighting*. (2004).
34. Garg, J., Arik, M., Weaver, S., and Saddoughi, S., Micro fluidic jets for thermal management of electronics, Proc. ASME Heat Transfer/Fluids Engineering Summer Conference, Charlotte, North Carolina, FED F-346, July 11–15, 2004.
35. Incropera, Frank P., DeWitt, David P., "Introduction to Heat Transfer," Third Edition, John Wiley and Sons.
36. Arik, M. and Icoz, T., "Predicting heat transfer from unsteady synthetic jets", ASME Journal of Heat and Mass Transfer, Vol 134, 2012.

CHAPTER 7

RECENT ADVANCE IN THERMOELECTRIC DEVICES FOR ELECTRONICS COOLING

Peng Wang

Qualcomm Inc., San Diego, CA 92121
wangpeng@qti.qualcomm.com

Thermal management of on-chip hot spot, with a heat flux of around 1000 W/cm2, has become one of the major challenges in the development of next-generation microprocessors. Solid state thermoelectric cooler (TEC) offers great promise for hot spot thermal management because of their compact structure, fast response, high reliability, localized cooling, and high flux removal capability. To date TEC has received great attentions in electronics cooling community as one of the potential hot spot cooling solutions. In this paper, recent development and application of hot spot cooling strategies based on micro thermoelectric technologies will be reviewed and discussed, three hot spot cooling concepts, including thin-film thermoelectric cooling, mini-contact cooling, and semiconductor self-cooling in silicon substrate and germanium substrate will be discussed. The advantages and disadvantages of these on-chip cooling solutions for high flux hot spots will be evaluated.

1. Introduction

The Moore's law progression in semiconductor technology, including shrinking feature size, increasing transistor density, and faster circuit speeds, is leading to ever higher chip power dissipation and heat flux. In addition, device miniaturization and design complexity are causing highly non-uniform power distribution on the microprocessors and producing high heat flux hot spots. These hot spot regions have a heat flux several times higher than the chip average[1] and results in extremely high local temperatures that can dramatically degrade microprocessor performance and system reliability. Figure 1 shows a typical power map and temperature map on the silicon chip, which are characterized by substantial non-uniformities in power dissipation, resulting in localized, high heat flux hot spot with a very large on-chip temperature gradient[2]. Since chip-level thermal management must ensure the junction temperature of the microprocessors will not exceed an application-driven maximum temperature, typically in the range of 85 to 120°C, it is often these hot spots, not the entire chip power dissipation, that drive the thermal design and system reliability[3]. This leads to two undesirable consequences for electronics cooling: (1) non-uniform heat generation limits the total heat dissipation that can be managed by a conventional thermal solution and, thus, a much more aggressive thermal solution than would be required for uniform heating, is required, and (2) the focus on controlling the hot spot temperature can lead to over-design of the microprocessor cooling solution.

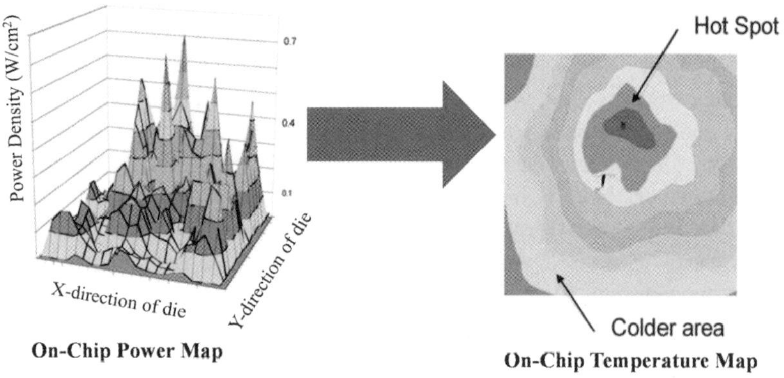

Fig. 1. Typical power map (a) and temperature map (b) on the silicon chip[2].

Therefore, beyond the challenge of continuously enhancing available thermal management techniques for high heat flux chips, the need to develop cooling approaches that can specifically handle such micro-scale high flux hot spots has become a primary roadblock to design of high performance electronic devices. Thus new and novel cooling technologies with the ability to selectively cool micro-scale hot spots are in urgent demand. Advanced liquid cooling techniques, such as micro-channel cold plates, liquid immersion cooling, spray cooling, and jet impingement cooling provide very high heat transfer coefficients and could be used to meet these requirements. However, the reliability, complexity, volume, weight and cost are major barriers to successful commercial implementation of these approaches.

Thermoelectric cooler is a solid-state pump which utilizes energy conversion from electricity to thermal energy to implement electronics cooling. Recent developments in solid state thermoelectric coolers have shown great promises for hot spot thermal management. In this paper, recent development and application of hot spot cooling strategies based on micro thermoelectric technologies will be reviewed and discussed, and three hot spot cooling concepts, including thin-film thermoelectric cooling, mini-contact cooling, and semiconductor self-cooling in silicon substrate and germanium substrate will be discussed.

2. Thin-Film Thermoelectric Cooling

2.1. *Hot Spot Cooling Concept*

TEC is a solid-state pump which utilizes energy conversion from electricity to thermal energy to implement electronics cooling. The maximum cooling heat flux, or the maximum pumping capability, for a TEC is given by:

$$q''_{\max} = \frac{S^2 T_c^2}{2\rho L} \tag{1}$$

where S is the Seebeck coefficient, ρ the electrical resistivity, L the thickness of the thermoelectric element, and T_c the cold-junction temperature. It suggests that an increase of Seebeck coefficient and cold-junction temperature or decrease of electrical resistivity and element thickness will improve the maximum achievable pumping capability.

Since the pumping capability is inversely proportional to the thickness of thermoelectric element (TE) as indicated in Equation (1), the main advantage of going to thin-film thermoelectric cooler (TFTEC) is the dramatic increase in cooling heat flux. For a TE thickness in an order of 10 μm, a theoretical maximum cooling heat flux of 1000 W/cm^2 can be achieved for Bi_2Te_3 materials. Thus, the TEC designed for hot spot cooling has to be very thin in order to remove a high heat flux from the electronic components. It should be noted that although the coefficient of performance (COP) is still pretty low for today's state-of-the-art TFTEC making it unrealistic to apply to cooling the entire chip, it is a very ideal solution for local cooling of small hot spot on the chip. For example, if a TFTEC with a COP of 1 is targeted to cool 100 W chip, the thermal package has to provide an additional 100 W to activate the TFTEC, which would cause significant problems in power delivery, substrate capability, and as well as more aggressive global cooling solution[4]. Therefore, application of TFTEC to cool the entire high-power chip is not a realistic solution. On the hand other, if TFTEC is locally applied to remove a 1.0 W high flux hot spot, the input power to operate the TFTEC would be 1.0 W and the thermal package needs only to remove an additional 1.0 W to cool the hot spot, which makes TFTEC a very attractive thermal solution for hot spot cooling.

Thermoelectric hot spot cooling solution can suppress hot spot temperature, isothermalize chip surface, and reduce the junction-to-case thermal resistance. Relative to other electronics cooling technologies, TFTEC also has a compact structure, no moving parts, fast response, and high reliability. In addition, TFTEC can be used with other cooling technologies such as heat sinks, heat pipes, vapor chambers, and liquid cooled cold plates, to develop hybrid cooing solution to enhance the cooling capability and flexibility[5].

There are four possible configurations which can be employed to integrate a TFTEC into the silicon package for hot spot cooling[4]:

(1) *Configuration #1*: TFTEC is attached to the backside of the chip and embedded inside the thermal interface material (TIM) (Figure 2(a)).
(2) *Configuration #2*: TFTEC is attached to the integrated heat spreader (IHS) and embedded inside the thermal interface material (TIM) (Figure 2(b)).
(3) *Configuration #3*: TFTEC is embedded in the silicon chip (Figure. 2(c)); and
(4) *Configuration #4*: TFTEC is embedded in the integrated heat spreader (IHS) (Figure 2(d)).

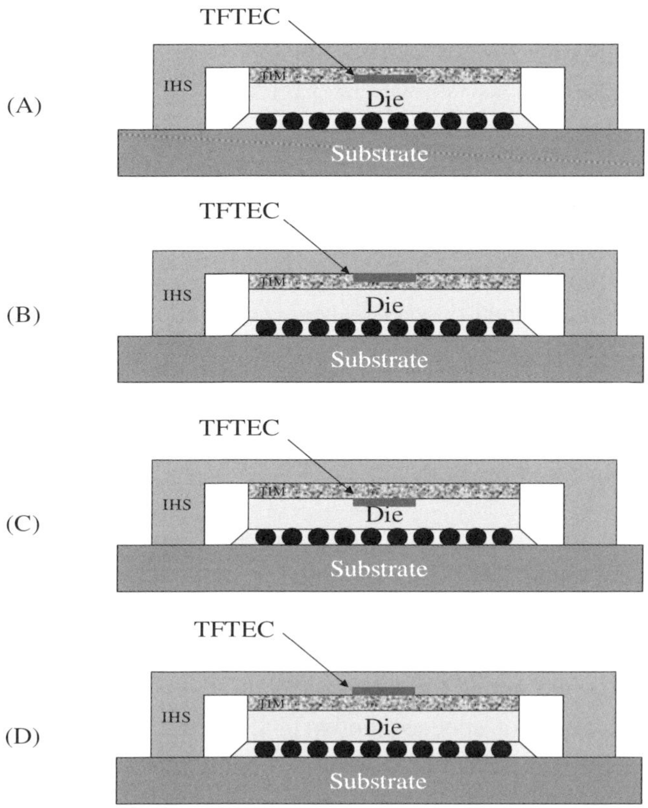

Fig. 2. TFTEC assembling configuration in the silicon package for hot spot cooling: (a) TFTEC is attached on the backside of silicon chip, (b) TFTEC is attached on the IHS; (c) TFTEC is embedded in the cavity side of the silicon chip, and (d) TFTEC is embedded in the cavity side of the IHS[4].

Chau *et al.* examined the effect of TFTEC assembling configurations on hot spot cooling performance[4]. In his analysis, the silicon chip testing vehicle, with the dimension of 11 m × 13 mm × 780 μm, has a uniform power density of 70 W/cm^2 and a 400 μm × 400 μm, 1875 W/cm^2 hot spot located on the center of the chip. The Bi_2Te_3-based TFTEC, with a dimension of 2.5 mm × 2.5 mm × 100 μm, is positioned over the hot spot on the silicon chip. His results suggest that TFTEC integration configurations have little effect on hot spot cooling and the cooling performance of all the configurations is very similar (typically within

2°C). This gives a great deal of flexibility in integrating a TFTEC in electronic package for hot spot cooling. For example, the benefit of the *Configuration #2*, as shown in Figure 2(b), is its simplicity in manufacturing and assembling because the fabrication of TFTEC for this configuration is completely separate from the silicon chip so that it does not incur any post-silicon process constraints imposed on it. This configuration is also less complicated in fabrication in comparison to other assembling configurations. However, care should be taken to reduce various thermal contact resistances introduced by the integration of TFTEC into the silicon package, which could bring in a significant cooling degradation as discussed in the sequent sections.

2.2. *Hot Spot Cooling Analysis*

Ramanathan *et al.* conducted the theoretical analysis to explore the possibility of using TFTEC for on-chip hot spot cooling[6]. The test vehicle is a 100 W/cm^2, 10 m × 10 mm silicon chip with an 800 W/cm^2, 500 μm × 500 μm hot spot located on the center of the chip. The TFTEC with a footprint area of 3 mm × 3 mm is attached on the backside of the chip and directly over the hot spot as shown in Figure 2(a). Packaging density, which is defined as the total footprint area of all thermoelectric elements to the TFTEC footprint area, is 80%. The thickness of thermoelectric element is varied from 5 to 30 μm to determine the optimum cooling capability. Since a filler material is typically needed between the thermoelectric elements to provide mechanical stability, the analysis includes the parasitic losses due to back heat conduction through the filler material. The simulation shows that the hot spot temperature is 102.5°C if there is no TFTEC integrated in the silicon package as shown in Figure 3. When the TFTEC is attached on the chip and powered on at an input power of 9.0 W, the hot spot temperature drops by 7.3°C if electrical contact resistance or back-conduction through the filler is not included. If the impact of Joule heat at electrical contact is included with an electrical contact resistance of 1×10^{-7} Ω-cm^2, the hot spot cooling is reduced to 6.6°C. From mechanical stability and processing point of view, it is important to have a filler material. However, this degrades the hot spot cooling performance as shown in Figure 3. From the above analysis, it is clear that in order to obtain the maximum hot spot temperature reduction, one would have to optimize processing to minimize electrical contact resistances and minimize back heat conduction through the filler material by choosing low thermal conductivity filler.

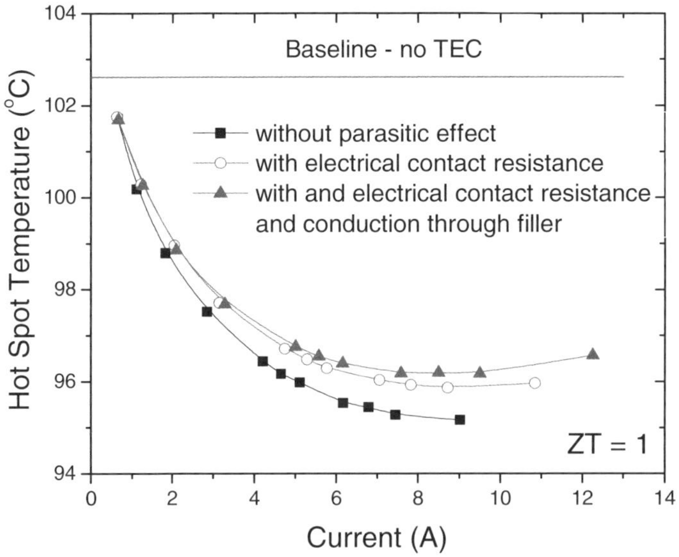

Fig. 3. Hot spot temperature as a function of the input current on TFTEC [5].

More detailed package-level thermal analyses of TFTEC for hot spot cooling are discussed by Narasimhan *et al.*[7,8]. The hot spot cooling performance is evaluated using a 11 m × 13 mm × 780 μm silicon chip with a 400 μm × 400 μm hot spot located on the center of the chip. The power dissipation is 20 to 40 W on the chip and 0.5 to 2 W on the hot spot. Figure 4(a) shows the temperature map on the chip without TFTEC integrated in the package, indicating a highly non-uniform temperature distribution with a peak hot spot temperature close to 104°C at an ambient temperature of 38°C. Figure 4(b) shows the temperature map of the chip cooled by a 4.25 mm × 4.25 mm TFTEC. The hot spot temperature is 98°C corresponding to a temperature suppression of 6°C at the hot spot. From Figure 4(b) it can be seen that when a TFTEC is integrated in the silicon package at an applied current of 6.5A, a new hot spot located outside the TFTEC shadow region occurs.

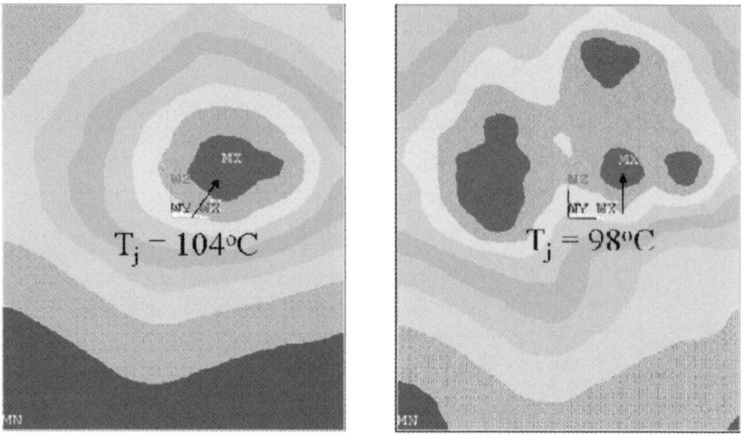

(A) No TFTEC (B) TFTEC at 6.5A

Fig. 4. The temperature contours on the bottom of silicon chip: (A) without TFTEC, and (B) with TFTEC integrated on the top of silicon chip[6].

2.3. *Hot Spot Cooling Demonstration*

The first experimental demonstration of TFTEC for electronics cooling was reported by Chowdhury[9]. The superlattice thermoelectric materials used in the TFTEC are p-type Bi_2Te_3/Sb_2Te_3 and n-type $Bi_2Te_3/Sb_2Te_{2.83}Se_{0.17}$ grown by metal–organic chemical vapor deposition on GaAs substrates. The TFTEC has a dimension of 3.5 mm × 3.5 mm × 100 μm. The measured Seebeck coefficient is 301 μV/K, electrical resistivity is 10.8 μΩ-m, and thermal conductivity is 1.2 W/m-K, which gives an intrinsic *ZT* value of 2.1 at 300K. The integration of the TFTEC into the package follows *Configuration* #1 as shown in Figure 2. The testing chip has a uniform background heat flux of 42.7 W/cm^2. On the center of the chip there is a 400 μm × 400 μm hot spot with a heat flux of 1250 W/cm^2 corresponding to 2 W local heating. Thermal grease with the thermal conductivity of 1.5 W/m-K is used as the thermal interface material (TIM) between the backside of the chip and the IHS to reduce the thermal contact resistance.

The impact of thermoelectric cooling on the hot spot temperature is shown in Figure 5. It can be seen that if there is no TFTEC embedded into the IHS, the hot spot temperature reaches 124.5°C. When the TFTEC is integrated into the package, the hot spot temperature reduces to 116.9°C even without powering on the TFTEC. This 7.6°C of "passive cooling" on the hot spot is attributed to the fact that the effective thermal conductivity of the TEC module, which is a combination of low-thermal-conductive thermoelectric elements and two layers of high-thermal-conductive metal contact, is larger than that of thermal grease. When the TFTEC is activated, the hot spot temperature decreases as the applied current increases and reaches a minimum hot spot temperature of 109.6°C at 3.0A. The temperature reduction of 7.3°C at the hot spot is referred to as the "active cooling" which is attributed to thermoelectric cooling effect. Therefore, a total of 14.9°C hot spot temperature reduction, combined with "passive cooling" and "active cooling", has been demonstrated in the experiment. A total of 12 TFTECs were tested and the repeatability of thermoelectric active cooling is shown in Figure 5(b). It can be seen that a maximum active cooling of 9.6°C, a minimum active cooling of 5.4°C and an average value of active cooling of approximately 6.0°C is observed across these units. This large variation in active cooling was attributed by the authors to unit-to-unit variability in TFTEC fabrication and package assembly, which results in variation in thermoelectric properties, electrical contact resistance, and thermal contact resistance.

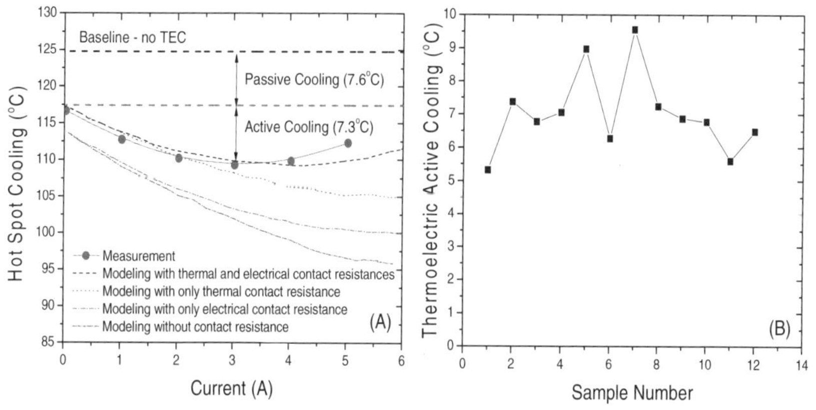

Fig. 5. Hot spot cooling performance of the TFTEC: (a) hot spot temperature as a function of applied current, and (b) hot spot cooling repeatability testing[8].

The electrical and thermal contact resistances of the solder interface between the IHS and the TEC were estimated to be about 10^{-10} Ω-m^2 and 8×10^{-6} m^2-K/W.[10] The simulation results shown in Figure 5(a) indicate that, in the absence of parasitic effects of electrical contact resistance and thermal contact resistance, the TFTEC is able to reduce the hot spot temperature approximately by 23.0°C at an applied current of 3.0A and 30.0°C at an applied current of 6.0A. However, with an electrical contact resistance of 1×10^{-10} Ω-m^2, the hot spot cooling is reduced to 22.0°C at an applied current of 3.0A; if there is a thermal contact resistance of 8×10^{-6} m^2-K/W, the hot spot cooling is reduced to 17.0°C at an applied current of 3.0A. Since the temperature rise on the hot spot is proportional to the thermal contact resistance and hot spot heat flux, it is expected that the hot spot cooling performance will degrade severely with thermal contact resistance, as the hot spot heat flux increases.

3. Mini-Contact Enhanced Cooling

3.1. *Hot Spot Concept*

TEC has unique characteristics making it suitable for thermal management of electronics in comparison to other cooling solutions. However, the relatively low cooling heat flux of conventional TEC, at a level of 5 ~ 10 W/cm^2, has seriously limited it to hot spot cooling[11]. Recently, a novel use of a mini-contact pad, which connects the TEC and the silicon chip, was proposed to improve hot spot cooling performance[12,13].

Figure 6 illustrates the physical phenomenon underpinning the use of mini-contact enhanced thermoelectric coolers. The packaging structure of the TEC without mini-contact is shown in Figure 6.1(b) where the TEC is embedded in the TIM while Figure 6.1(c) illustrate the mini-contact enhanced TEC embedded in the TIM. It can be expected that the cooling heat flux at the mini-contact will increase as the mini-contact cross-sectional area reduces, since it is approximately inversely proportional to the cross-sectional area of the mini-contact tip. In addition, introduction of the mini-contact provides a unique benefit that allows the TEC to act locally on the hot spot heat load. Since the TEC does not have to pump the additional background heat load, the power requirement to activate TEC is reduced or the hot spot can be cooled further for the same amount of input power.

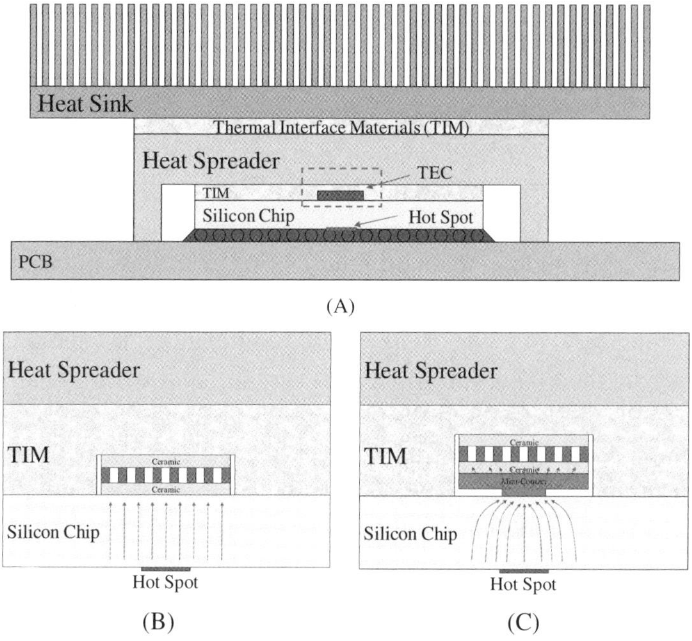

Fig. 6. (A) Schematic of the silicon chip package with an embedded TEC for hot spot cooling. (B) TEC embedded in the TIM without mini-contact, (C) TEC embedded in the heat spreader with mini-contact. The arrows indicate the heat flow pattern in silicon chip and mini-contact. The detailed chip package is not shown in this figure[13].

3.2. *Hot Spot Cooling Analysis*

Wang and Yang developed a three-dimensional modeling to explore hot spot cooling using mini-contact TEC[12,13]. The TEC is integrated with a copper mini-contact pad, attached on the top of the silicon chip, and then embedded inside the thermal interface materials as shown in Figure 6. The TEC consists of a 4 × 4 array of 400 μm × 400 μm × 20 μm thermoelectric elements sandwiched between two 50 μm thick ceramic substrates. The mini-contact pad features a 2.4 mm × 2.4 mm × 100 μm base to facilitate heat spreading, and a 50 μm-thick mini-contact tip with various cross-sectional areas, from 600 μm × 600 μm to 2400 μm × 2400 μm, to concentrate the thermoelectric cooling capability.

The beneficial effect of mini-contact on hot spot cooling could be limited by the heat spreading resistance inside the pad as well as inside the silicon chip. Consequently, care should be taken to optimize the geometric configuration to achieve the maximum hot spot cooling performance. Figure 7 shows the temperature profiles on the bottom of the silicon chip, with and without the embedded TEC, revealing the characteristic "W" shaped temperature profile generated by the mini-contact-enhanced TEC. For this case study, the thermoelectric element thickness is 20 μm and the input power is 10 W to activate the TEC. It is seen that if there is no hot spot on the chip and no TEC embedded inside the package, the peak chip temperature is about 109°C. However, if there is a 400 μm × 400 μm hot spot with a heat flux of 1250 W/cm^2, the peak chip temperature will increase to 137°C. So, the concentrated hot spot heat flux leads to about a 28°C peak temperature rise on the chip. If the TEC is activated with a 10 W and enhanced with a 600 μm × 600 μm mini-contact pad, the temperature profile around the high flux region shows a 9°C reduction in the hot spot temperature (to 128°C), a low temperature ring with temperatures below the uniform heat dissipation values, and a warm outer ring with slowly decaying temperatures in the radial direction. If the mini-contact tip grows to 1250 μm × 1250 μm, the hot spot temperature will reduce further, down to 120°C, resulting in 17°C maximum cooling at the hot spot. However, if we expand the mini-contact size further to 2400 μm × 2400 μm, the hot spot cooling is limited to just 12°C. Obviously there exists an optimum mini-contact size for each configuration. The observed temperature increase in the outer ring or "side lobes" of the profile, as well as a very modest increase in the average chip temperature is due to the additional power dissipation associated with the operation of the TEC device.

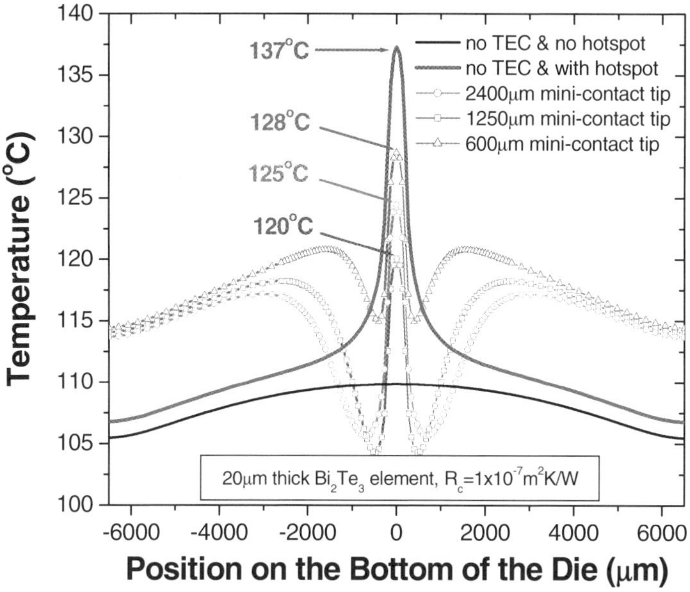

Fig. 7. Effect of mini-contact size on TEC-induced temperature profile on the active side of silicon chip [12].

Low thermal contact resistance are critical to mini-contact enhanced hot spot cooling, since the thermal resistance at the mini-contact/chip interface — where the cooling flux is highest — will reduce the effectiveness of the mini-contact enhancement. Moreover, a high thermal resistance interface between the TEC and the TIM2 will impede the conduction of thermoelectric heat and Joule heat into the heat spreader and then into the heat sink and the ambient. Figure 8 displays the interplay between the thermal contact resistance and the achievable hot spot cooling, with the assumption of equal thermal contact resistance at the two interfaces (e.g. $R_{c1} = R_{c2} = R_c$), and reveals that with increasing thermal contact resistance at both interfaces, the net cooling achievable on the hot spot diminishes. It may be seen that if the thermal contact resistance is 1×10^{-5} m^2-K/W or higher, the hot spot temperature will exceed 140°C and the embedded TEC will raise rather than lower the hot spot temperature.

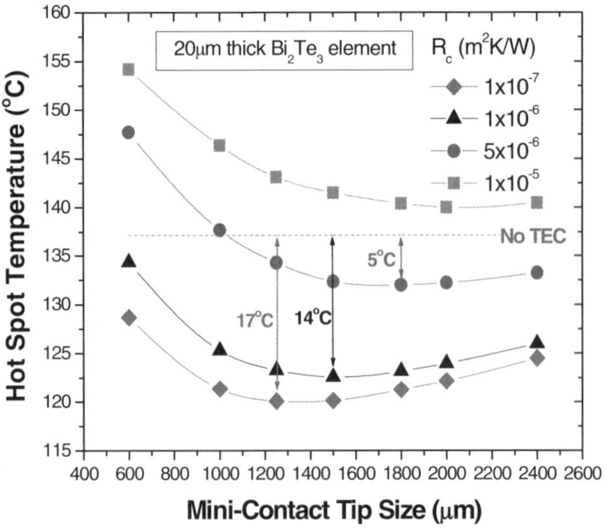

Fig. 8. Influence of thermal contact resistance on hot spot cooling performance[13].

The thermal contact resistance also has an impact on optimized mini-contact size. As shown in Figure 8, with increasing thermal contact resistance, the optimized mini-contact increases from 1250 μm × 1250 μm for the thermal contact resistance of 1×10^{-7} m^2-K/W, representing a nearly-perfect interface, to 2000 μm × 2000 μm for a thermal contact resistance of 1×10^{-5} m^2-K/W, typical of thermal grease interfaces. It should be noted that if the thermal contact resistance is 5×10^{-5} m^2-K/W or lower, use of an optimized mini-contact tip always provides a lower hot spot temperature than achieved without the mini-contact. However, the mini-contact is seen to provide diminishing returns as the contact resistances increase and to elevate the hot spot temperatures for thermal contact resistances of 1×10^{-5} m^2-K/W or higher.

3.3. *Hot Spot Cooling Demonstration*

Wang *et al.* developed a prototype to quantify the spot cooling improvement by use of mini-contact pad, and experimentally determined its relationship to the TEC input power and power dissipation on the silicon chip[13]. In their proof of concept experiment, the miniaturized TECs have a dimension of 3.6 mm × 3.6 mm × 1.6 mm. The copper mini-contact pad with various mini-contact tip sizes, from 0.8 mm × 0.8 mm to 3.6 mm × 3.6 mm, was bonded onto the silicon chip using an indium-based solder. The mini-contact pad and the TEC are accommodated inside the copper spacer. Above the copper spacer and the TEC, the copper heat spreader was attached using thermal grease. The heat spreader was then attached to the air-cooled copper heat sink, as illustrated in Figure 9(A).

The experimentally observed effect of the mini-contact size on the temperature reduction at the targeted spot is displayed in Figure 9(B). For the three different power dissipations and a 500 μm thick silicon chip, the maximum spot cooling is seen to display a parabolic dependence on the mini-contact tip size, showing a favorable improvement as the mini-contact tip size decreases in area from the "full coverage" limit, but ultimately reversing direction as the tip size shrinks below an optimum value and approaches point contact. The presence of an optimum tip size reflects the competing effects of the favorable cooling flux concentration and the parasitic spreading resistance in the mini-contact tip. As shown in Figure 9(B), for the case of no power dissipation on the silicon chip, if the mini-contact is of the same size as the TEC base, the measured maximum spot cooling is about 3.3°C. However, if a 1.8 mm × 1.8 mm copper mini-contact is integrated onto the TEC, 7.1°C maximum spot cooling can be obtained which results in 115% improvement on spot cooling performance. Similarly, spot cooling performance can be improved by 100% and 80% if the power dissipation of the silicon chip is 30 W and 67 W, respectively.

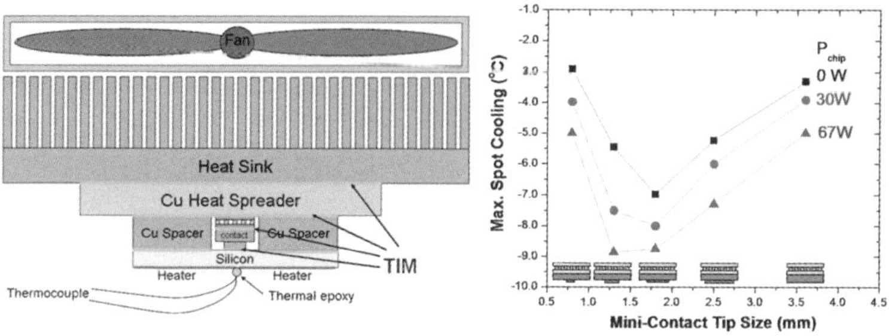

(A) (B)

Fig. 9. (A) Schematic of proof of concept experiment for mini-contact enhanced TEC, and (B) Variation of measured spot cooling performance with copper mini-contact size[13].

4. Semiconductor Self-Cooling

Semiconductor materials, such as silicon (Si), germanium (Ge), silicon germanium (SiGe), silicon carbide (SiC), gallium arsenide (GaAs), and gallium nitride (GaN), have been the heart of microelectronics for decades, but their thermoelectric natures[14,15,16,17,18,19] which can be utilized for on-chip self-cooling have received little attention. Wang and Bar-Cohen have systematically investigated semiconductor self-cooling phenomena and applied it to hot spot cooling on germanium chip[20], silicon chip[21,22] and silicon carbide chip[23], and demonstrated that self-cooling on semiconductor substrate could be a viable on-chip cooling solution due to its simple structure, ease of fabrication, and elimination of thermal contact resistance compared with traditional cooling technologies.

4.1. *Germanium Self-Cooling Concept*

Germanium is an important semiconductor material, mainly used in transistors and integrated circuits. A metal oxide semiconductor field effect transistor (MOSFET) with germanium channel currently receives much attention as a possible candidate for next-generation high mobility devices because germanium could offer two-fold higher mobility for electrons and four-fold higher mobility for holes in comparison to silicon. While germanium solid-state devices provide outstanding electrical benefits, they also offer significant challenges in thermal management and raise a major concern over the effect of on chip hot spots on their reliability and performance.

The germanium self-cooling structure is illustrated in Figure 10, where the Peltier cooling, Peltier heating and Joule heating effects are identified. The metal contact is integrated onto the germanium substrate and developed as a thermoelectric micro-cooler. A chip-level thermal-electric coupling model with a germanium micro-cooler integrated on the top of 12 mm × 12 mm germanium chip is developed. The heat generation from these components is modeled as a uniform 70 W/cm^2 background heat flux on the bottom of the germanium chip and a 70 μm × 70 μm, 700 W/cm^2 hot spot located on the center of the chip. The backside of the chip experiences an effective heat transfer coefficient of 8700 W/m^2-K in reference to an ambient temperature of 25°C.

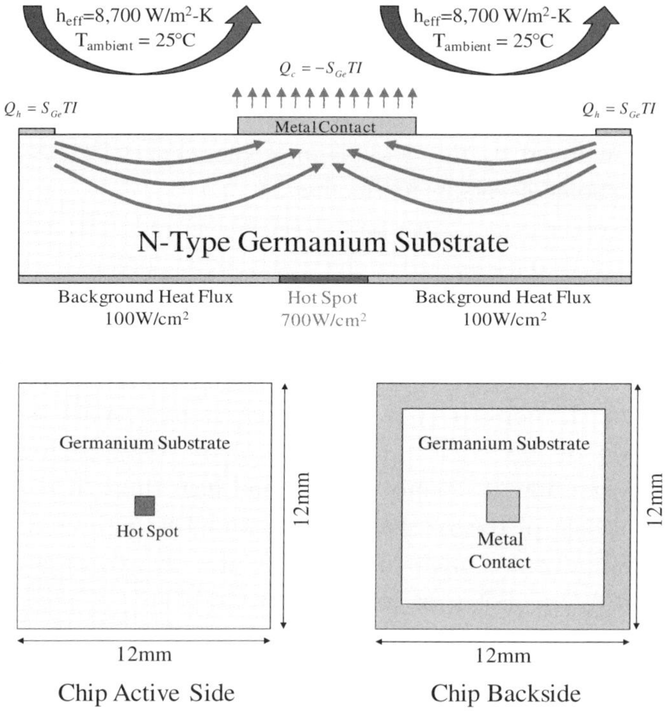

Fig. 10. Structure of on-chip germanium self-cooling for hot spot thermal management. The arrows indicate the direction for electrical current[20].

The hot spot cooling capability of germanium coolers can be characterized using the following metrics:

(1) ΔT - the temperature reduction anywhere in the studied domain that is achievable by activating the germanium cooler. This metric characterizes the intrinsic thermoelectric cooling capability of the germanium cooler.

$$\Delta T = T_{cooler,on} - T_{cooler,off} \tag{2}$$

(2) $\Delta T_{hotspot}^{*}$ - the ratio of the temperature change at the hot spot due to activating the germanium cooler to the temperature rise engendered by the hot spot. This metric quantifies the hot spot cooling effectiveness of the germanium cooler.

$$\Delta T^*_{hotspot} = \frac{T_{hotspot\_on,cooler\_off} - T_{hotspot\_on,cooler\_on}}{T_{hotspot\_on,cooler\_off} - T_{hotspot\_off,cooler\_off}} \tag{3}$$

For $\Delta T_{hotspot}^* = 1$, the temperature rise engendered by the hot spot can be completely removed by the germanium cooler. For $\Delta T_{hotspot}^* = 0$, the germanium cooler is totally ineffective and for $0 < \Delta T_{hotspot}^* < 1$, the germanium cooler can achieve partial success in reducing the hot spot temperature. For $\Delta T_{hotspot}^* > 1$, the germanium cooler is capable of "overcooling" the hot spot relative to the base temperature of the germanium chip.

Figure 11 shows the temperature profiles along the center line of a 100 μm-thick, 12 mm × 12 mm germanium chip at an arsenic doping concentration of 2.8×10^{18} cm^{-3} and with a 70 μm × 70 μm, 700 W/cm^2 hot spot and 70 W/cm^2 background heat flux. It is observed that the presence of the 70 W/cm^2 background heat flux on the chip produces a uniform temperature profile of 106.3°C. If a 70 μm × 70 μm, 700 W/cm^2 hot spot is added to the center of the chip, the peak temperature increases to 111.3°C resulting in 5.0°C peak temperature rise on the chip. When a 600 μm × 600 μm micro-cooler is integrated onto the backside of the chip and activated with a current of 0.5A (the input power = 0.063 W), the peak temperature is reduced to 108.5°C, bringing in a 2.8°C hot spot cooling. If the current is increased to 1.25A (the input power = 0.39 W), the peak temperature is further reduced to 106.8°C and a 4.5°C hot spot temperature reduction is achieved. However, if the current is increased further to 2.0A (the input power = 1.0 W), the peak temperature reaches 108.1°C and only 3.2°C hot spot cooling is obtained. It should be noted that use of the micro-cooler results in a locally-complex temperature variation, including a cool ring at the edge of the hot spot and a slightly elevated temperature in a second ring surrounding the hot spot. For the stated conditions, germanium self-cooling yields 90% (4.5°C) removal of the 5.0°C local temperature rise produced by the 700 W/cm^2 hot spot but consumes 0.39 W to operate the thermoelectric cooling.

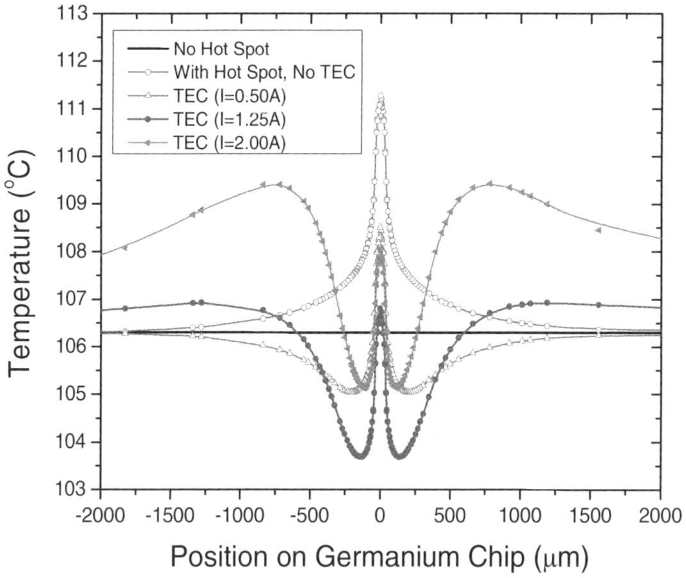

Fig. 11. Temperature profiles on the front side of the germanium chip with 700 W/cm^2, 70 μm × 70 μm hot spot and 70 W/cm^2 background heat flux. The germanium chip thickness is 100 μm and the arsenic doping concentration is 2.3×10^{18} cm^{-3} in the germanium substrate[20].

Figure 12 shows the temperature distribution, heat flux distribution, and current density distribution inside the germanium chip when the self-cooling is activated. It can be seen that if the self-cooling is activated, the heat flux leaving the hot spot is directed towards the germanium cooler and the majority of heat from the hot spot is absorbed by the cooler. Due to the use of an oversized germanium cooler, the self-cooling not only cools the hot spot but also cools the surrounding germanium. Figure 12(C) illustrates that the current density and thus joule heating is highly non-uniform in the chip. The current density reaches its maximum value around the metal contact but fully spreads inside the chip. On the other hand, the current density reaches its minimum value around the hot spot thereby bringing in least joule heating at the high heat flux region. These thermal-electric coupling phenomena have an implication that, due to three dimensional current spreading, the proposed self-cooling concept can offer great advantages

— not only reducing the Joule heating around the micro-cooler but also minimizing the Joule heating at the hot spot — and effectively improve the thermoelectric cooling performance in comparison to traditional 1-D configuration.

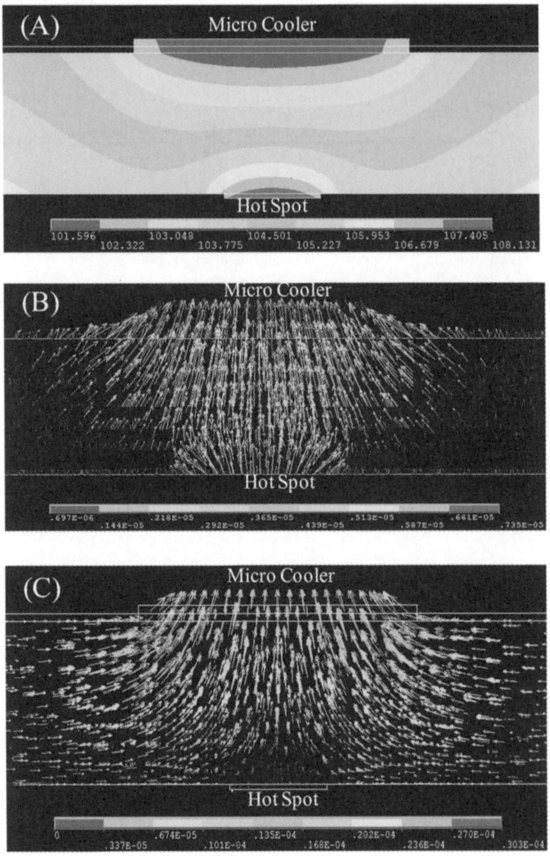

Fig. 12. (A) Temperature distribution, (B) heat flux distribution, (C) and current density distribution inside the germanium chip with self-cooling activated. A 200 μm × 200 μm micro-cooler located on the top of 100 μm-thick germanium chip and operated at an applied current of 0.75A; The germanium substrate has an arsenic doping concentration of 2.3×10^{18} cm^{-3} with a 700 W/cm^2, 70 μm × 70 μm hot spot and 70 W/cm^2 background heat flux[20].

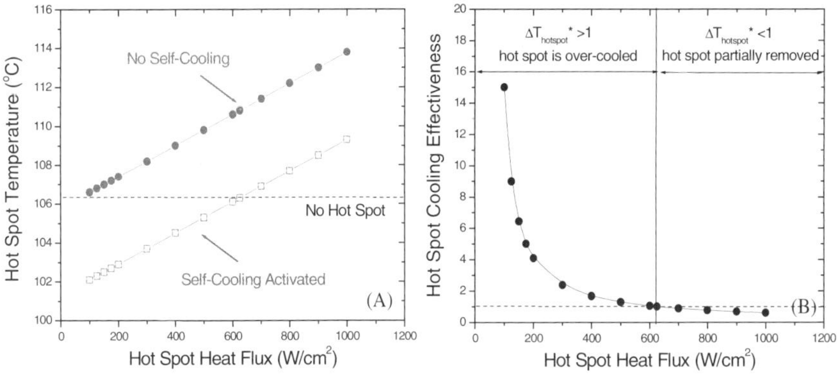

Fig. 13. (A) Hot spot temperature and (B) hot spot cooling effectiveness as a function of hot spot heat flux. 100 μm-thick germanium substrate has a 70 μm × 70 μm hot spot and 70 W/cm^2 background heat flux at an arsenic doping concentration of 2.3 × 10^{18}cm^{-3}. The micro-cooler size is 600 μm × is 600 μm[20].

Figure 13 shows the effect of hot spot heat flux on hot spot cooling and hot spot cooling effectiveness when the hot spot size and micro-cooler size is fixed at 70 μm × 70 μm and 600 μm × 600 μm, respectively. It can be seen that an increase of hot spot heat flux from 70 W/cm^2 to 1000 W/cm^2 has little effects on the hot spot cooling but significantly reduces the hot spot cooling effectiveness. As indicated in Figure 14(B), the hot spot cooling effectiveness is substantially improved with decreasing hot spot heat flux. A 600 μm × 600 μm micro-cooler operated at the optimized current of 1.25A with a cooling power consumption of ~0.4 W can completely remove a 620 W/cm^2 hot spot and reach a value of $\Delta T_{hotspot}^* = 1.0$. The micro-cooler will overcool the hotspot and reach a value of $\Delta T_{hotspot}^* > 1.0$ if the heat flux is less than 620 W/cm^2. When the hot spot heat flux is more than 620 W/cm^2, the micro-cooler reaches a value of $\Delta T_{hotspot}^* < 1.0$ and thus partially suppress the hot spot temperature.

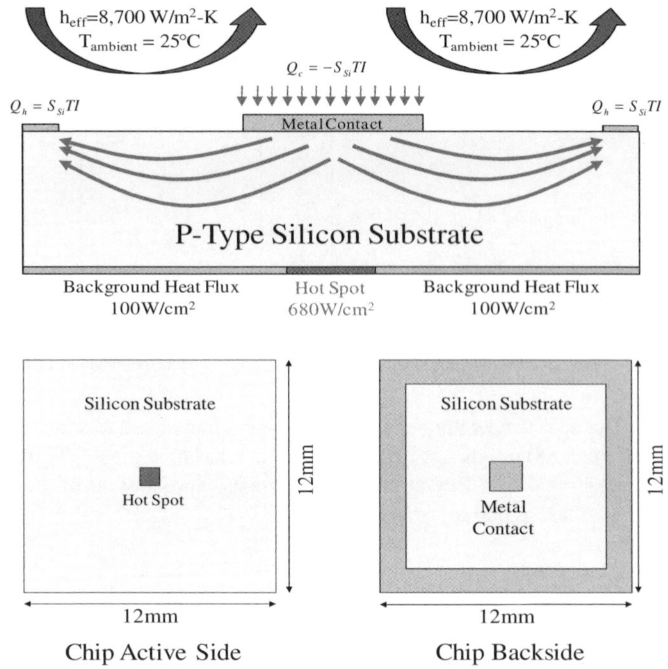

Fig. 14. Structure of on-chip silicon self-cooling for hot spot thermal management. The arrows indicate the direction for electrical current[21].

4.2. *Silicon Self-cooling Concept*

The concept of silicon self-cooling for on-chip hotspot cooling, fabricated on the back of the silicon chip, is illustrated in Figure 14[21]. Single-crystal silicon has a high power factor when its doping concentration is optimized. High thermal conductivity of silicon aids in the flow of heat from the hot spot to the micro-cooler. In this design, the lateral electrical current flow makes it possible to dissipate Peltier heating at the edge of the chip, far from the location of the hot spot, thus substantially reducing its detrimental effect on the temperature of active circuitry. In Wang's analysis, a 12 mm × 12 mm silicon chip with 70 W/cm^2 background heat flux and 70 μm × 70 μm hotspot, with a heat flux of 680 W/cm^2, located at the center of the active side of the chip is used as the test vehicle to explore the hot spot cooling potential[24,25,26].

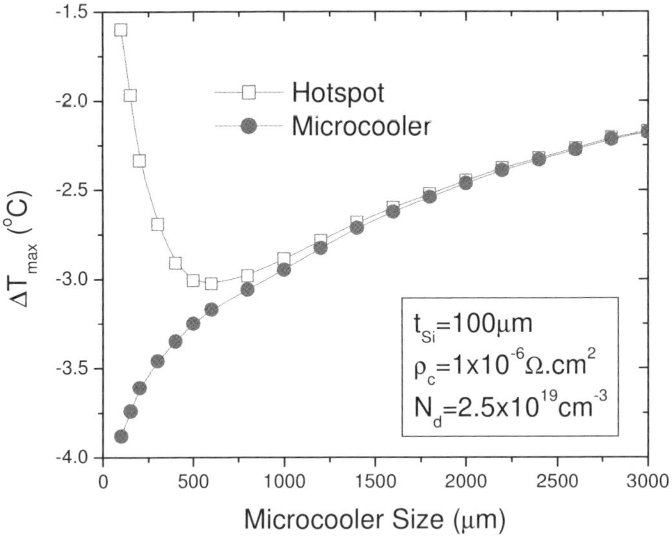

Fig. 15. Variation of temperature reductions at the hotspot and the micro-cooler with micro-cooler size[21].

The effect of silicon cooler size on cooling performance involves the interplay of thermoelectric cooling by the silicon cooler and thermal diffusion from the hot spot to the silicon cooler. With decreasing cooler size, the effective cooling flux and thus the temperature reduction at the cooler increases, while the thermal resistance between the hot spot and the silicon cooler also increases. Consequently, this larger cooling flux at smaller coolers cannot effectively translate into larger temperature reduction at the hotspot. On the other hand, with smaller thermal resistances between the hot spot and the silicon cooler, the more modest cooling flux on larger coolers can be projected effectively onto the hot spot, narrowing the temperature difference between the hot spot and the silicon cooler. However, the modest cooling flux achievable on the larger coolers reduces the beneficial temperature reduction at both the hotspot and the micro-cooler. The competition between these two effects results in an optimum micro-cooler size.

Figure 15 displays this behavior and shows the temperature reductions at the hot spot and the silicon cooler for a wide range of micro-cooler sizes for 100 μm thick chip operating under the background and hot spot heat fluxes of 70 W/cm^2 and 680 W/cm^2, respectively. It is seen that the temperature reduction at the hot spot first increases with silicon cooler size and, after reaching the maximum value of 3.0°C for 600 μm × 600 μm silicon cooler, decreases with a further increase in cooler size. Interestingly, the temperature reduction at the silicon cooler varies monotonically with cooler size, yielding progressively larger temperature reductions, to as much as 3.9°C, as the silicon cooler dimension shrinks to 100 μm × 100 μm which, however, only provides 1.6°C temperature reduction at the hot spot.

In the application of silicon coolers to hot spot remediation, the silicon chip plays multiple roles, functioning as a thermoelectric material, to provide on-chip cooling and, at the same time, as an electrical conductor to transfer electrons from the ground electrode to the micro-cooler, and as a thermal conductor to provide a diffusion path for the heat generated in the chip to the ambient. Therefore, the chip thickness influences Joule heating distribution inside the chip, heat spreading from the hotspot, heat diffusion from the hotspot to the micro-cooler, and heat diffusion from the ground electrode, where Peltier heating occurs, to the hotspot. As the chip becomes thinner, the thermal resistance between the micro-cooler and the hotspot decreases, allowing the micro-cooler to achieve greater hotspot temperature reductions, e.g., 2.05°C to 3.03°C as the chip thickness decreases from 500 μm to 100 μm, for the conditions of Figure 16. However, due to the smaller heat spreading effect in thinner chips, the temperature rise engendered by the hotspot is also higher for thinner chips and increases with decreasing chip thickness from 2.2°C for a 500 μm thick chip to 2.9°C for 100 μm-thick chip. These two trends compete with each other, yielding the maximum hotspot cooling effectiveness at the chip thickness of 200 μm, with $\Delta T_{hotspot}^* = 1.2$ as shown in Figure 16. At this chip thickness, the silicon micro-cooler is, thus, capable of reducing the hotspot temperature below the baseline temperature of the chip by approximately 0.5°C. Moreover, for the present 70 μm × 70 μm hotspot with a heat flux of 680 W/cm^2, the silicon micro-cooler is capable of completely suppressing or over-cooling the hotspot, with $\Delta T^*_{hotspot} \geq 1$, for the chip thicknesses between 100 μm and 475 μm.

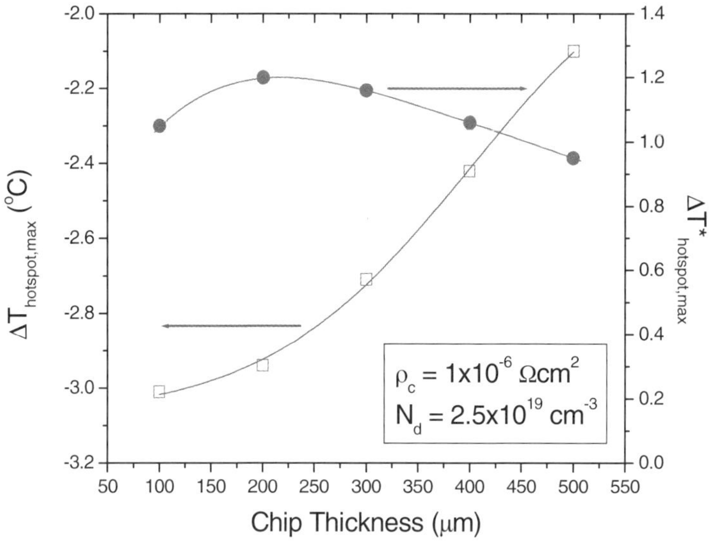

Fig. 16. Hotspot cooling and hotspot cooling effectiveness as a function of chip thickness. The hotspot is 70 μm × 70 μm with a heat flux of 680 W/cm^2 [21].

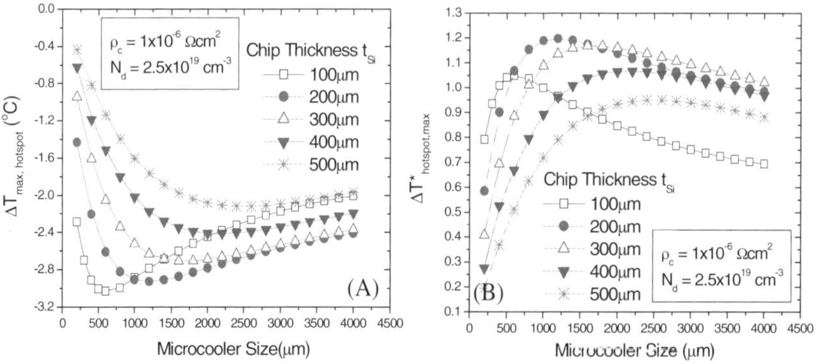

Fig. 17. (A) Hot spot cooling and (B) hot spot cooling effectiveness as a function of silicon cooler size for various chip thicknesses[21].

It is interesting to find that the chip thickness also influences the optimum micro-cooler size. As shown in Figure 17, the thicker the silicon chip the larger the optimized cooler size. For a 100 μm thick chip, the maximum hot spot cooling of 3.0°C for the conditions studied, is achieved with a 600 μm × 600 μm silicon cooler, while for 500 μm thick chip, 2500 μm × 2500 μm silicon cooler is required in order to attain the maximum hot spot cooling of 2.1°C. It should be noted that the optimum ratio of cooler size to chip thickness is approximately 5.5, with a modest sensitivity to chip thickness, reaching 6.0 for 100 μm, 200 μm and 300 μm thicknesses, dropping to 5.5 for the 400 μm thick chip, and to 5.0 for the 500 μm chip. This decreasing ratio can be related to the growing contributions of silicon Joule heating and Peltier heating to the hot spot temperature as the optimized current increases.

5. Conclusions

This paper reviews the recent advances in developments and applications of thermoelectric devices for electronics cooling with a major focus on hot spot thermal management. Three thermoelectric cooling concepts, including thin-film thermoelectric cooling, mini-contact cooling, and semiconductor self-cooling in silicon substrate and germanium substrate, are discussed, the advantages and disadvantages of these on-chip cooling solutions for high flux hot spots are evaluated, and the physical phenomena underpinning these on-chip thermal management approaches are discussed. It is theoretically demonstrated that, with proper thermal, electrical, and geometric optimizations, the semiconductor self-cooling can provide 3 ~ 5°C temperature reduction on a 700 W/cm^2 70 μm × 70 μm hot spot on the silicon or germanium chip, the mini-contact cooling technology can suppress at least 17°C at a 1250 W/cm^2 400 μm × 400 μm hot spot on the silicon chip, and the thin-film thermoelectric cooling, when integrated into air-cooled heat sink, can achieve more than 23C°C. Experimentally, it is identified that around 9.0°C spot cooling can be achieved on the silicon chip using mini-contact cooling technology and more than 15°C temperature reductions at a 1250 W/cm^2 400 μm × 400 μm hot spot on an actual silicon chip package can be obtained using TFTEC cooling solution. These recent progresses suggest that thermoelectric cooling could a very promising active cooling solution for hot spot thermal management.

References

1. A. Shakouri, "Nano-scale thermal transport and microrefrigerators on a chip", *Proceedings of IEEE*, **94**, 1613 (2006).
2. R. Mahajan, C. Chiu, and G. Chrysler, "Cooling a microprocessor chip", *Proceedings of IEEE*, **94**, 1476 (2006).
3. A. Majumdar, "Thermoelectric devices: Helping chips to keep their cool", *Nature Nanotechnology* 4, 214 (2009).
4. D. S. Chau, G. Chrysler, S. Narasimhan, D. Ganapathy, and K. Lofgreen, "Feasibility Study of Using Solid State Refrigeration Technologies for Electronic Cooling", *Proceedings of The Tenth Intersociety Conference on Thermal and Thermomechanical Phenomena in Electronics Systems (ITHERM '06)* (U.S.A, 2006), pp. 464–469.
5. P. Wang, P. McCluskey, and A. Bar-Cohen, "Hybrid Solid- and Liquid-Cooling Solution for Isothermalization of Insulated Gate Bipolar Transistor Power Electronic Devices", *IEEE Transactions on Components, Packaging and Manufacturing Technology,* **3**, 601 (2013).
6. S. Ramanathan, and G. M. Chrysler, "Solid-State Refrigeration for Cooling Microprocessors", *IEEE Transactions on Components and Packaging Technologies* **29**, 179 (2006).
7. S. Narasimhan, D. Ganapathy, G. Chrysler, D. Chau, and K. Lofgreen, "Use-condition thermal metrics for characterization of thin film TEC modules on electronic packages", *Proceedings of The Tenth Intersociety Conference on Thermal and Thermomechanical Phenomena in Electronics Systems (ITHERM '06)* (U.S.A, 2006), pp. 476–0482.
8. S. Narasimhan, K. Lofgreen, D. Chau, and G. Chrysler, "Thin Film Thermoelectric Cooler Thermal Validation and Product Thermal Performance Estimation," *Proceedings of The Tenth Intersociety Conference on Thermal and Thermomechanical Phenomena in Electronics Systems (ITHERM '06)* (U.S.A, 2006), pp. 470–475.
9. I. Chowdhury, R. Prasher, K. Lofgreen, G. Chrysler, S. Narasimhan, R. Mahajan, D. Koester, R. Alley, and R. Venkatasubramanian, "On-chip Cooling by Superlattice-Based Thin-film Thermoelectrics", *Nature Nanotechnology*, 25 January 2009.
10. I. Chowdhury, R. Prasher, K. Lofgreen, S. Narasimhan, and R. Mahajan, "Site-specific and on-demand high heat flux cooling using superlattice based thin-film thermoelectrics," *Proceedings of the ASME InterPACK Conference*(San Francisco, U.S.A, 2009), Paper No: InterPACK2009-89268.
11. A. Shakouri, and Y. Zhang; "On-chip solid-state cooling for integrated circuits using thin-film microrefrigerators", IEEE Transactions on Components and Packaging Technologies, **A28**, 65 (2005).
12. B. Yang, P. Wang, and A. Bar-Cohen, "Mini-contact enhanced thermoelectric cooling of hot spot in high power devices", *IEEE Transactions on Components and Packaging Technologies* **30**, 432 (2007).

13. P. Wanga, B. Yang, and A. Bar-Cohen, "Mini-Contact Enhanced Thermoelectric Coolers for On-Chip Hot Spot Cooling", *Heat Transfer Engineering* **30**, 736 (2009).
14. T. H Geballe., and G. W. Hull, "Seebeck effect in silicon", *Phys. Rev.* **98**, 940 (1955).
15. T. H Geballe., and G. W. Hull, "Seebeck effect in germanium", Phys. Rev. 94, 1134 (1954).
16. S. K. Sutradhar, and D. Chattpoadhyay, "Thermoelectric power of n-GaAs", *J Physics C: Solid State Physics* **12**, 1693 (1979).
17. M. S. Brandt, P. Herbst, H. Angerer, O. Ambacher, and M. Stutzmann, "Thermopower investigation of n- and p-type GaN", *Phys. Rev.* **B 58**, 7786 (1998).
18. A. Sztein, H. Ohta, J. Sonoda, A. Ramu, J.E. Bowers, S. P. DenBaars and S. Nakamura1, "GaN-Based Integrated Lateral Thermoelectric Device for Micro-Power Generation", *Appl. Phys. Express* **2**, 111003 (2009).
19. S. Fukuda, T. Kato, Y. Okamoto, H. Nakatsugawa, H.Kitagawa, and S. Yamaguchi, "Thermoelectric Properties of Single-Crystalline SiC and Dense Sintered SiC for Self-Cooling Devices", *Jap. J. Appl. Phys.* **50**, 031301 (2001).
20. P. Wang and A. Bar-Cohen, "Self-Cooling on Germanium Chip," *IEEE Transactions on Components, Packaging, and Manufacturing Technology*, **1**, 705 (2011).
21. P. Wang, and A. Bar-Cohen, "On-chip hot spot cooling using silicon thermoelectric microcoolers," Journal of Applied Physics, **102**,034503(2007).
22. P. Wang, A. Bar-Cohen, B. Yang, *et al.*, "Analytical modeling of silicon thermoelectric microcooler," *Journal of Applied Physics*, **100**, 14501 (2006).
23. P. Wang, M. Manno, and A. Bar-Cohen, "Quantum-Well Si/SiC Self-Cooling for Thermal Management of High Heat Flux GaN HEMT Semiconductor Devices," *Proceedings of 3rd ASME Micro/Nanoscale Heat & Mass Transfer International Conference*, March 3-6, 2012, March 3-6, 2012, Paper Number: MNHMT2012-75290.
24. V. Litvinovitch, P. Wang, and A. Bar-Cohen, "Impact of Integrated Superlattice micro-TEC Structures on Hot Spot Remediation," *IEEE Transactions on Components and Packaging Technologies*, **33**, 229 (2010).
25. A. Bar-Cohen and P. Wang, "Thermal Management of On-Chip Hot Spot," *ASME Journal of Heat Transfer*, **134**, 051017 (2012).
26. P. Wang, A. Bar-Cohen, B. Yang, G. L. Solbrekken, Y. Zhang, and A. Shakouri, "Thermoelectric microcooler for hotspot thermal management", *Proceedings of InterPACK'05*, San Francisco, U.S.A, 2005), Paper No: 2005-7324.

CHAPTER 8

ENERGY EFFICIENT SOLID-STATE COOLING FOR HOT SPOT REMOVAL

Kazuaki Yazawa[1], Andrei Fedorov[2], Yogendra Joshi[2], Ali Shakouri[1]

[1]*Birck Nanotechnology Center, Purdue University*
1205 W. State St., West Lafayette, IN 47906, USA
kyazawa@purdue.edu
[2] *G.W. Woodruff School of Mechanical Engineering*
Georgia Institute of Technology

In this chapter, modeling and analysis of a hybrid scheme of a thermoelectric microcooler and a microchannel single-phase heat sink is discussed for a hotspot cooling. Following the introduction, the hybrid scheme concept is described. The Section 3 describes thermoelectric materials and fabrication of the solid-state microcoolers to give the necessary information for the thermal modeling, analysis, and the optimization of thermoelectric element in Section 4. Microchannel geometry and the pump power are discussed in Section 5 with an analytic model, and then the heat sink design itself is designed to optimum for lowest power used for the required cooling performance. Integrated cooling power for an integrated circuit (IC) with a hotspot as a function of heat flux is demonstrated. Section 6 summarizes the energy efficient cooling performance by the discussed hybrid scheme. To make technological challenges clear, concept of a new packaging approach for this integration is illustrated in Section 7 followed by the conclusions.

1. Introduction

As an application of heat transfer, electronics cooling imposes a wide variety of challenges. History of electronics thermal packaging is found elsewhere, but a comprehensive description of the transition of thermal packaging will be found in [1]. A hot spot in general has existed since the light bulb was invented by Thomas Edison [2]. It has not become a major concern, however, until the 21st century. The heat dissipation per unit area or per unit volume of semiconductor devices has been sufficiently moderate to enable design of an air-cooled heat sink to manage the heat dissipation and to maintain the maximum allowable temperature of the heat source. The temperature limitations originate from several factors for deep submicron CMOS ICs. One factor is, the electron mobility is a function of temperature for high frequency operation. Another factor is, the leakage current at the transistor exponentially increases with increasing temperature.

Recent progress in precision technology enables dimensional features on a nanometer scale for mass produced silicon chips. These technology developments following Moore's law [3] have also created inconvenient parasitic concerns. Due to over a billion transistors on a single microprocessor, and the huge complexity of recently developed integrated circuits, localization of multiple hot spots has been accelerated both in space and in time. An industrial research report [4] projects that due to the small features and high clock frequencies, the localized heat flux may reach 1000 W/cm^2.

Many approaches have been studied for cooling this super high heat flux including: single phase microchannel [5], two-phase microchannel [6], jet impingement [7], mist cooling [8], and emission cooling [9]. Most of these prior studies have focused on a uniform high heat flux in a relatively small area. The non-uniform distribution of heat flux across the chip area due to co-location of different functional blocks is significantly increasing. The following discussion in this chapter assumes a 100 W/cm^2 of uniform background heating, along with a 400-1,000 W/cm^2 of heat flux on a hotspot located at the center of the chip.

Additionally, to meet the needs for compact electronic packaging, thermal management solutions for the localized heat fluxes must fit into a very low profile package. The use of a solid-sate thermoelectric (TE) device is one of the solutions for local high heat flux removal. A TE device works as a heat pump. A typical TE module consists of solid semiconductor TE elements in an array within a sandwich structure between electrodes and ceramic substrates. A TE cooler is a relatively moderate energy conversion device compared to a mechanical thermodynamic refrigerator, due to its internal Joule heating and parasitic heat leakage. TE is currently the only active refrigeration solution available to locate directly on a hotspot, which is less than 1 mm. Recent thin film TE devices can be fabricated as thin as tens of microns [10]. The device has no moving parts and, with direct electric drive and a small thermal mass, can respond very quickly to pulsed operation. Studies of embedded thermoelectric cooling have been reported [11], however, an important advantage of this solid-state device has not yet been presented. Thermoelectric cooler performance is maximized by designing the thermoelement (leg) dimensions, as well as the drive current. Therefore, material development and systematic device design are critical challenges for cooling high heat flux hot spots.

In this chapter, the unique challenges and technology development of microcoolers are discussed with emphasis on silicon-based superlattice thermoelectric microcoolers in view of their applicability to monolithic integration onto the silicon substrate. The proposed Hybrid Cooling Scheme combining solid-state and fluidic cooling aims to minimize the cooling power consumption for hotspot heat removal while also addressing entire heat removal [12]. The superlattice microcooler is fabricated by nanoscale layer-by-layer growth with thin film deposition. This monolithic fabrication provides an excellent contact to the silicon substrate, compatible with CMOS ICs.

2. Hybrid Cooling Scheme

The Hybrid Cooling Scheme combines micro-fluidic and solid-state cooling techniques to exploit its unique advantages, and overcomes challenges associated with each method. It uses a liquid cooled microchannel heat sink to dissipate the background heat flux and solid state devices to remove high heat flux from the localized hotspots. The microchannel heat sink is designed to primarily remove the uniform background heat, not the heat from the hotspots. The superlattice cooler (SLC) is strategically located right at the hotspot locations. By using the small heat pump exclusively on the hot spot, both the uniform back ground heat and the hot spot heat are removed without significantly increasing cooling power for the microchannel heat sink.

Figure 2-1 shows a schematic of the Hybrid Cooling Scheme. It consists of a microchannel heat sink, modified by fabricating an array of superlattice coolers on the back side. SLCs are silicon micro-fabrication compatible and hence can be directly fabricated on the back of the silicon microchannel heat sink. SLCs are positioned at the expected locations of the hotspots. An array of superlattice coolers can be used to manage multiple hotspots, or dynamically shifting hotspots.

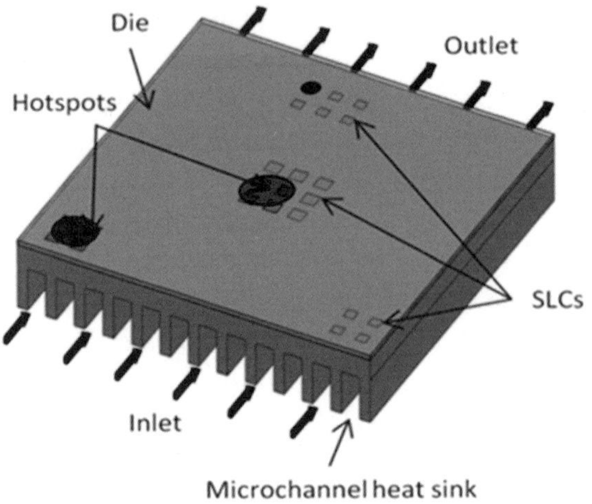

Fig. 2-1. Schematic of the proposed Hybrid Cooling Scheme. Small rectangular features show the location of superlattice coolers (green and blue features represent non-activated and activated SLCs respectively), transferred from [13].

The hot electrodes of SLCs are placed away from the active region of the die. This modified microchannel heat sink is bonded to the die. The working fluid enters the microchannel heat sink through the inlet port, picks up the background heat dissipated by the die, and exits the microchannel through the outlet port. If a hotspot is present, SLCs located in the proximity of the hotspot are activated. This provides a localized cold spot that absorbs the heat dissipated by the hotspot. The heat generated at the hot electrode of the SLC is dissipated to the working fluid.

The key advantage of the above Hybrid Cooling Scheme is that it is able to handle the non-uniform heat flux without overdesigning the microchannel heat sink. In the presence of the hotspot, the heat sink has to be designed based on the power dissipated by the hotspot. However, since the hotspot characteristic size is typically a few to hundreds of microns compared to the die size of ~ 1 × 1 cm, the heat sink will have to be significantly overdesigned.

Some of the other key features of the Hybrid Cooling Scheme are:

- It provides an energy efficient solution to address non-uniform heat dissipation from the die.
- It can dynamically adapt to address the microprocessor power dissipation map in real time, due to the small size of the solid-state elements.
- It is compatible with silicon micro-fabrication techniques.
- With proper design, it can be integrated into 2D as well as 3D ICs.

Through modeling, device fabrication, and testing of the above Hybrid Cooling Scheme, it was found that the optimum current of the SLC is a function of the superlattice size, ambient temperature, ground electrode location, as well as the thermal and interface resistance [14]. The optimum current increases with the superlattice size and ambient temperature. However, it decreases with interface and thermal resistance as the parasitic heat transfer from the ground electrode to the superlattice increases. Furthermore, the superlattice cooler should always be operated below its optimum current to minimize the operating power.

Two configurations fabricated and evaluated for the investigation include one with an interface layer between the SLC ground electrode and the heat sink, and another with microchannels on the same wafer as the SLC [15]. The interface layer configuration resulted in ~400% reduction in cooling obtained at the superlattice cooler. Thermal resistance between the ground electrode and the superlattice cooler is another critical factor affecting the performance of the Hybrid Cooling Scheme. For the off-chip microchannel configuration, thermal resistance between the superlattice and ground electrode reduced the maximum temperature drop achieved by 25%. However, for the on-chip microchannel configuration, the heat sink is effective enough to remove the heat, and the cooling performance is not impacted. For both cases, the power required to operate the superlattice cooler increases considerably as the ground electrode is moved farther away from the superlattice cooler due to the increase in the substrate electrical resistance.

In the actual implementation of such a scheme, an array of SLCs is utilized. Thermal coupling between SLCs was found by Sahu [16] to reduce the individual performance compared to an isolated SLC by up to 87%. Thermal coupling is affected by the geometric and operating parameters. As the separation between the ground electrode and the superlattice cooler is reduced from 1000 μm to 100 μm, thermal coupling increases from 10% to 60%. Reducing the convective heat transfer coefficient from 250,000 W/m^2K to 50,000 W/m^2K increases the thermal coupling by 87%. Even in the ideal case corresponding to an infinite heat transfer coefficient, thermal coupling reduces the performance of the SLC by 4% compared to an isolated SLC. Experiments confirmed the significant thermal coupling between the SLCs. Trends observed with experiments agreed with the computational model.

Due to their fast response time, SLCs are well suited to handle transient hot spots. Response time of the SLC was found to be independent of the operating current, as well as its size. The measured time constant of the SLC was 35 μs exhibiting good agreement with test device characterizations [17], [18]. Experiments revealed that; to reduce the temperature fluctuation at the hotspot, the SLC should be activated as soon as a hotspot is detected. Since the temperature at the hotspot increases exponentially, even a few microseconds delay in the activation of the SLC can result in a significant temperature rise at the hotspot. Increasing the ZT of the superlattice cooler dramatically improves its performance. ZT is a dimensionless figure-of-merit of thermoelectric materials. For a 100 μm × 100 μm superlattice cooler, cooling power density as high as 1700 W/cm^2 and a temperature drop of 20 K were predicted if the ZT is increased to 1.0.

3. Microcoolers, materials, and characterization

3.1. *Superlattice thermoelectric microcoolers*

Recent developments in thermoelectric materials are summarized in [19]. This section discusses the superlattice microcoolers. Unlike commercially available Peltier cooler modules, the superlattice microcooler is a monolithic solid device with a nanostructure containing very thin layers [20]. The device consists of hundreds to thousands of layers of two different materials as shown in Figure 3-1, transcribed from [21]. The dark layer made of Si0.7Ge0.3 is a barrier for hot electrons excited by the bias such that only the electrons of certain energy level pass through the barrier. The current flow induced by an external bias brings the heat energy from the heat source to the opposite side.

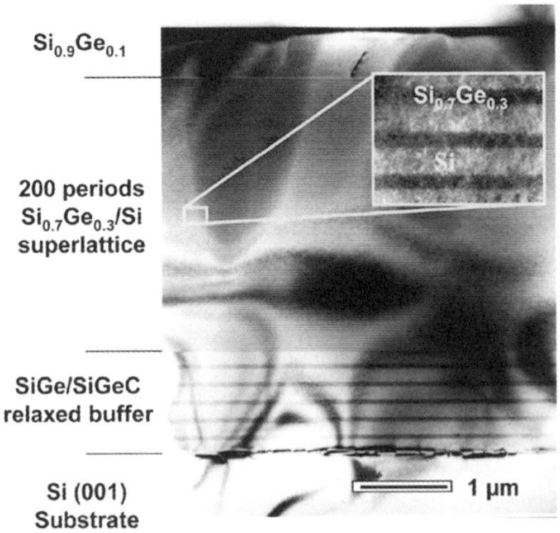

Fig. 3-1. Cross sectional transmission electron microscopy image of SiGe/Si superlattice thermoelectric device [21].

3.2. *Nanoscale heat transfer*

There are several approaches to enhance the dimensionless figure-of-merit (ZT) of the semiconductor thermoelectric materials defined as Eq. (1), where S is the Seebeck coefficient [V/K], σ is the electrical conductivity [1/ohm.m], and $\overline{T}$ [K] is the mean temperature of the material being used. A low dimension structure is one of the methods employed to reduce the bulk thermal conductivity [W/m.K] by reducing the phonon contribution β_p of the heat transport. β_e is the electron contribution of the heat transport. Thermoelectric materials of thinfilms [10], nanowires [22], and embedded nano particles [23] are studied as techniques to reduce the dimension smaller than the phonon mean free path (MFP) such that the phonon scattering increases [24],[25]. The phonon MFP is material dependent since it is directly related to the lattice structure. Manipulating lattice structure allows us to design for reduced thermal conductivity for ZT enhancement, while σ and S do not significantly change.

$$ZT = \frac{\sigma S^2}{\beta_p + \beta_e}\bar{T} \tag{1}$$

3.3. *Power factor enhancement*

The application for hot spot cooling, unlike refrigeration, requires the hot spot temperature to be no less than the surrounding area temperature but still pump the extremely high heat flux. This suggests that the back heat flow, following Fourier's law, caused by the reverse temperature gradient does not apply to this case. Therefore, thermal conductivity will not play a significant role. The material development for the enhancement of the power factor, σS^2, for this application then becomes more important. This electron transport related characteristic is hard to improve since the two components are inversely related. Doping intensity, relating to the free electron concentration, is the major factor influencing to optimize the power factor. Recent research [19] revealed that embedded nano particles also enhance the power factor, as well as reduce thermal conductivity. Methods to tune the band gap of the material have also been studied [26].

Detailed device modeling and analysis on micro refrigerators are reported in [27] including the impact of contact interfaces. Further analysis, based on Quadrupoles, was discussed comparing Si/SiGe superlattice and bulk SiGe [28].

3.4. *Fabrication of superlattice coolers*

Superlattice microcoolers and micro heaters are fabricated in a test chip for the characterization of thermoelectric performance. The test chip includes all functional units such as localized heating, localized cooling, and background heating as shown in Figure 3-1. To prevent the processing failure of bad insulation between the heaters and the coolers, a Plasma-Enhanced Chemical Vapor Deposition (PECVD) SiN_x insulation layer is deposited on top of the metals. A DC-sputtered TiN thin layer on top of the metals is better for the deposition of an insulation layer. Multiple PECVD SiO_2 layers are then deposited as insulation layers. A SiO_2 layer performed better than SiN_x for insulation. Pt heater lines are used instead of gold heaters on top of the superlattice mesas for higher power density for the same current input (Figure 3-3). Three different sizes of coolers are fabricated for characterization of the cooler performance as a function of size. Figure 3-3 shows the performance data of the test devices.

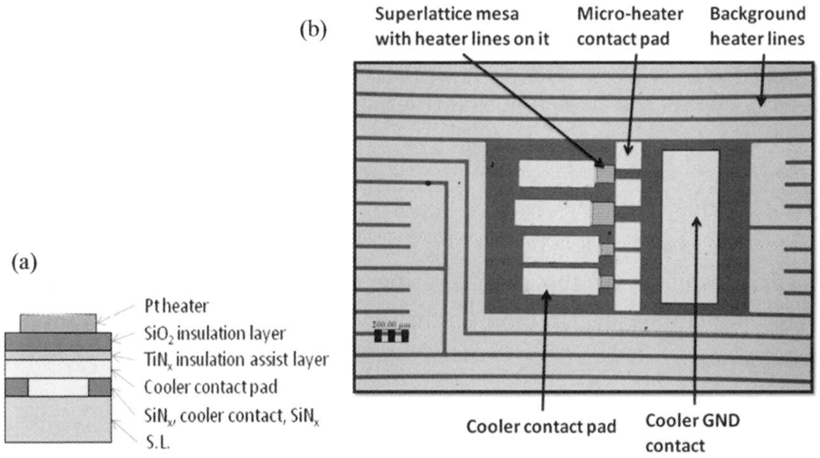

Fig. 3-2. (a) Schematic of the superlattice micro-cooler cross-section, and (b) picture of the hot-spot micro-heaters and micro-coolers surrounded by background heater lines on fabricated test chip.

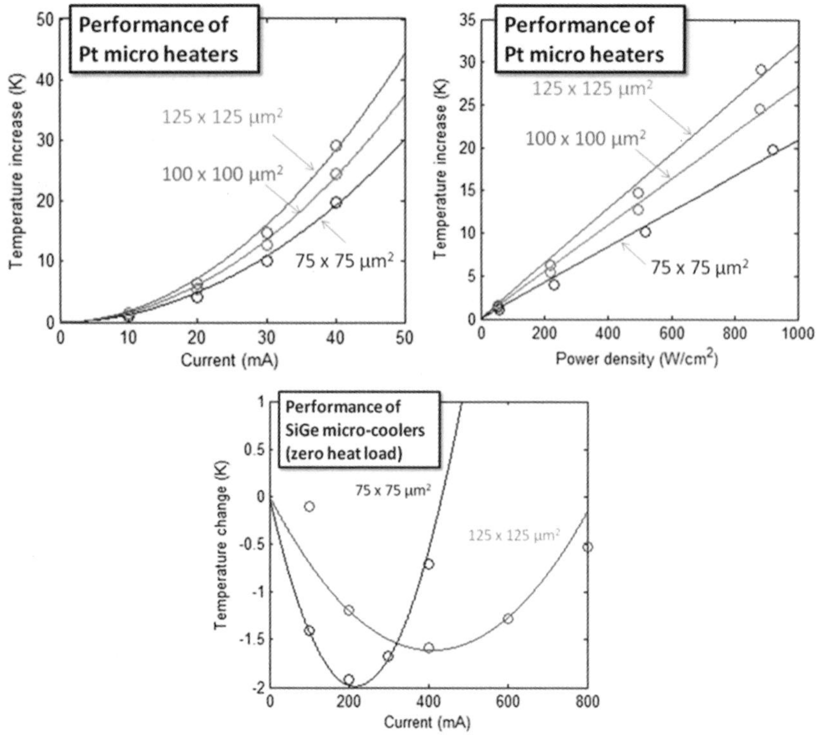

Fig. 3-3. Heating performance of the Pt micro-heaters (a) as a function of current input, and (b) as a function of power density. (c) Cooling performance of the SiGe superlattice micro-coolers with zero heat load.

Characterizing the superlattice cooler is also a challenge. Micro refrigerator is not much bigger than the smallest micro-thermocouples. Thermal imaging of the device using thermoreflectance showed the actual temperature distribution on the device [29] and very fast transient phenomena.

4. Modeling and optimization of hot spot coolers

4.1. *Analytic 1-D model*

Fukutani *et al.* obtained an analytical model for the optimum thickness of the micro-refrigerator for a given hot spot cooling application [30]. The following extensive study optimizes the device thickness along with the driving current. A 500 μm × 500 μm size solid-state microcooler is located underneath a similar sized hotspot. The entire chip uniformly dissipates a total of 100 W for 1 cm × 1 cm chip area. Figure 4-1 shows the thermal network diagram of the hotspot cooling portion of the system. Temperature, Tm, in Figure 4-1 is the base temperature of the chip where the chip is cooled by the liquid cooling microchannel not shown in this diagram.

a)

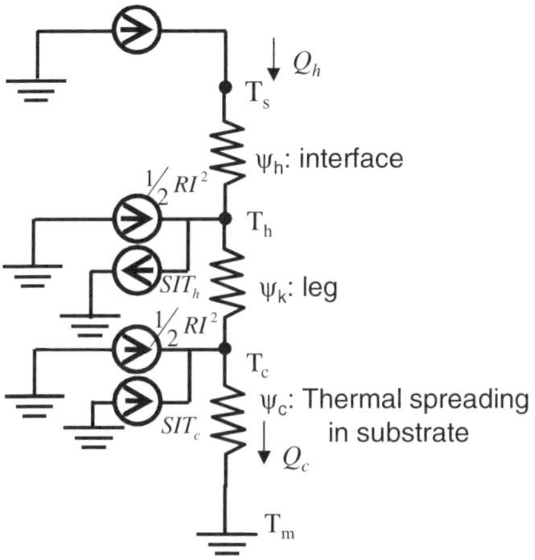

b)

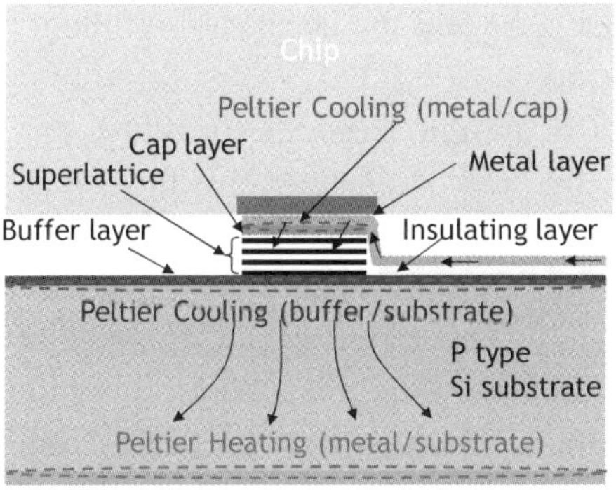

Fig. 4-1. a) Thermal resistance network diagram of the proposed model with a superlattice cooler on a silicon substrate and b) the corresponding physical structure diagram, transferred from [12].

The following four equations express the energy balance in the system shown in Figure 4-1.

$$Q_h = \frac{(T_s - T_h)}{\psi_h} \tag{2}$$

$$Q_h - ST_hI + \frac{R}{2}I^2 = K(T_h - T_c) \tag{3}$$

$$K(T_h - T_c) + ST_cI + \frac{R}{2}I^2 = Q_c \tag{4}$$

$$Q_c = \frac{T_c - T_m}{\psi_c} \tag{5}$$

where,

$$K = \frac{\beta FA}{d} \text{ and } \psi_k = 1/K \tag{6}$$

and

$$R = \frac{d}{\sigma FA} \tag{7}$$

In the equations, I is the drive current for pumping the heat and d is the thickness of the superlattice cooler. In this study, the superlattice cooler is treated as an effective bulk material. The model also applies to a bulk material such as Bi_2Te_3 or even a polymer thin film material, as long as the fabrication technique can yield the necessary optimum thicknesses of the thermoelectric material.

Assuming there is design freedom to adjust the thickness of the superlattice cooler, the key parameters are; the thickness and the drive current. The hotspot side thermal resistance, ψ_h, defines the temperature difference between the hotspot (junction), T_s, and the source side of the superlattice cooler, T_h. The term, ψ_h; in Eq. (2) is, in practice, the interface thermal resistance, thus it is always linear to the heat flow and the value is quite small. Therefore this parameter can be ignored in the optimization process.

First, T_h as a function of the thickness and the drive current, I, are found. From Eqs. (3) and (4),

$$Q_h - Q_c = -RI^2 + SI(T_h - T_c) \tag{8}$$

Substituting Eq. (5) into Eq. (8),

$$Q_h - \frac{(T_c - T_m)}{\psi_c} = -RI^2 + SI(T_h - T_c) \tag{9}$$

T_h is found by substituting Eq. (9) to Eq. (3),

$$T_h = \frac{-\frac{\psi_c RS}{2} I^3 + \left(\psi_c K + \frac{1}{2}\right) RI^2 - \psi_c Q_h SI + (\psi_c K + 1) Q_h + KT_m}{\left(K + SI - \psi_c S^2 I^2\right)} \tag{10}$$

Substituting Eqs. (6) and (7) into Eq. (10), T_h is found as a function of the design parameters, d and I as,

$$T_h = \frac{-\frac{\psi_c S}{2\sigma FA} d^2 I^3 + \left(\beta FA \psi_c + \frac{1}{2} d\right) \frac{d}{\sigma FA} I^2 - \psi_c Q_h SdI + Q_h d + \beta FA(\psi_c Q_h + T_m)}{\left(\beta FA + dSI - \psi_c dS^2 I^2\right)} \tag{11}$$

To test the correctness of the equation, we apply $I \rightarrow 0$ and $d \rightarrow 0$. Then, Eq. (11) yields the following formula, which exactly matches the case without the superlattice cooler.

$$T_h = \psi_c Q_h + T_m \tag{12}$$

4.2. *Spreading thermal resistance in the silicon substrate*

Since the hotspot area, (A_s), is much smaller than the chip substrate area (A_p), a spreading thermal resistance from the hotspot must contribute to the heat path for the hotspot cooling. A closed-form spreading resistance by Lee *et al.* [31] is commonly used in the thermal packaging community. Based on that, Eq. (13) gives the spreading thermal resistance, ψ_S, instead of solving the diffusion equations [32].

$$\psi_S = \frac{\sqrt{A_P} - \sqrt{A_S}}{\beta\sqrt{\pi A_P A_S}} \frac{\phi\beta A_P \psi_E + \tanh(\phi t)}{1 + \phi\beta A_P \psi_E \tanh(\phi t)} \tag{13}$$

where ψ_E is the thermal resistance applied to the dissipation side using the effective uniform heat transfer coefficient h, ϕ is a function of the area of chip A_p and the area of hotspot A_s.

$$\psi_E = \frac{1}{hA_p} \text{ and } \phi = \frac{\pi^{1.5}}{\sqrt{A_P}} + \frac{1}{\sqrt{A_S}} \tag{14}$$

4.3. *Spreading Joule heating in silicon substrate*

The configuration in which the superlattice cooler is directly attached to the silicon substrate enables the substrate to be used as an electrode. The electrical current can spread to the substrate as heat flow spreads. The model that Sahu *et al.* reported [14] could be used in this model, however, we leave this factor as a perfect electrode (zero resistance) to avoid the complexity. Current density through the silicon substrate is much lower than through the SLC device. This is a crude assumption but should be valid as long as this impact is sufficiently small.

4.4. *Joule heating on the metal top layer*

Due to a finite electrical resistance, the current flow through the top metal layer induces Joule heating. Unlike the spreading above, this impact is not very predictable, since the edge of the device has a singularity in the electrical current flow. Rather than making assumptions, we feel it is reasonable to ignore this impact in the model.

4.5. *Optimization of the superlattice cooler*

Numerical investigation revealed a characteristic for the hotspot side temperature, T_h. Figure 4-2 shows a 3D plot for T_h (Z-axis) based on the thickness, d, and the drive current, I. The 3D surface represents Eq. (11) in the I, d, and T_h spaces. In the region of near minimum temperature, the temperature, $T_{h,}$ corresponds to the drive current, I (X-axis, axis of d) in a smooth quadratic shape. On the other hand, the shape caused by changing thickness, d (Y-axis, axis of I) exhibits quite different characteristics. For a smaller (thinner) d, the 3D surface changes gradually. For a larger (thicker) d, after hitting the minimum temperature, the temperature steeply increases as d increases. This may cause some difficulty for optimization.

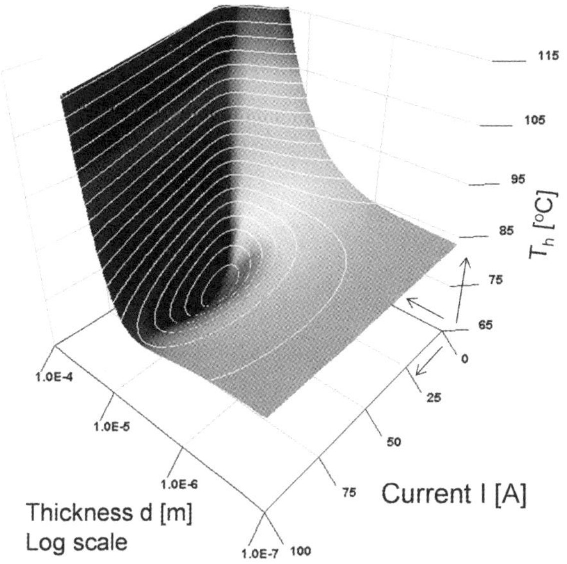

Fig. 4-2. T_h as a function of both TE cooler thickness, d, and drive current, I. ZT = 0.5 (thermal conductivity: 3.0 W/mK, electrical conductivity: 5.5×10^4 1/Ohm.m, Seebeck coefficient: 2.8×10^{-4} V/K). T_a = 35 °C, transferred from [12].

Based on Figure 4-2, if the system is optimized for a specific fixed temperature, T_h, there will be many selectable combinations of d and I. The solution is along one of the curves shown as contoured white lines in the figure. If T_h is fixed at 84 °C, considering the interface resistance to the hotspot, there is an option to choose either the minimum drive current, I_{min} or the minimum thickness, d_{min}. The I_{min} is the optimum for an energy-sensitive solution and the latter, d_{min}, is the optimum for a cost-sensitive solution when using expensive materials, which may also require a relatively complex fabrication process.

4.6. *Mathematical optimum*

To find the mathematical expression for I_{min}, the following process has been carried out. The relationship between d and I for a specific T_h may exist in a circular or a U-shaped function on a particular T_h plane. Reformulating Eq. (11), I as a function of $d$is obtained, i.e. $I = f(d)$. The function has three solutions including an imaginary part. However, in a narrow range the minimum current may be found for the real part only. Setting the derivative dI/dd equal to zero for a given temperature T_h, should yield the optimum thickness, d_{opt}. Substituting this d_{opt} into $I = f(d)$, would then yield I_{min}. However, obtaining the formula that only includes d_{opt}, is still too complex to solve. Finding the mathematical expression for the optimum design is reserved for future work.

4.7. *Numerical simulation*

A series of 3D numerical simulations solves the junction temperature, T_j, while the active hotspot cooling portion of the hybrid system is considered. The results compared with the analytic 1D model show a difference, which is based on the 3D temperature profile. The model is shown in Figure 4-3 and the example temperature contours are shown in Figure 4-4. The comparison for 600 W/cm^2 as an example is shown in Figures 4-5 and 4-6. The optimum current is ~50 A for a 500 μm × 500 μm hot spot. This requires special low resistance contacts and wirings. By putting several n- and p-type thermoelectric legs of the same thickness electrically in series and thermally in parallel, one can tradeoff current and voltage and significantly reduce the optimum current.

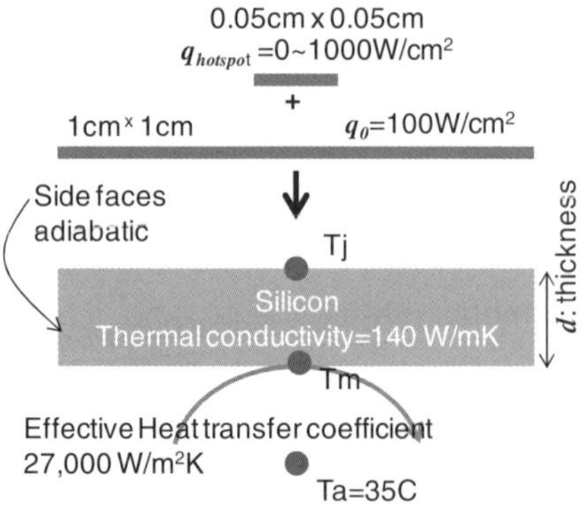

Fig. 4-3. Schematic of the model used for the simulations. A 500 μm × 500 μm hot spot with different heat fluxes is applied at the center of a 1 cm × 1 cm chip with a uniform heat dissipation of 100 W/cm^2, transferred from [12].

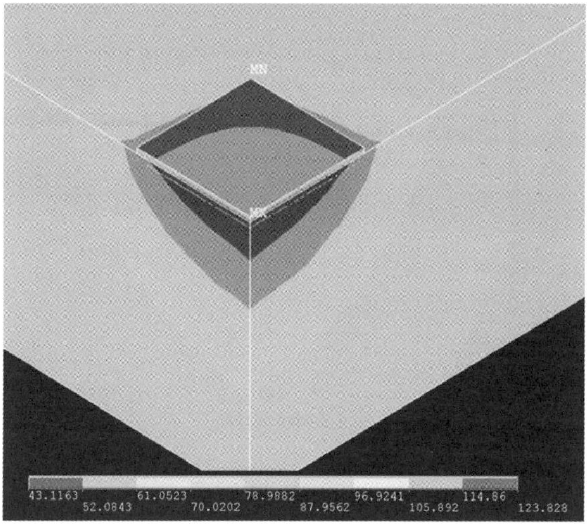

Fig. 4-4. Simulation result - temperature contour of a quarter of the model. The heat flux is 600 W/cm^2 at the hotspot located at the center of the chip, transferred from [12].

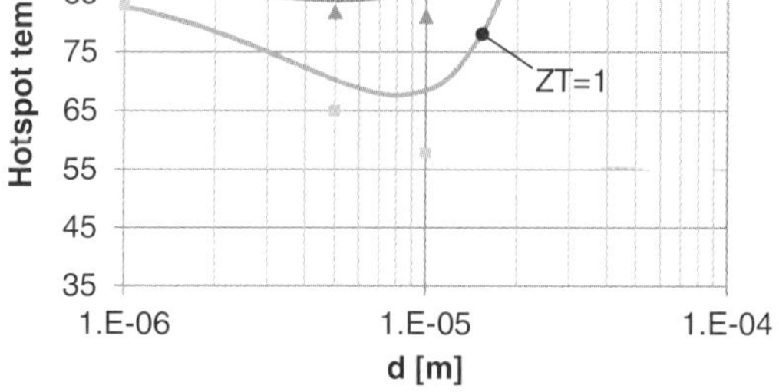

Fig. 4-5. Hotspot temperature vs. superlattice thickness. Drive current is 60 A. The hotspot heat flux is 600 W/cm^2 on a 500 μm × 500 μm area. Solid line is for analytical model and symbols are ANSYS simulation results, transferred from [12].

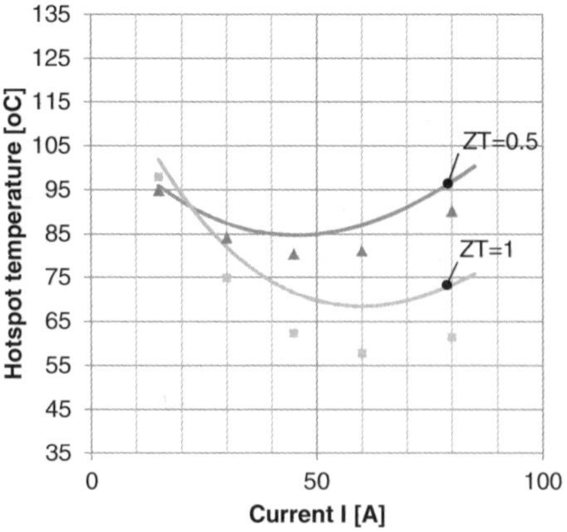

Fig. 4-6. Hotspot temperature versus drive current. The superlattice thickness is 10 μm. The other conditions are the same as Figure 8. Solid line is for analytical model and symbols are ANSYS simulation results, transferred from [12].

The thickness of the silicon substrate has an impact on the thermal spreading resistance. The spreading resistance is tested for some specific thicknesses and obtained from Eq. (13) along with 3D numerical simulations (see Figure 4-7). The optimum substrate thickness in this case is 478 μm. If one desires a thinner substrate for packaging height reduction purposes, the minimum substrate thickness can be found as 290 μm, defined as the thickness for 110% of the minimum thermal resistance.

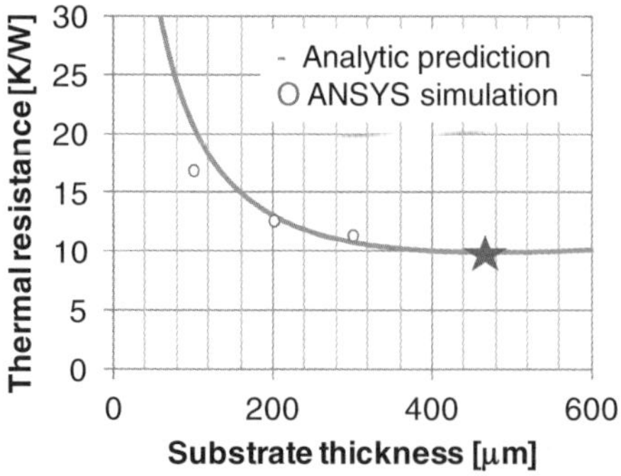

Fig. 4-7. Spreading thermal resistance vs. substrate thickness. A 500 μm × 500 μm hotspot with a heat flux of 400 W/cm^2 is applied with uniform heating of 100 W/cm^2. Substrate is silicon with thermal conductivity of 140 W/mK, transferred from [12].

4.8. *Parametric analysis*

The following Figures 4-8, 4-9, 4-10, and 4-11 show the optimum thickness or the optimum current density for several heat flux cases at 400, 600, and 800 W/cm^2 on a hotspot with uniform background heating of 100 W/cm^2 for the entire chip. The bulk ZT factor varies based on the thermal conductivity, which was changed to investigate the impact of the cooler characteristics. Figures 4-8 and 4-9 are based on a 500 μm thick substrate while Figures 4-10 and 4-11 are based on a 200 μm thick substrate for comparison. The 200 μm substrate case follows the major characteristics of the 500 μm case. However, the cooling temperature difference, $\Delta T = T_j$-T_m, is significantly reduced compared to the thicker substrate, while the required cooling power is lower for the optimum design.

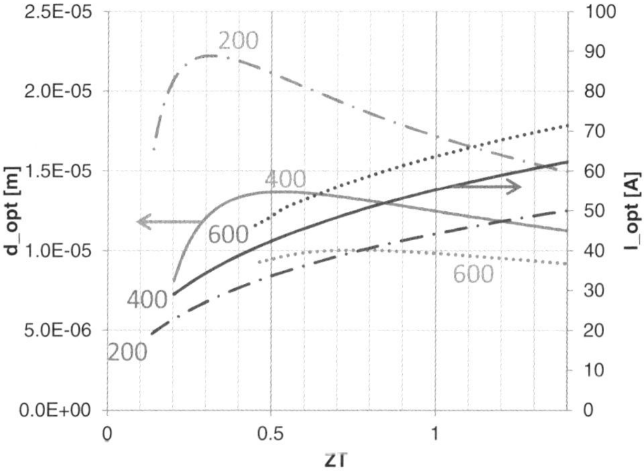

Fig. 4-8. Optimum leg thickness and the optimum drive current vs. ZT value of the material. Substrate thickness is 500 μm. Dashed, solid, and dotted lines correspond to different hot spot heat fluxes of 200, 400 and 600 W/cm^2, respectively - transferred from [12].

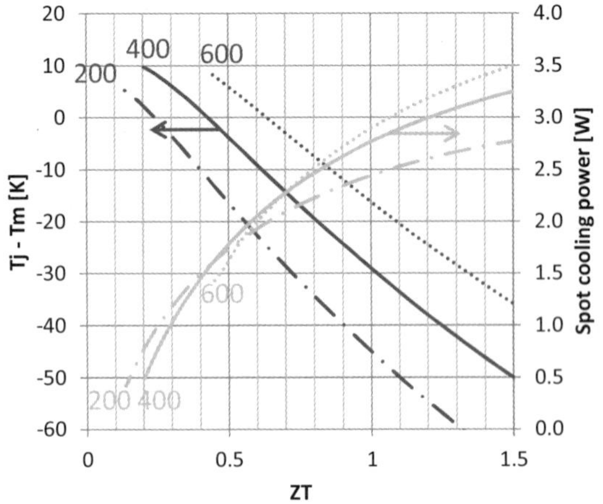

Fig. 4-9. The cooling temperature difference, T_m-T_j and the cooling power used for a superlattice cooler vs. ZT value at the optimum TE leg thickness design in Figure 4-8, transferred from [12].

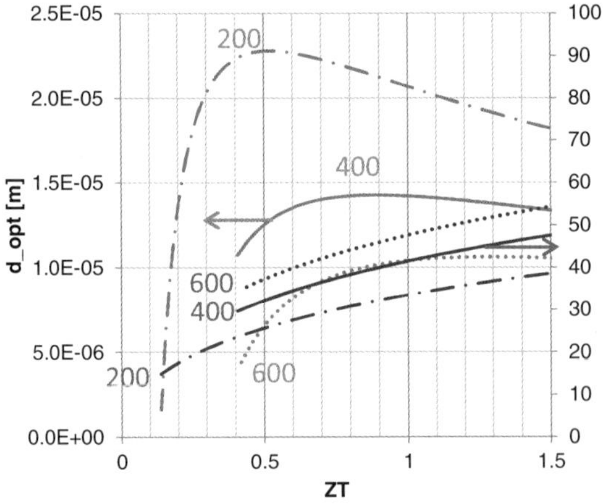

Fig. 4-10. Optimum leg thickness and the optimum drive current vs. ZT value of the material. Substrate thickness is 200 μm. Dashed, solid, and dotted lines correspond to different hot spot heat fluxes of 200, 400 and 600 W/cm^2, respectively - transferred from [12].

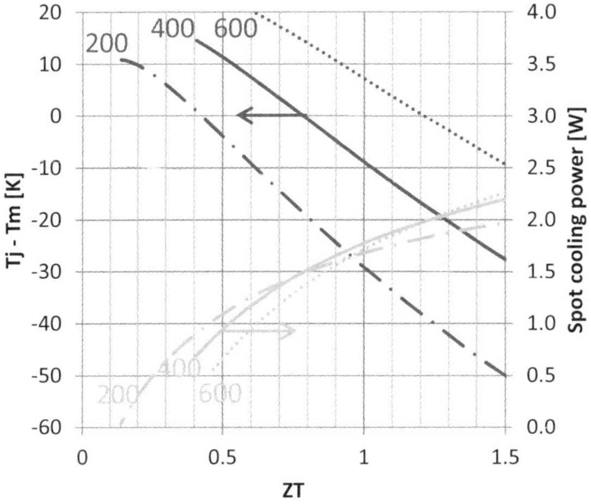

Fig. 4-11. The cooling temperature difference T_m-T_j and the cooling power used for superlattice cooler vs material ZT value at the optimum TE leg thickness design shown in Figure 4-10, transferred from [12].

5. Modeling of liquid pump power and integration

The integration of heat sink performance and the optimized superlattice microcooler is discussed in this section with a focus on quantitating the liquid pump power for overall heat removal.

5.1. *Microchannel heat sink*

The purpose in modeling the microchannel in this case is to find the electrical power that matches a specific thermal resistance of the microchannel for a unit area (1 cm × 1 cm) of foot print. Note that the size impact on the cooling performance is non-linear. The microchannel length across this square area affects the performance by a power of approximately 0.2.

Figure 5-1 shows the cross section of a channel. A mid length fluid temperature, T_{fm}, defines the temperature point to which a convective heat transfer from the channel wall is considered. Total thermal resistance has two components, sensible heat transport by heat capacitance of the fluid, and wall-to-liquid heat convection. Both of these components act as limiting factors in heat transfer through the heat sink, and an optimum balance should exist that leads to the lowest thermal resistance, ψ_{hs} [33]. The hydraulic diameter is based on the perimeter of the cross section of the channel. Since the channel is long enough compared to the diameter, the fully developed approximation for the entire channel length should be reasonable.

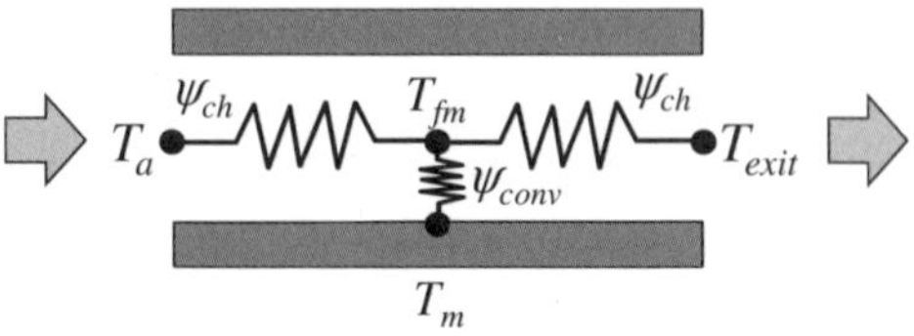

Fig. 5-1. Temperatures at various locations and thermal resistances in the microchannel. Fluid flow is shown by the arrow, transferred from [12].

$$\psi_{hs} = \psi_{ch} + \psi_{conv} = \frac{1}{2\dot{m}C_p} + \frac{1}{hA_{ch}} \tag{15}$$

where, A_{ch} is the internal surface area of the channel and mass flow rate is

$$\dot{m} = \rho u \delta H \tag{16}$$

and

$$h = \frac{N_u}{D_h}\beta_f \tag{17}$$

where, β_f is the thermal conductivity of the fluid. Hydraulic diameter, D_h, is found as

$$D_h = \frac{4\delta H}{2(\delta + H)} \tag{18}$$

By following Shah and London [34], for a fully developed and laminar flow in a rectangular channel, the Nusselt number is a function of the channel cross sectional aspect ratio α given by

$$N_u = 8.235(1 - 2.0241\alpha + 3.0853\alpha^2 - 2.4765\alpha^3 + 1.0578\alpha^4 - 0.1861\alpha^5) \tag{19}$$

Fluid dynamic power for channel flow, the inlet contraction, outlet expansion, and wall friction are considered following Munson *et al.* [35], for total flow pressure drop Eq. (20) based on Kays and London [36]. The friction flow in a channel is also considered as a fully developed flow for the entire channel length and its friction factor is $f = 96/Re$.

$$\Delta P = \Delta P_{ent} + \Delta P_{ch} + \Delta P_{exit} = K_{ent}\frac{\rho u^2}{2} + f\left(\frac{L}{D_h}\right)\frac{\rho u^2}{2} + K_{exit}\frac{\rho u^2}{2} \quad (20)$$

while

$$K_{ent} = -0.3861r^5 + 1.3345r^4 - 0.8864r^3 - 0.4105r^2 - 0.0308r + 0.378 \quad (21)$$

and

$$K_{exit} = \left(1 - \frac{\delta}{b+\delta}\right)^2 \quad (22)$$

The pump power should include conversion efficiency, η, from the electric power to the fluid-dynamic power as in Eq. (23). Based on some industrial information, the efficiency is about 60% similar to the 60%-72% used by Daud *et al.* for the electrical to mechanical conversion efficiency [37]. In total, the number of channel $N \sim W/(b + \delta)$ while $\delta << W$.

$$W_{pp} = \frac{\Delta P(N\delta Hu)}{\eta} \quad (23)$$

The following shows the relationship between the thermal resistance and the electric pump power using the channel design shown in Figure 5-2. The thermal resistance does not include fin efficiency in the solid, since silicon used for the substrate of the chip is a thermally conductive material (140 W/mK) with good isothermal characteristics as observed for the optimum design based on Eq. (15).

Due to the impact of flow speed, which is nearly square function of the heat transfer performance, the progression curve in the chart can be fitted as Eq. (24) with $C_1 = 0.0482$ WK, and $\psi_\infty = 0.329$ K/W is the thermal resistance limit at pumping power → infinity. Figure 5-3 shows the thermal resistance as a function of the pumping power. It is important to note that for a very large pump power, thermal resistance tends to converge to a specific value, which is practically the lowest limit. When a required thermal resistance is close to this value, the required pump power steeply increases as thermal resistance needs to decrease.

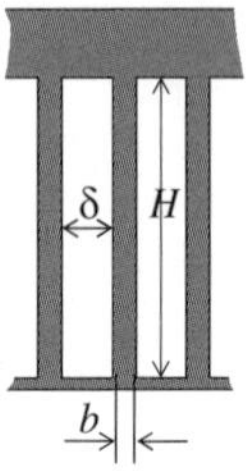

Fig. 5-2. Microchannel configurations. In cross section, H is channel height, b is channel wall thickness, and δ is channel spacing, transferred from [12].

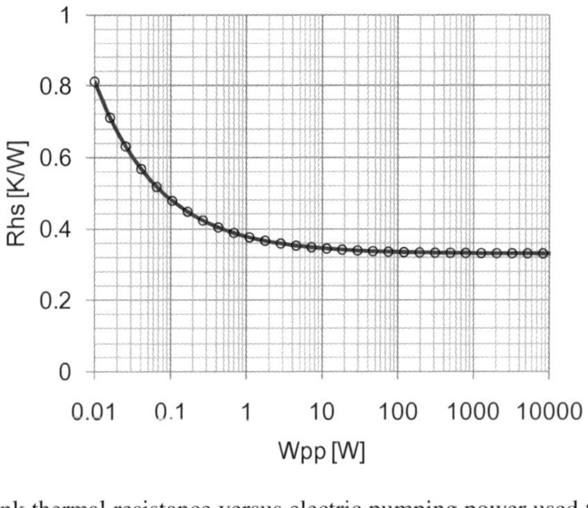

Fig. 5-3. Heat sink thermal resistance versus electric pumping power used to run the fluid in the microchannels. The regression curve is based on Eq. (24). Transferred from [12].

$$W_{pp.elec} \cong \frac{1}{\eta}\sqrt{\frac{C_1}{\psi_0 - \psi_\infty}} \tag{24}$$

5.2. *Integrated system optimization*

Overall integration of the hybrid cooling components in this model allows us to predict the minimum pumping power needed for hotspot cooling.

Before analyzing the hybrid cooling integration, we describe the challenge of hot spot cooling with only a single phase micro channel solution. Figure 15 shows the cooling power versus the total chip power without the hotspot cooler(s). For this case, the cooling power includes the electrical power, as described in Eq. (24), for only pumping liquid. The curve for the uniform heating without any hotspots shows a smooth transition of cooling power with increasing uniform power. In contrast, if a single hotspot is on the same chip, the required cooling power significantly and steeply increases even if the hotspot dissipation is a small fraction of the remaining background uniform heating. Both curves are based on Eq. (24) but use a different constant and ψ_∞ due to the spreading thermal resistance for the hot spot case.

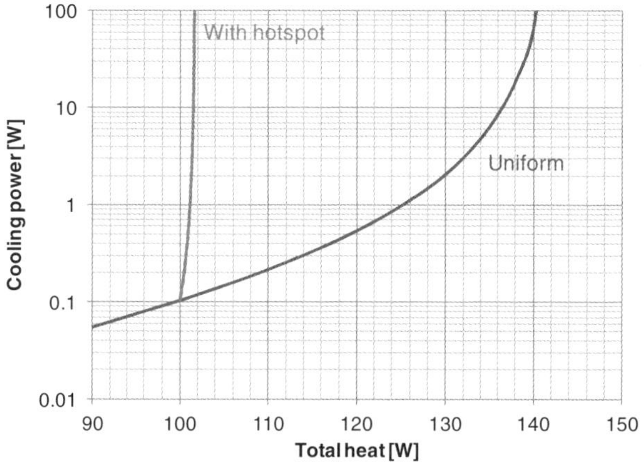

Fig. 5-4. Cooling power required to keep the hottest point in the chip below 85 °C versus total power dissipation on the chip. The case without hotspots (i.e. "uniform" power dissipation) is shown with the violet curve and that including a single hotspot is shown with the blue curve. Transferred from [12].

In Figure 5-5 the effects of using a superlattice cooler for the hotspot as the hybrid solution is shown. The cooling power is defined by $W_{cooling} = W_{pp} + W_{SLC}$, where the W_{pp} is found by Eq. (24) and $W_{SLC} = I^2R$.

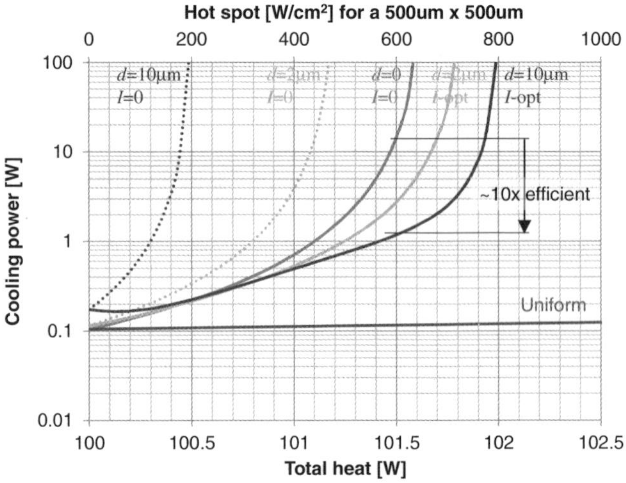

Fig. 5-5. Cooling power required to keep the hottest point in the chip below 85 °C versus total power dissipation for a 1 cm × 1 cm chip with a single 500 × 500 μm^2 hotspot. ZT = 0.5 is assumed and power dissipated at the hot spot in shown on the top horizontal axis, transferred from [12].

Figure 5-5 shows two cases for *d; d* = 2 μm and *d* = 10 μm. The drive current for each design is optimized for the system. The dashed curves show the case for no current and the solid curves are for the case with optimum drive current. As an example, for *d* = 10 μm, which is near optimum thickness, if we pick 600 W/cm^2 for the hotspot, the total cooling power required for the hybrid cooling system is about 1.2 W. This is about an order of magnitude improvement in energy efficiency compared to the 12 W cooling power for the non-spot-cooler operation. Based on this nearly 10 fold performance, if the same superlattice coolers are assumed, up to 10 hotspot locations of the same size on a single chip could be addressed with comparable results.

6. Summary of analysis

A superlattice micro cooler, as a solid-state device, was modeled with spreading resistance through the silicon substrate with a liquid cooling microchannel. The investigation went quantitatively through the hybrid cooling system for hotspots with uniform background heating on chip.

The minimum achievable hotspot temperature is found based on the optimization. The optimum thickness of the superlattice cooler is in the range of 5 μm up to 25 μm for practical hotspots that dissipate hundreds of watts per square centimeter. The drive current needs to be optimized for the system cooling COP point of view. For a 500 μm × 500 μm hot spot, the optimum current is approximately 50 A, thus requiring special low resistance contacts and wirings. By putting several n- and p-type thermoelectric legs of the same thickness electrically in series and thermally in parallel, one can tradeoff current and voltage and significantly reduce the optimum current. Since the optimum shows a 10 micron range thickness for this particular condition, it would be important to take into account the parasitic contact resistance both for Joule heating as well as the interface resistance, considering the Wiedemann-Franz law.

For the cooling application of hotspot(s) on a chip, however, the junction temperature may not always need to be at the minimum. If both T_j and the ambient temperature T_a are fixed, there are multiple design options for the thickness and the current. The unique optimum shape is demonstrated in a 3D plot and the optimum design can be found on the cross cut of the 3D surface in the plot for the specific temperature for the superlattice cooler. The heat spreading resistance, inherently the ΔT across the substrate due to the spreading effect, plays a significant role in the cooling power requirement. For spot cooling purposes, thicker substrates such as 500 μm or even 750 μm yields better cooling COP, so it is not worth thinning the die too much. The hotspot heat dissipation has a big influence on pumping power, even though it is a smaller amount of power compared to the uniform chip heating. This Hybrid Cooling Scheme removes the localized hotspot and allows the microchannel liquid cooling to be much more energy efficient. As an example, when using a moderate ZT (~0.5) material, when the leg thickness is designed to minimize the driving current, this Hybrid Cooling Scheme provides a *10 fold* energy efficient cooling improvement, for the case of a 600 W/cm^2 heat hotspot, in addition to 100 W/cm^2 background heat flux dissipating on a chip. This is due to the fact that the maximum allowable junction temperature on a chip should be lower than certain prescribed limits, even though a very small fraction of the chip reaches this limit.

7. Prospects of hot spot cooling

The Hybrid Cooling Scheme with TE microcooler(s) may provide solutions to address additional challenges for thermal packaging. As an example, recent three dimensional stacked die packages (3D-Pkg) have challenges in removing the heat from the hotspots on the intermediate chips. This hybrid scheme could provide the parallel cooling path for two opposite face of the chips for each stack as shown in Figure 7-1. It adds complexity to the packaging structure, but remains the cooling COP scalable to the single die solution.

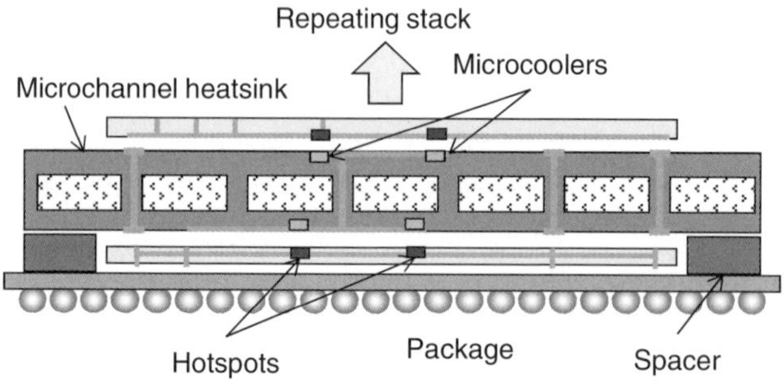

Fig. 7-1. Multi stack chip package cooling concept.

Another extensive technology is dynamic spot cooling. TE microcoolers can be fabricated in an array on a chip. Thanks to the nature of high speed response of electrons and small thermal mass, the thin film microcoolers would work very quickly; in microseconds to tens of microseconds [18]. The Hybrid Cooling Scheme could provide dynamic hotspot cooling while maintaining a stable die mean temperature. The TE device is also able to work as a thermal sensor. A closed loop dynamic thermal control of the hotspot will be an interesting approach for future enhancements to this concept.

8. Conclusions

A Hybrid Cooling Scheme of solid-state and single phase microchannel cooling for a chip with hotspot(s) was addressed in this chapter. Solid-state microcoolers possess a number of unique characteristics - scalable, no moving parts, and quick response time. An important design factor for microcoolers is the leg thickness to yield the maximum performance, a factor that has been rarely considered in electronics cooling. The Hybrid Cooling Scheme hot spot cooling devices with microchannels and superlattice microcoolers were fabricated to demonstrate the capability of a monolithic solution. The performance of the hotspot cooler has been characterized by sophisticated metrology without contacting a small area ranging from 40 to 200 microns square. The analysis based on the developed analytic model revealed a unique performance curve as a function of drive current and leg thickness. A minimum temperature point (maximum cooling ΔT) was observed. For most electronics devices, this will likely be an overcooling. The solution for a given maximum allowable temperature T_j_max may be found at the minimum drive current on cross cut curve in the T_j_max plane in the plot. The Hybrid Cooling Scheme provides a balanced combination of flattening of the hotspot temperature and heat removal from the entire chip. The Hybrid Cooling Scheme leads to a minimum power to pump sufficient refrigerant liquid through the microchannel to remove the overall heat and the scheme results in minimum power to pump the hotspot heat by minimizing the drive current to the superlattice microcooler. The Hybrid Cooling Scheme thus provides a significant improvement in energy efficient cooling of the chip with hotspot(s) compared to conventional uniform cooling schemes.

Acknowledgements

The authors acknowledge the contribution of Dr. Vivek Sahu on his efforts on analytic modeling and the microchannel fabrication, system integration, and characterization. Also authors acknowledge the contribution of Dr. Xi Wang and Dr. Je-Hyeong Bahk for their valuable discussions and efforts on the superlattice cooler design, cooler device fabrication, and testing, and Mr. Amirkoushyar Ziabari for his work on 3-dimensional numerical analysis.

Nomenclature

A	area, m^2
a	source length, m
b	channel wall thickness, m
C_p	specific heat, kJ/kg.s
d	thickness of superlattice cooler, m
F	fractional area coverage, %
H	channel height, m
h	heat transfer coefficient, $W/m^2.K$
I	current, A
K	thermal conductance, W/K and friction coefficient
m	mass flow rate, kg/s
P	pressure, Pa
Q	heat flow, W
q	heat flux, W/m^2
R	electrical resistance, ohm
T	temperature, K
t	thickness, m
u	velocity, m/s
W	power, W

Greek symbols

β	thermal conductivity, W/mK
δ	channel spacing, m
η	efficiency, %
ϕ	geometric factor, 1/m
ρ	density, kg/m^3
σ	electrical conductivity 1/Ohm.m
ψ	thermal resistance, K/W

Subscripts

a	ambient
c	cold side

ch	channel section
conv	convective term
ent	entrance section
exit	exit section
h	source side
m	microchannel side of silicon substrate
P	chip surface
pp	pumping
S	hotspot surface

References

1. A. Bar-Cohen ed., *Encyclopedia of Thermal Packaging: Thermal Packaging Technique*, World Scientific, (2012).
2. T. Edison, Electric Lamp, U.S. patent 223898, (1880).
3. G. E. Moor, Electronics Magazine, Vol. 38, No. 8, 4–6, (1965).
4. International Technology Roadmap for Semiconductors (ITRS) 2011 Edition, Semiconductor Industry Association, (2011)
5. D. B. Tuckerman, R.F.W., Pease, IEEE Electron Device Letters-2, No.5, 126–9, (1981).
6. R. L. Webb, N. H. Kim eds., *Principles of Enhanced Heat Transfer – Second Edition*, Taylor and Francis, 738–741, (1994).
7. E. Bergles, C. F. Ma, Proc. the Symposium - Heat transfer in electronic equipment, ASME, 5–12, (1983).
8. M. Ohadi, J. Qi, J. Lawler, Microscale Heat Transfer Fundamentals and Applications, 321–328, (2005).
9. T. S. Fisher, D. G. Walker, Proc. ASME International Mechanical Engineering Congress & Exposition, (2000).
10. H. Böttner, J. Nurnus, A. Gavrikov, G. Kühner, M. Jägle, C. Künzel, D. Eberhard, G. Plescher, A. Schubert, K. Schlereth, J. Microelectromechanical Systems, Vol. 13, No. 3, 414–420, (2004).
11. J. Snyder, M. Soto, R. Alley, D. Koester, B. Conner, Proc. 22nd IEEE SEMI-THERM Symposium, 135–143, (2006).
12. K. Yazawa, A. Ziabari, Y. R. Koh, A. Shakouri, V. Sahu, A. G. Fedorov, Y.K. Joshi, Proc. IEEE ITherm2012 conference, (2012).
13. V. Sahu, Hybrid Solid-State/Fluidic Cooling for Thermal Management of Electronic Components, Ph.D thesis, (2012).
14. V. Sahu, Y.K. Joshi, A. G. Fedorov, Nanoscale and Microscale Thermophysical Engineering, Vol. 13, No.3, 135–150, (2009).
15. V. Sahu, Y. K. Joshi, and A. G. Fedorov, Proc. the 12th IEEE ITherm, (2010).

16. V. Sahu, A. G. Fedorov, Y.K. Joshi, K. Yazawa, A. Ziabari, A. Shakouri, Proc. the 28th Semi-Therm symposium, (2012).
17. Y. Ezzahri, J. Christofferson, K. Maize, A. Shakouri, MRS Proceedings, 1166, (2009).
18. B. Vermeerschal, J. Bahk, J. Christofferson, A. Shakouri, MRS Proceedings, 1329, (2011).
19. A. Shakouri, Annual Review of Materials Research, Vol. 41, 399–431, (2001).
20. X. Fan, G. Zeng, C. LaBounty, J. E. Bowers, E. Croke, S. Huxtable, A. Majumdar, A. Shakouri, Appl. Phys. Lett., Vol. 78, No.11, 1580–1582, (2001).
21. A. Shakouri, Proc. IEEE, 94(8): 1613–1638, (2006).
22. A. I. Boukai, Y. Bunimovich, J. Tahir-Kheli, J. Yu, W. A. Goddard III, J. R. Heath, Nature 451, 168–171, (2008).
23. M. Zebarjadi1, K. Esfarjani, A. Shakouri, J. Bahk, Z. Bian, G. Zeng, J. Bowers, H. Lu, J. Zide, A. Gossard, Appl. Phys. Lett., Vol. 94, 202105, (2009).
24. L. D. Hicks, M. S. Dresselhaus, Phys. Review B, (1993).
25. G. Chen, J. Nanoparticle Research, Vol. 2, Issue 2, 199–204, (2000).
26. J. O. Sofo, G. D. Mahan, Phys. Review B, Vol. 49, no. 7, (1994).
27. D. Vashaee, J. Christofferson, Y. Zhang, A. Shakouri, G. Zeng, C. LaBounty, X. Fan, J. Piprek, J. E. Bowers, E. Croke, Microscale Thermophysical Engineering, Vol. 9, 99–118, (2005).
28. Y. Ezzahri, G. Zeng, K. Fukutani, Z. Bian, A. Shakouri, Microelectronics Journal 39, pp. 981–991, (2008).
29. Y. Ezzahri, S. Dilhaire, S. Grauby, J. M. Rampnoux, W. Claeys, Y. Zhang, G. Zeng, A. Shakouri, Appl. Phys. Lett., Vol. 87, 103506 1–3, (2005).
30. K. Fukutani and A. Shakouri, IEEE Transaction on Component and Packaging Technologies, Vol. 29, No. 4, 750–757, (2006).
31. S. Lee, S. Song, V. Au, and K. P. Moran, Proc. the 4th ASME/JSME Thermal Engineering Joint Conference, Vol. 4, 199–206, (1995).
32. M. M. Yovanovich, Y. S. Muztchaka, and J. R. Culham, J. Thermophysics and Heat Transfer, **13**(4), 495–500, (1999).
33. K. Yazawa, G. L. Solbrekken, and A. Bar-Cohen, IEEE Transactions on Advanced Packaging, Vol. 28, No. 2, 231–239, (2005)
34. R. K Shah and A. L. London, *Advances in Heat Transfer* Supplement 1, Academic Press, New York, (1978).
35. B. R. Munson, D. F. Young, and T. H. Okiishi, *Fundamentals of Fluid Mechanics*, John Wiley & Sons, New York, NY, (1990).
36. W. M. Kays and A. L. London, *Compact Heat Exchangers*, 3rd edition, Kreiger Publishing, Florida, (1984).
37. A. Dauda, M. M. Mahmoud, Renewable Energy 30, 701–714, (2005).

CHAPTER 9

AN OVERVIEW OF THE USE OF PHASE CHANGE MATERIALS FOR THE THERMAL MANAGEMENT OF TRANSIENT PORTABLE ELECTRONICS: BENEFITS AND CHALLENGES

Amy S. Fleischer

Professor, Department of Mechanical Engineering, Villanova University
Villanova, PA, USA
amy.fleischer@villanova.edu

This paper provides an overview of the implementation of phase change materials to absorb thermal energy in transient electronics. A thorough review of existing literature is included with an emphasis on the evolution of the technology, its potential benefits and the remaining challenges to its implementation.

1. Introduction

The thermal management of portable and transient use electronics presents a unique challenge to the thermal engineer. Although the heat flux and power loads for portable electronics tend overall to be lower than comparable desktop or workstation electronics, the portability requirement imposes weight, size, orientation and reliability constraints on the thermal management solution. The demand for additional add-on casings, whether as part of the original design or as consumer aftermarket additions, that withstand the abuse of everyday use while also allowing users to customize their electronics, also adds to the thermal complexity. The use of these aftermarket casings for portable electronics makes it difficult to effectively implement the possibility of using the exterior surfaces as high conductivity heat spreaders to dissipate the heat to the ambient, which has been a popular solution to date. The size and weight limits often rule out the use of large fans and heat sinks, while the requirement for high reliability and the need to operate in various orientations eliminates the choices of thermosyphons, pumped liquid cooling and most liquid-vapor to phase systems. Thus, for portable electronics, the number of possible solutions is limited.

Much of the current design emphasis is on reducing power demands through control algorithms that prevent or minimize hot spot formation and have the additional benefit of extending battery life, a critical issue in portable electronics. However, the insatiable demand for increased computing power at one's fingertips is driving power needs up and up, and the resultant waste heat is a growing problem. One possible solution which meets the need for a reliable system in a small form factor package is the use of solid-liquid phase change energy storage. In phase change thermal management systems, the waste heat from the electronics is absorbed by the material as it changes phase from solid to liquid. As the phase change material (PCM) melts, it absorbs heat at a rate equivalent to its latent heat, without increasing in temperature. The system is completely passive, dependent only on the thermophysical properties of the PCM, and thus is highly reliable.

2. Background

The goal of PCM thermal management is to absorb heat and maintain a constant temperature at the melt point. The ideal situation is where the temperature distribution throughout the medium is constant as melting (or solidifying) occurs and is described by a PDE boundary value problem where the phase boundary is a function of time. This is referred to as the classical Stefan problem [1]. The problem is challenging to solve and there are exact solutions available only under certain circumstances. The most common simplifications are one-dimensional, semi-infinite solid with a uniform initial temperature, a constant surface temperature and a sharp interface between the phases as shown in Figure 1.

Figure 1 depicts the 1D solidification process, where the medium is initially entirely liquid. At time t = 0, the base temperature (T_b) is lowered to below the melt/solidification temperature (T_m) and solidification of the medium begins from the left hand side, progressing towards the right. The instantaneous location of the interface between the solid and liquid is shown as S(t). This results in a set of partial differential equations, coupled at the melt front location, S(t). The temperature distribution in the solid phase is given by Eq. (1), and the temperature distribution in the liquid phase is given by Eq. 2. The initial condition for these equations is given by Eq. 3, the left hand boundary condition by Eq. 4 and the right hand semi-infinite boundary condition by Eq. 5.

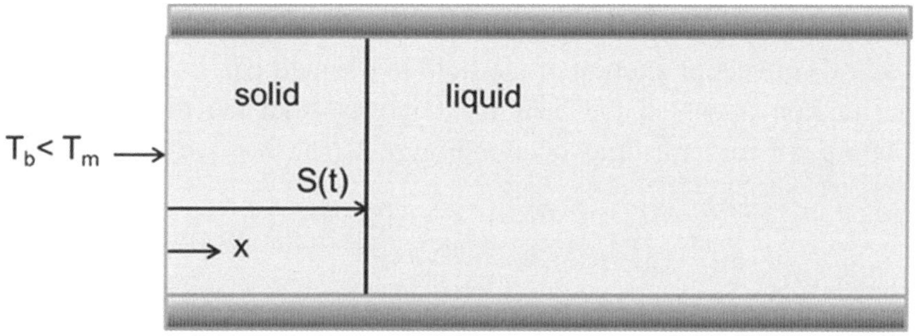

Fig. 1. Classical Stefan solidification problem with solidification progressing from left to right.

$$\left(\rho c_p\right)_s \frac{\partial T_S}{\partial t} = k_s \frac{\partial^2 T_S}{\partial x^2} \; where \; 0 \le x \le S(t) \tag{1}$$

$$\left(\rho c_p\right)_L \frac{\partial T_L}{\partial t} + \left(\rho c_p\right)_L u_x \frac{\partial T_L}{\partial x} = k_L \frac{\partial^2 T_L}{\partial x^2} \; where \; S(t) \le x \le \infty \tag{2}$$

$$T_L(x,0) = T_M \quad at \; t = 0 \tag{3}$$

$$T_s\left(0,t\right) = T_b < T_M \tag{4}$$

$$T_L\left(x,t\right)_{x \to \infty} = T_m \tag{5}$$

These equations are coupled at the phase change boundary through the latent heat of fusion (h_{fg}) and the melt temperature as given in Eq. 6

and 7. While the classical Stefan problem is typically depicted in terms of the solidification of a material, in thermal management of electronics, the primary interest is the energy absorption and energy absorption during the melting process. The equation formulation is similar but the boundary condition changes to reflect that $T_b > T_m$. As noted above, this formulation of the ideal problem assumes a sharp interface between phases and a 1D motion of the solidification front. In reality however, the phase change actually occurs over a region known as the "mushy zone" and convection in the liquid phase makes the melt front motion two-dimensional. As shown in experimental and numerical simulations by Pal and Joshi [2] using tricontane PCM ($C_{30}H_{62}$) in a tall side heated enclosure with an aspect ratio of 10, the natural convection motion of the fluid in the melt phase actually creates a situation in which the melt front moves from the top right to the bottom left corner as illustrated in Figure 2.

$$k_S \frac{\partial T_S}{\partial x} - k_L \frac{\partial T_L}{\partial x} = h_{fg} \rho_s \frac{dS}{dt} \quad at\ x = X(t) \tag{6}$$

$$T(x,t)_L = T_s(x,t) = T_{melt}\ at\ x = S(t) \tag{7}$$

$$u_x = \frac{dS}{dt} \tag{8}$$

One of the key aspects of PCM implementation is thus making sure that the heat penetrates effectively into the medium, and is not isolated at the source. In the event that the thermal resistance into the solid is high, the result is a superhcated liquid layer circulating near the source as in Figure 2. This liquid layer can heat rapidly as the heat fails to penetrate effectively into the solid layer at the melt front, and the result is overheating of the source device despite an ample supply of solid PCM. The thermal diffusivity and thermal conductivity of the selected PCM thus become important parameters in the problem and the PCM material must be carefully selected.

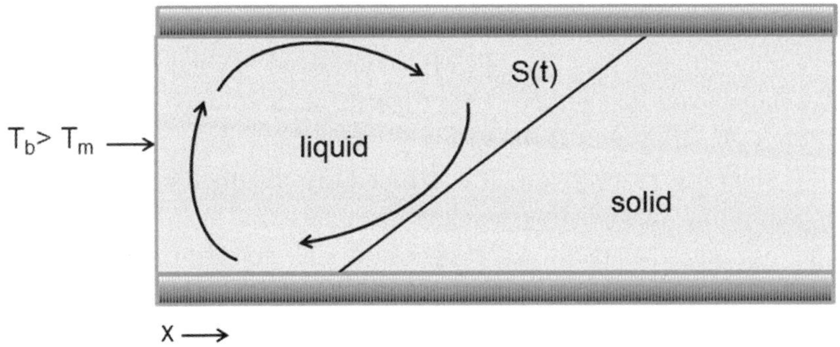

Fig. 2. Melt front motion with convective influence.

3. Material Selection

The design of the PCM system must be carefully matched to the source so that when the electronics are in operation, the heat flows directly into the PCM and is absorbed during the melt process. The length of the melt process must be matched to the expected duration of the transient use cycle, as the system is only effective as long as the PCM continues to melt. Once the use cycle ends, the energy stored in the PCM is dissipated to the ambient and the PCM re-solidifies. PCMs have been shown to be very effective in the thermal abatement of electronics by slowing the rate of temperature increase during transient operation [3].

The feasibility of using a PCM for transient thermal management of a cellular phone was investigated by Hodes *et al.* [3]. In this project, the transient response of a mock handheld phone was determined using two types of PCM: tricosane, a common paraffin wax, and Thermasorb-122, a commercially available PCM encapsulated in a pellet-like plastic shell. It was determined that for low power loads even small masses of PCM can substantially increase the operational time for a handheld phone before overheat occurs. For a heat load of 3 W and a defined steady-state temperature of 50°C the presence of a PCM doubled overall usage time of the device.

Krishnan *et al.* [4] performed a theoretical analysis which compared the thermal performance of PCMs to a conventional copper heat sink. The analysis considered the effects of conduction and phase change through the different materials when subjected to large heat loads of between 300-600W. The PCMs considered included triacontane, aluminum foam impregnated with triacontane, and two metallic PCMs a Bi/Pb/Sn/In alloy and Bi/In/Sn alloy which have melt temperatures of 57°C and 60°C respectively. It was found that that the PCMs were better at controlling the junction temperature (chip-heat sink/PCM container junction) than the traditional copper heat sink and consistently delayed steady state longer than just the copper heat sink.

There are various types of phase change materials in use for thermal management systems, including organics, inorganics and metallics. The criterion used to select a phase change material is first and foremost its phase change temperature. This temperature must of course be below the maximum allowable operating temperature for the system, and above the ambient temperature in which the electronics will operate.

The general guideline for selecting a melting point is to pick the material with the highest possible melting point that is still lower than the electronics maximum temperature. The advantage to using a high melting point material is that it leads to a longer melting time, and thus a longer effective duty cycle. The higher melting point means that the system dissipates more heat through sensible heating as the system transitions from the initial temperature to the melting temperature, and that system dissipates more heat to the ambient throughout the melt process due to the larger ΔT, again extending the melt time. A study by Leoni and Amon [5] showed that changing the PCM from one with a melt temperature of 36°C to one with a melt temperature of 44°C with all other variables held constant, extended the transient melting time, and thus the effective operation of the system, by 50%. The higher melting point is also beneficial to the system during the recharge process when the larger ΔT from the hot liquid to the ambient drives a faster solidification process.

Secondary criteria include the need for the material to have a high latent heat to maximize storage potential, and to be chemically and physically stable over repeated melting and solidification cycles. Finally, it is desirable for the PCM to have a high thermal conductivity to prevent thermal bottlenecking at the source.

Of the three main classifications of phase change materials, metallics have the highest thermal conductivities (100-200 W/mK) but the lowest latent heats (30-40 kJ/kg) [6]. Their low level of energy storage, combined with their high densities, typically makes them unsuitable for portable electronics due to weight concerns. Inorganics, such as molten salts, have much higher latent heats (180-260 kJ/kg) than metallic PCMs, but tend to break down under repeated thermal cycling, which is unsuitable for use electronics. Furthermore, their melt temperatures tend to be extremely high (300-400°C), again making them unsuitable for portable electronics, although they are commonly used in solar energy applications [7].

Among the most commonly studied PCMs for electronics thermal management are the organics, specifically paraffin waxes. Paraffin waxes have among the highest latent heats of common PCMs (240-260 kJ/kg), allowing them to store a higher amount of energy in a smaller mass. Even better, paraffins exhibit melting points that make them well-matched for energy storage in electronics applications, ranging from as low as 35°C (meaning that ambient temperature may be a limiting factor) to as high as 70°C depending on the specific paraffin type or blend. Additionally, paraffin is inexpensive and widely available, thermally and chemically stable, lighter than metallics and compatible with a wide variety of casing materials.

However, the low thermal conductivity of most organic PCMs (~0.2 W/m·K for most paraffin waxes) decreases their effectiveness. The low thermal conductivity leads to bottlenecking of the heat flow and isolation of the melt process near the heat source. Therefore, much research work has focused on addressing the challenge of increasing the thermal diffusivity and thermal conductivity of phase change materials by using embedded high conductivity structures of various designs.

A comparison of the sample properties for three main classifications of PCMs can be seen in Table 1. The organic, tricontane, has the highest latent heat by far, this allows energy storage in a significantly lighter system, or conversely, if mass is held constant, a much smaller system. The inorganic, KNO_3, is suitable for higher operating temperatures, as in concentrating solar systems, while the metallic system has a high thermal conductivity, but is significantly heavier.

Table 1: Energy Storage Capability of Various PCMs.

PCM	Latent heat (kJ/kg)	Melt Temp (°C)	Thermal Conductivity (W/mK)	Volume (cm^3)	Mass (kg)	Total Energy Storage (kJ)
Tricontane (organic)	251	65	0.23	100	0.081	20.3
KNO_3 (inorganic)	95	335	0.5	100	0.210	19.95
Bi/In/Sn (metallic)	29.5	57	21.9	100	0.864	25.5

4. Use of Fins and Foams for Performance Enhancements

Many different techniques have been implemented to improve the thermal response of paraffins and other organics. These techniques typically involve embedding a high thermal conductivity matrix material within the low thermal conductivity paraffin to provide a low thermal resistance heat flow path into the material. In application, the PCM is usually contained within a sealed container (module) located adjacent to the heat source. Thus, as the PCM melts it stays within the containment unit. Researchers have embedded heat sinks of various designs in this container and then filled the container with paraffin between the fins [8-10]. This provides a heat flow path along the high conductivity fins and then into the paraffin.

For example, Krishnan *et al.* [8] studied a hybrid heat sink/paraffin combination for use in electronics cooling applications. The hybrid heat sink featured a plate fin heat sink with the fin tips immersed in a PCM. The exposed area of the fins dissipated heat during periods when high convective cooling was available. When the air cooling was reduced, the heat was absorbed by the PCM. The fin heat transfer rate increased with an increase in fin thickness and amount of PCM and decreasing melting temperature of the PCMs. The heat sink was found to facilitate more effective heat flow into the PCM and it was concluded that paraffin alone was unsuitable for transient heating applications due to its low thermal diffusivity. In fact, it was found that a metallic PCM performed better than the organic PCM owing to its superior thermal diffusivity, although weight concerns and low latent heats usually eliminate metallic PCMs from application.

Nayak *et al.* [9] used a numerical simulation to study the performance of various embedded plate fin and pin fin heat sink designs to enhance the performance of PCM thermal management systems for electronics cooling. The effect of melt convection was found to be significant as it strongly affected the melt front motion and played an important role in establishing temperature uniformity within the melt phase. In terms of overall performance, it was found that thinner fins and pin fins were more effective than plate fins at maintaining uniformity of the temperature within the PCM which resulted in lower chip temperatures. Interestingly, it was found that the use of fins led to initiation of melting of the PCM at multiple locations, such as along the fin faces, and that more fins led to better temperature control. Unfortunately, each additional fin displaces more PCM, thus the thermal engineer must balance the need for increased energy storage with increased thermal diffusivity.

Rather than considering finned heat sinks, some researchers investigated metallic foams with the thought that the reticulated foam ligaments could provide heat transfer pathways without displacing as much PCM as a heat sink [9, 11-13]. Pal and Joshi [13] considered a horizontally mounted heat source in a low form factor enclosure intended to mimic a notebook computer application. Three cases were considered, a situation with no thermal management at all, a situation with a paraffin PCM module used near the heat source, and a situation with an aluminum foam module filled with paraffin PCM near the heat source. It was found that with no thermal management at all, the heat source simply increased in temperature due to sensible heating through the transient process and the limit of 100°C was reached in just over 25 minutes. When PCM was used to absorb the dissipated heat through latent heat effects, the melting process slowed the temperature rise, and it took more than four times as long (110 minutes) to reach the maximum temperature. However, when aluminum foam was used in the module design to provide a heat flow path into the system and spread the heat throughout the module, the system reached a steady-state condition at only 90°C, 10°C below the maximum allowable temperature. The clear benefit of using a high conductivity filler material in this case is a more uniform temperature throughout the system creating a situation in which the ultimate operating temperature is suppressed even below the maximum allowable.

Nayak *et al.* [9] numerically simulated PCM and high conductivity foams and found that inserting an aluminum matrix into paraffin offered an order-of-magnitude increase in thermal conductivity and melting rate. However, it was also found that with too high of a volume fraction of foam, the convection within the melt phase was inhibited. Therefore, conduction was enhanced but convection was decreased. They found that convection in the melt region leads to quicker heat transfer from the base plate to the upper regions of the heat sink and more uniform temperatures throughout the system. But as the volume fraction of the metal foam increased, the metal network provided more resistance, resulting in a conduction dominated heat transfer and significant temperature stratifications in the system.

So while there are clear benefits to embedded metal finned heat sinks and metallic foams there are also several significant disadvantages, including added weight, displaced PCM and inhibited convection in the melt.

5. Use of Carbon Based Materials for Performance Enhancements

Several researchers have turned to carbon based materials for thermal diffusivity enhancement in order to take advantage of their lower weights and higher thermal conductivities when compared to aluminum or copper. Many of these works have focused on carbon fibers and expanded exfoliated graphite materials. For instance, Fukai *et al.* [14] used embedded carbon fibers in either a random orientation or a specifically designed brush type configuration where the fibers are oriented in the direction of heat flow within the paraffin PCM. The fibers were microscale in diameter but up to 200 mm long. The fiber volume fraction in the paraffin was restricted to less than 2% to avoid displacing an excessive amount of PCM. However, even at these loading levels the fibers were found to suppress natural convection in the melt phase. Ultimately there was no significant effect of the convection suppression, because the increased conductivity of the carbon matrix fully compensated for the elimination of convection. Overall, the effective thermal conductivity of the brush type carbon fiber/PCM composite with 2% loading levels was increased by a factor of six over the base paraffin and the brush type orientation was found to be significantly more effective than the randomly oriented fibers.

Other researchers, including Py *et al.* [15] used expanded graphite rather than carbon. Py *et al.* [15] used commercial expanded graphite powder which was poured into a cubic mold and compressed into a graphite matrix. This matrix was submerged in liquid paraffin which then infiltrated the matrix through capillary forces. This design was successful in improving the bulk conductivity of the system with measured thermal conductivities of between 4-70 W/m•K for the different mixtures studied, in comparison to the 0.24 W/m•K of the base paraffin. However, this design was less successful in optimizing the energy storage potential of the material. The PCM had difficulties in fully impregnating the expanded graphite matrix and the resulting system had between 20-65% open air space, reducing effectiveness.

Carbon foams are similar in many ways to expanded and compressed graphite. Carbon foam is an open pore material with low density and high thermal conductivity [16]. The effects of porosity on the thermal properties of carbon foams filled with PCM has been investigated by among other, Osama *et al.* [17], and Chintakrinda *et al.* [18], both numerically and experimentally. The results indicate that the higher the porosity of the foam, the longer the latent heat absorption and the more effective the module becomes [17], but also that the primary effect of the PCM saturated foams is to act as a heat sink not an energy storage device as the PCM within the system melted rapidly, without exhibiting any significant delay to steady-state [18]. The graphite foams also acted to inhibit convection, just as the metallic foams did [18].

Additionally there are inherent difficulties in effectively filling small cell graphite materials with PCM. A study by Warzoha *et al.* [19] showed that the commonly used method of submerging the foam or expanded graphite in melted PCM and letting it wick into the solid is ineffective. Repeated tests showed that at least 25% of the reported open space within the foam was left unfilled using this method. It was shown that for effective use of these materials the foam must be fully evacuated and then filled under vacuum conditions [19].

6. Use of Embedded Nano Materials for Performance Enhancements

The use of metallic and graphite foams is thus found to be effective by some measures, but also to eliminate convective motion creating a melt regime dominated by conduction and significant temperature stratifications; and to lead to an incomplete melt of the PCM and ineffective thermal storage. Some researchers have therefore turned to the idea of embedding nanoparticles of various materials and shapes into the PCM [18, 20-27]. The majority of these studies to date have focused on quantifying the material properties of these nanoenhanced PCMs such as thermal conductivity, latent heat and melt temperature, without yet quantifying the material behavior and overall system performance in repeated thermal cycling.

Different base PCMs, nanoparticle shapes, and loading levels have all been found to influence the thermal properties, and it is reported that the addition of nanoparticles to commonly used PCMs can increase the base thermal conductivity from 20% to 320% [20-27]. This enhancement in thermal properties is expected to have a significant effect on thermal performance. Recent work suggests that nanoparticle influences on the specific heat and overall thermal diffusivity may be even more significant than on the thermal conductivity [27]. While the benefit of nanoenhancement is seen during melting [18, 20] the benefit may be even greater during solidification as the conduction-dominated solidification takes many times longer than the convection-dominated melting phase [21]. The enhanced thermal diffusivity of the solid phase is particularly advantageous during this phase transition.

However, at high loading levels, the nanofibers can exhibit characteristics similar to those of foams or compressed graphite. Chintakrinda *et al.* [18] investigated the use of graphite nanofibers at 11 wt% embedded in paraffin. It was seen that while standard paraffin consistently showed the development of convection currents in the melt region, with the addition of 11wt% graphite nanofibers the convection currents in the melt were suppressed just as with foams; and the temperature stratification through the conduction dominated system led to an incomplete melt.

The effect of graphite nanofiber loading level was investigated in a separate study by Chintakrinda *et al.* [20] in which the nanofibers were loaded at 5.5 wt% and 11 wt% in paraffin. The 5.5 wt% sample had improved heat spreading when compared to the paraffin and convection in the liquid layer was not suppressed while the 11wt% sample did exhibit suppressed convection [20].

In application, the suspension of these nanoparticles can be challenging. Most PCMs exhibit a significant density change during the transition from solid to liquid. For example, in paraffin the liquid density is 10-15% lower than the solid density. The significant contraction during solidification exerts forces on the nanoparticles that act to bring them out of suspension over repeated thermal cycles [28]. Research is ongoing to determine methods to keep the nanofibers in suspension during the phase transition using techniques such as shape stabilization to reduce or eliminate contraction [28].

7. Conclusions and Future Challenges

It is clear that energy storage in phase change materials can be extremely effective for thermal management in portable electronics but that significant design challenges still remain before full implementation. Full characterization of the various promising nanoenhanced materials needs to be completed under repeated thermal cycling conditions, focusing on overall system performance and implementation issues, not just on thermal property enhancement. Optimization of these materials needs to be completed to minimize filler material and maximize energy storage so that performance in enhanced in both the solid and liquid phases, during melting and solidification. Finally, the suspension of these nanomaterials needs to be consistent within the solid and liquid phases without settling over thousands of cycles. Ongoing research in each of these areas is expected to solve these remaining challenges and lead to effective and reliable thermal management using PCMs.

References

1. J. Stefan, *Annalender Physik und Chemie*, 42, 269–286 (1891).
2. D. Pal, and Y.K. Joshi *International Journal of Heat and Mass Transfer*, 44, pp. 375–387 (2001).
3. M. Hodes, R.D. Weinstein, S.J. Pence, J.M. Piccini, L. Manzione, and C. Chen, *Journal of Electronic Packaging,* 124, pp. 419–426 (2002).
4. S. Krishnan and S.V. Garimella, *Journal of Electronic Packaging,* 126(3), pp. 308–316 (2004).
5. N. Leoni, and C.H. Amon, ASME HTD-Vol. 343, ASME Proceedings of the 32nd National Heat Transfer Conference, Baltimore, MD, August 8-12, Vol. 5, pp. 49–56 (1997).
6. W.R. Humphries, and E.I. Griggs, NASA Technical Paper 1074, NASA (1977).
7. L.A. Chidambaram, A.S. Ramana, G. Kamaraj, and R. Velraj, *Renewable and Sustainable Energy Reviews,* 15(6), pp. 3220–3228 (2001).
8. S. Krishnan, S.V Garimella,, and S.S. Kang, *IEEE Transactions on Components and Packaging Technologies*, 28, pp. 281–289 (2005).
9. K.C. Nayak, S.K. Saha, K., Srinivasan, and P. Dutta, P, *International Journal of Heat and Mass Transfer*, 49(11-12), pp. 1833–1844 (2006).
10. S.K. Saha, K., Srinivasan, and P. Dutta,, *Journal of Heat Transfer*, 130(3), pp. 034505-1-4, 2008.
11. X. Tong, J. Khan, and M.R. Amin, *Numerical Heat Transfer Part A: Applications*, 30(2), pp. 125–141 (1996).
12. A. Siahpush, J. O'Brien, and J. Crepeau, *Journal of Heat Transfer*, 130, pp. 082301–1-11 (2008).
13. D. Pal and Y.K. Joshi, *Proceedings of InterPACK '99*, ASME EEP Vol. 26, 1999, pp. 1625–1630 (1999)
14. J. Fukai, Y. Hamada, Y. Morozumi, and O. Miyatake, *International Journal of Heat and Mass Transfer,* 45(24), pp. 4781–4792 (2002).
15. X. Py, R. Olives, and S. Mauran, *Int. J. Heat and Mass Transfer,* 44, pp. 2727–2737 (2001).
16. N.C. Gallego, and J.W. Klett, International Seminar on Advanced Applications for Carbon Materials. Jeju Island, Korea (2002).
17. M. Osama, K. Lafdi, and A. Efgafy, *Carbon*, 44(10), pp. 2080–2088 (2005).
18. K. Chintakrinda, R.D. Weinstein, and A.S. Fleischer, *International Journal of Thermal Sciences*, 50(9), pp. 1639–1647 (2011).
19. R. Warzoha, O. Sanusi, B. Mcmanus, and A.S. Fleischer, ASME Journal of Solar Energy Engineering, 135(2), 021007, (2013).
20. K. Chintakrinda, R. Warzoha, R.D. Weinstein, and A.S. Fleischer, *Journal of Heat Transfer*, 134(7), pp. 071901 (2012).

21. O. Sanusi, R. Warzoha, and A.S. Fleischer, *International Journal of Heat and Mass Transfer*, 54(19/20), pp. 4442–4436 (2011).
22. R. Zheng, J. Gao, J. Wang, and G. Chen, *Nature Communications*, 2:289, DOI: 10.1038 (2011).
23. Y. Zhang, J. Ding, X. Wang, R. Yang, and K. Lin, *Solar Energy Materials and Solar Cells*, 90, pp. 1692–1702 (2006).
24. C.J. Ho, and J.Y. Gao, International Communications in Heat and Mass Transfer, 36, pp. 467–470 (2009).
25. S. Kim, and L.T. Drazal, *Solar Energy Materials and Solar Cells*, 93, pp. 136–142 (2009).
26. J.L. Zeng, Z. Cao, D.W. Yang, F. Xu, L.X. Sun, X.F. Zhang, and L. Zhang, *Journal of Thermal Analysis and Calorimetry*, 95(2), pp. 507–512 (2009).
27. R. Warzoha, R.D. Weinstein, and A.S. Fleischer, *ASME Summer Heat Transfer Conference*, Paper HT2012_58043, Puerto Rico, July 8-12, (2012).
28. R. Ehid, R.D. Weinstein, and A.S. Fleischer, *Energy Conversion and Management*, 57, pp. 60–67 (2012).

CHAPTER 10

ESTIMATION OF COOLING PERFORMANCE OF PHASE CHANGE MATERIAL (PCM) MODULE

Masaru Ishizuka and Tomoyuki Hatakeyama

Department of Mechanical Systems Engineering,
Toyama Prefectural University
5180 Kurokawa, Imizu, Toyama, Japan
ishizuka@pu-toyama.ac.jp

This chapter describes estimations of cooling performance of two kind of phase changing materials (PCMs) which would help develop high density packaging. A low cost alloy, composed of Bi/Pb/Sn/In in which melting points of individual elements are less than 373 Kelvin, and a kind of paraffin were used as the PCMs. Thermal experiments using the above-mentioned materials were carried out by the substrate back surface temperature could be fixed at the PCM's melting temperature for several minutes by thermal absorption, while the PCM phase changed from its original solid state into the liquid state. Also, in the present work, it has been confirmed that thermal network method is practically useful using a personal computer for the thermal design of a package with a PCM.

1. Introduction

Thermal design of electronic equipment is gaining in importance with the high density packaging of electronic equipment. Furthermore, in addition to steady cooling, a rapid and effective cooling system is needed to respond to temporary increases in heat generation, such as during peak use at electronic exchanges and other abnormal operating states. Temperature rise control using phase change materials (PCMs) has potential for use in high density equipment, and has already been the focus of several studies. Hodes[1] experimentally investigated the use of PCM for transient thermal management of a handset. They concluded that the use of PCM can increase the time for service for a handset. And they numerically calculated heat transfer rates from handset to ambient by natural convection and radiation. Krishnan and Garimella[2] carried out a transient analysis of a rectangular container including a PCM for electronics cooling application with pulsed power dissipation. Effects of frequency of pulses, location of heat sources and aspect ratio of the container were explored in their study. Alawadhi and Amon[3] investigated the effectiveness of a Thermal Control Unit (TCU) using PCM for portable electronic devices by performing experimental and numerical analysis. They confirmed that the TCU keeps the temperature of portable device almost constant at melting point of a PCM. Pal and Joshi[4] demonstrated computational study using finite volume method to predict the performance of passive cooling application using PCMs. n-Eicosene and eutectic alloy of Bi/Pb/Sn/In are considered as a PCMs in their study. Wirtz[5,6] explored the use of solid-solid dry PCM for electronics cooling application. They constructed semi empirical heat transfer model composed of elements and thermal resistances.

Organic PCMs generally have low thermal conductivity. To enhance apparent thermal conductivity of PCMs, several measures have been investigated. Weinstein *et al.*[7,8] embedded graphite nanofibers into a paraffin to improve the thermal conductivity of PCM. Graphite fiber types, fiber loading level, and PCM container size are explored in their study. Bauer and Wirts[9] measured the effective thermal conductivity of the PCM impregnated aluminum form. They had reported that an increase in aluminum metal form fraction enhance the effective thermal conductivity of PCM because most heat flow through the form. Lafdi [10] performed numerical analysis for heat sinks including form structures impregnated with PCMs. Form properties such as porosity, pore size, and thermal conductivity are investigated as design parameters.

Furthermore, a thermal analysis method must be provided for optimization of cooling devices. Commercially available Computational Fluid Dynamics (CFD) software are generally used for their design process, however, when conducting an analysis including phase changes, most CFD software allow in-depth analysis of temperature and flow distribution but require a fast computer and a great deal of time. Ishizuka[11-14] have reported some models for using the thermal network method for several thermal phenomena in electronic devices.

In this paper, 2 kind of models for the thermal network method to calculate phase change process in each package that contain PCM material are presented and it is shown that the thermal network method is an effective thermal design tool even when phase change process occurs.

2. Application of a Thermal Network Method to Thermal Analysis of Electronic Devices Using a Low Melting Point Alloy

In this section, application of a thermal network method to thermal analysis of a package using a low melting point alloy is described.

2.1. *Experiments*

The sectional view of the experimental model is presented in shown in Figure 1. Figure 1(a) shows the appearance of the experimental model without a PCM (Sample A). The appearance of the complete experimental model (Sample B) is shown in Figure 1(b). The alumina ceramic substrate dimensions for experimental model were 2 × 3 × 1.5 mm3. The height of kovar metal shell was 3 mm. The kovar metal is commonly used in the electronics-packaging field since the kovar metal has nearly the same thermal expansion rate as alumina ceramic substrate. The PCM properties are shown in Table 1.

2.1.1. *Experiment Procedure*

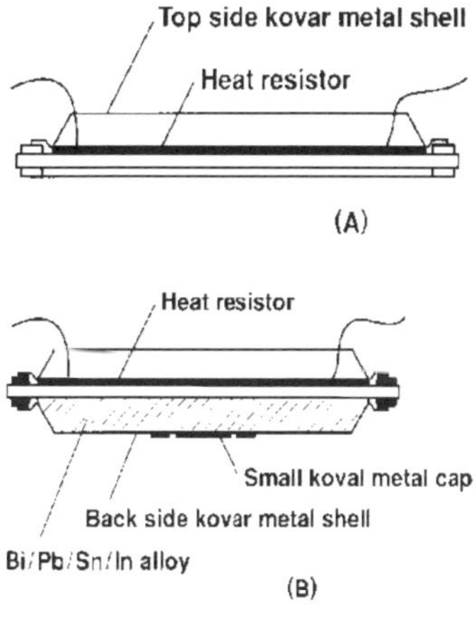

Fig. 1. Sample appearance.
Type A: top surface kovar metal shell.
Type B: A + back surface kovar metal shell with PCM.

2.1.2. *Experimental Procedure*

At first, thermal experiments for 2 kinds of samples were carried out, as shown in Figure 1.

Power dissipated by various mounted devices in the actual packages was simulated by attaching tungsten metallized heat resistor on the alumina ceramic substrate to serve the purpose of power dissipation sources. In the experiments, power dissipation was varied from 10 to 40 watts, with the intervals of 10 watts, which corresponds to 10 ohms resistance of the heat resistor. As soon as the voltage was supplied to the heat resistor, temperature values of the top and bottom surfaces of each sample package were measured by thermo-vision. Voltage application was stopped after the package temperature saturated and temperature drops were measured. The temperature measurement points (a-e) for each package are shown in Figure 2.

Table 1 PCM properties.

Material / Characteristics	Bi/Pb/Sn/In alloy		Substrate
	In solid state	In liquid state	Al_2O_3
Material density [kg/m^3]	9.06×10^3	8.22×10^3	3.6×10^3
Specitic heat [J/kgK]	323	721	879
Thermal conductivity [w/m°C]	33.2	10.6	16.7
Melting point [K]	330		
Heat of fusion [kJ/kg]	29.5		
Thickness [mm]	2.75		1.5

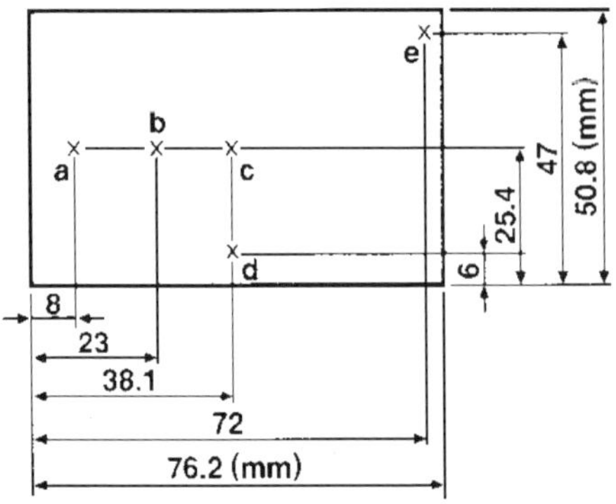

Fig. 2. Temperature measurement points for each package.

2.1.3. *Experimental Results and Discussion*

The experimental results obtained for the transient temperature rise and drop for Types A and B are individually shown in Figures 3 and 4. The power dissipation of the heat resistor was changed by varying the supplied voltage. The whole process of temperature variation with time presented in Figure 3, can be represented by three regions of prominent thermal behavior. Region 1 corresponds to the period during which the temperature starts rising and reaches the saturation temperature. Region 2 represents the period during which saturation temperature persists and this is the thermal equilibrium condition for the package, which was the thermal equilibrium condition for the package. Region 3 corresponds to the period during which the temperature starts decreasing as soon as the power supply is stopped.

The experimental results for Type A model presented in Figure 3 led to the following understanding;

Region 1: Region 1 is represented by the time period before saturation temperature is reached and this region is shortened with the increase in power dissipation for the package.

Region 2: In the Region 2, saturation temperature did not vary proportionally with the package power dissipation for the case of Type A model.

Region 3: Region 3 indicates that the higher the saturation temperature reached, the longer the cooling time is required. The cooling time is represented by the time between the start of the temperature drop and the attainment of ambient temperature.

Next, the experimental results for the Type B model presented in Figure 4 will be discussed. The results indicated seven regions of prominent change in thermal behavior.

Region 1: Region 1 corresponds to the period of time when the temperature started to rise and reached the melting point temperature of the PCM. In this region, the PCM was in solid state.

Region 2: Region 2 is represented by the time period during which the temperature of the ceramic substrate back surface remained constant to value equal to the melting point of PCM. And the phase of the PCM was changing from solid to liquid. In Region 2, the time required for the phase change of PCM was approximately inversely proportional to the package power dissipation. During this period, the substrate back surface temperature remained constant at the melting point of the PCM due to thermal absorption of latent heat.

Region 3: Region 3 corresponds to the period of time when the temperature started rising above the PCM melting point after the completion of the phase change process and reached the saturation temperature. The PCM in this Region was in liquid state.

Region 4: Region 4 corresponds to the period of time during which the temperature is in the state of saturation. This region indicates the thermal equilibrium condition for the package with a liquid state PCM.

Region 5: Region 5 corresponds to the period of time when the temperature drop as soon as the voltage supply was cut off to time the PCM reached its melting point. In this region the PCM was still in liquid state.

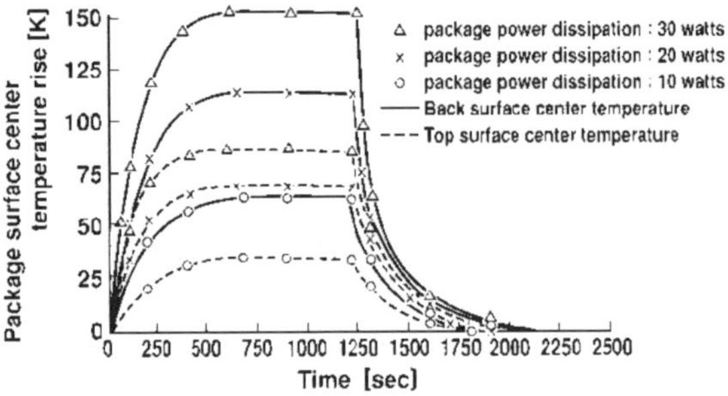

Fig. 3. Experimental results for transient temperature rise and drop for Types A model (ordinary).

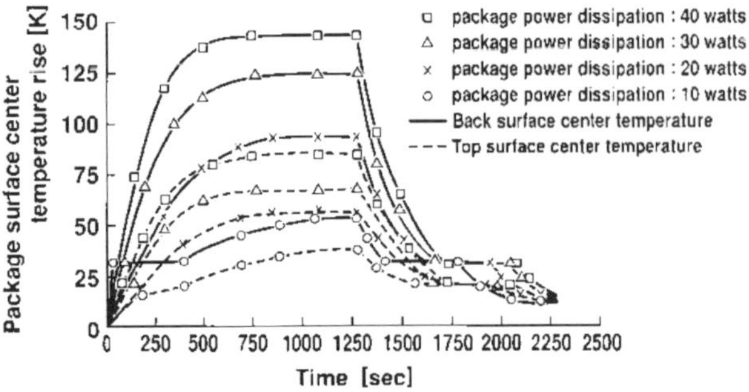

Fig. 4. Experimental results for transient temperature rise and drop for Types B model (ordinary).

Region 6: Region 6 corresponds to the period of time during which the ceramic substrate back surface temperature was kept constant at the melting point of the PCM. During this period PCM undergoes change in phase from liquid to solid. In this region, the time period during which the ceramic substrate back surface temperature was kept constant at the melting point of the low melting point alloy during the cooling stage was almost equal and was independent of saturation temperature. This period of time was approximately 360 seconds for the present PCM. Additionally, a hysterisis phenomenon was observed near the melting point of the PCM, which was somewhat lower (2-3 K) in region 6 than in Region 2. Region 7: Region 7 corresponds to the period of time after temperature began to drop from the melting point (334 K) for the PCM. During this period the PCM remained in solid state. In Region 1, the time required to reach the melting point temperature of the PCM decreases with the increase in package power dissipation. The package back surface temperature remains at a temperature equivalent to the melting point of the PCM. If the package is used for the back surface temperature having a value less than 348K Type A can be used for 120 seconds at 20 watts and Type B can be used for 250 seconds at 20 watts, as shown in Figures 3 and 4. The working time using the PCM is twice as that for the case without PCM.

2.2. *Thermal Analysis*

All the layers, including the heater, PCM and substrate, are divided into several sublayers, so that each sublayer has a uniform temperature distribution. Therefore, an individual temperature nodal point can represent each sublayer. The PCM area is divided into the liquid area, phase changing area, and solid area. Cap effects are neglected in this study. The nodal points are linked to each other by thermal resistance. Since the nodal point has thermal capacitance, thermal network is composed of thermal resistance R and thermal capacitance C, as shown in Figure 5.

2.2.1. *Equation Formulation*

Transient analysis was carried out by using a number of lumped RC tee sections, as shown in Figure 7, with subsequent employment of the nodal analysis method which is described, for instance, in the book of Kraus and Bar-Cohen (1983). The following equations, written here only for a single nodal point as examples, are

$$\begin{aligned} T0\text{-}T1 &= Q1R1 \text{ for } R1 \\ T_1\text{-}T_2 &= Q_2R_2 \text{ for } R_2 \end{aligned} \tag{1}$$

Where Q represents the heat flow value, which passes through thermal resistance R. The thermal resistance for thermal conduction is represented as follows;

$$R = d/(\lambda A) \tag{2}$$

And thermal capacitance is written as,

$$C = \rho V c_p \tag{3}$$

Where, ρ is density, V is volume and c_p is specific heat.

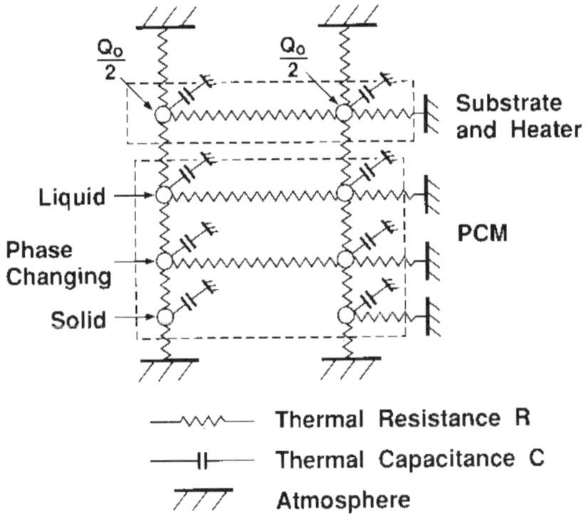

Fig. 5. Thermal network model composed of thermal resistance R and thermal capacitance C.

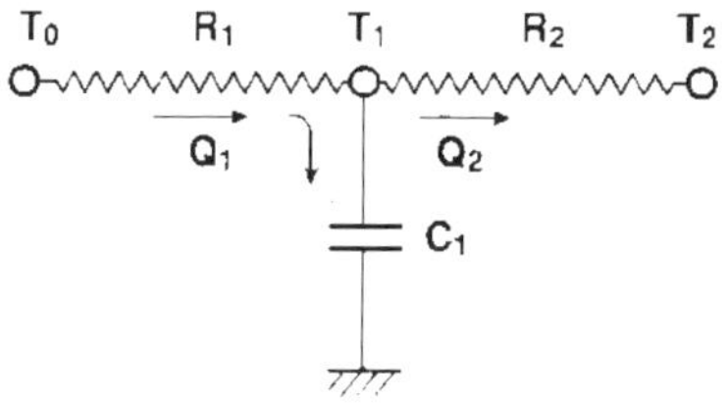

Fig. 6. RC Tee section.

Boundary conditions are specified by considering natural convective heat transfer and radiative heat. By defining α as, the natural convective heat transfer coefficient, its thermal resistance Rc is defined as,

$$Rc = 1/(\alpha A) = B(\Delta T/L)^{0.25} \tag{4}$$

where ΔT represents the temperature difference and B is the experimental constant. L represents the reference length and in this study, L is the axial length of the package. For radiative heat transfer between temperature values T_1 and T_2, resistance value Rr is defined as,

$$Rr=1/(FA\ (T_1^{\ 2} + T_2^{\ 2})(T_1 + T_2)) \tag{5}$$

where F is the shape factor and is taken as F=1.0 in this study.

The thermal characteristics of the nodes where phase change is occurring are as follows:

Originally, PCM is solid. It is assumed to be composed of several thin solid sublayers. In order to start the calculation, a very thin liquid layer is assumed to be associated with the solid. This solid is thought to be composed of several thin sublayers. In the present study it is 8. When the heat ΔQ from the heater is supplied to the solid sublayer (node) near the heater, this sublayer is heated up until the phase change temperature is reached. The change in thickness of the solid layer to liqui during the small time step Δt, as follows:

$$\Delta h = \Delta Q\ \Delta t/(\rho La\ A) \tag{6}$$

where, La is the latent heat associated with that node for the PCM. The following nodes, at which phase change is occurring, indicate that the temperature is set back to the phase change temperature. This process is repeated until the liquid thickness reaches the PCM thickness, leaving a very thin solid layer. Thereafter, the liquid temperature increases until the heat supply from the heater is cut off. When the heater is turned off, the reverse process takes place.

2.2.2. *Calculation Procedure*

A flow chart for the calculation technique is shown in Figure 7. Initially, the applied power Q and the PCM thickness are specified. Also, the temperature values T_0 at the initial time t_0, are given. Then, the thermal resistance values R_0 are obtained using Eqs. (4) and (5) on the basis of the above temperature values T and thermal capacitance C. The linearized system of equations is solved to yield the temperature values for the new time level $t = t + \Delta t$, based on the Runge-Kutta method. Thermal resistance value R is calculated on the basis of Eqs. (4) and (5) using the newly predicted temperature T and time level t advanced by Δt. In this procedure, when the temperature reaches the phase change temperature, calculation is carried out to obtain the change in liquid thickness Δh on the basis of Eq. (6) in a separate loop.

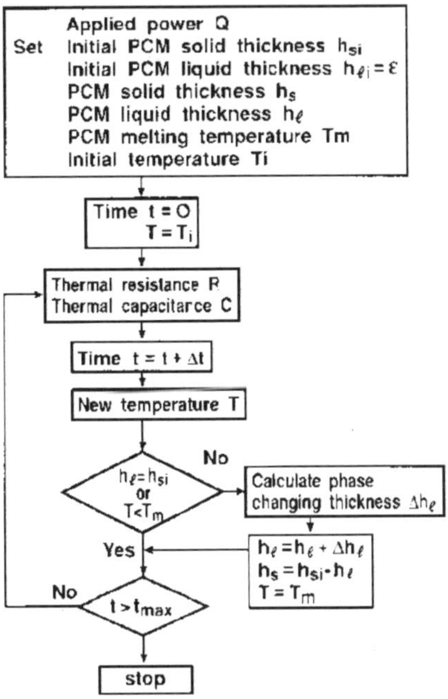

Fig. 7. A flow chart for the calculation procedure.

The results are fed back to the next time step. Also, when the solid temperature exceeds the melting temperature and all the PCM is not completely converted into liquid, computation for the change in liquid thickness is continued. The newly predicted temperature value replaced to the initial temperature and the procedure is repeated for the subsequent time level, until t = t_{max}. The liquid zone thickness Δh, which changes due to the melting of solid to liquid, is calculated using Eq. (6), when the temperature of the nodes near the heater reaches the phase change temperature. The procedure is repeated until entire the solid zone set melted leaving a very thin layer of solid for the convenience of computation.

2.2.3. *Calculated Results and Discussion*

Figure 8 shows a comparison between the measured and calculated results. The results of computation showed an excellent agreement with the measured data. In this calculation, the value of the coefficient B, which appeared in Eq. (3.4) for the thermal resistance due to the natural convection, was obtained as 10, on the basis of matching the computation results with the experimental value of the saturation temperature level for 20 watts. Furthermore, the applicability of this computational model was verified by using it to predict the measured surface temperature history of the Type B sample, which was already shown in Figure 4.

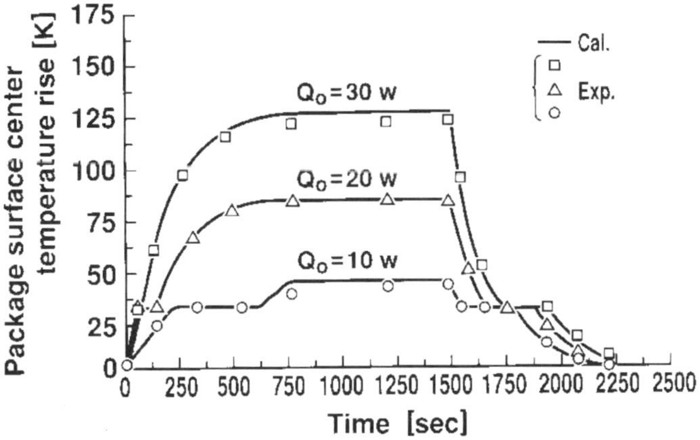

Fig. 8. A comparison between the measured and calculated results (ordinary).

Figure 9 shows a comparison between the measured and calculated temperature for the case of intermittent heat supply. Here also an excellent agreement between results of computation and measured data was observed. The calculated result was also in excellent agreement with the measured result. The thermal network method was found to be useful tool for thermal design of the package. However it should be mentioned here that the above-discussed experiments were accomplished using a uniformly heated package. In order to enhance the applicability of the present model to simulate the experimental result for some non-uniformly heated package (which generally occurs in practice), the present model may need to be slightly modified to take into account of non-uniformly thermal behavior.

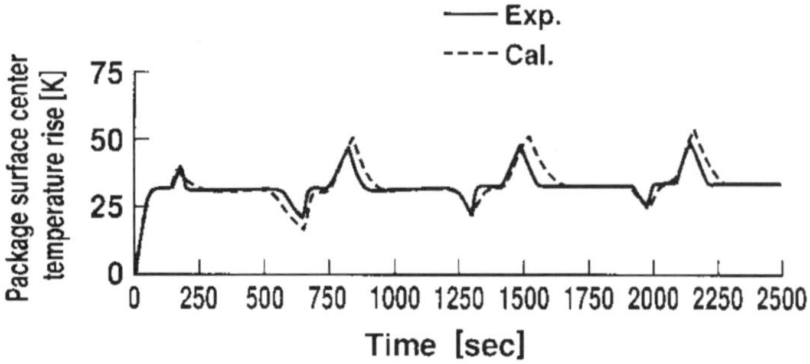

Fig. 9. Comparison between the measured and calculated results (intermittently).

3. Application of a Thermal Network Method to Thermal Analysis of Electronic Devices Using Paraffin

In this section, application of a thermal network method to thermal analysis of a device using paraffin is described.

3.1. *Experiments*

Figure 10 shows a schematic view of the experimental apparatus. It consists of a test section, a DC power supply to supply power to the heat source, a data logger to collect temperature data, a thermostatic chamber to eliminate variations in temperature.

An electronic module investigated as a test section in this study is shown in Figure 11. The module is consisted of a substrate and a heat sink. The 85 mm × 85 mm × 14 mm aluminum heat sink is mounted on the back of thc 2 mm–thick aluminum substrate on which two resistors are mounted as a heat sources. To reduce contact thermal resistance, silicone grease is applied between the substrate and the heat sink. The PCM are accommodated inside the heat sink. Cylindrical fins are provided inside the heat sink to enhance heat conduction to the PCM. Three fin diameters are employed as shown Figure 12: 2.5 mm fin × 180, 5 mm fin × 45, and 33.5 mm fin × 1.

Paraffin (Nippon Seiro's 155) was used as the PCM in our study. Table 2 lists the properties of the PCM and aluminum (ADC50). The specific heat and thermal conductivity in Table 2 used are that of general paraffin.

We conducted an experiment with heat dissipation P of 15, 20, and 30 W input to the resistors. The heat dissipation was varied by adjusting the supplied voltage. While applying power P/2 [W] to each of the two resistors on the substrate, Temperature of the module was measured every second using a 0.1 mm diameter K-type thermocouple. The uncertainty in temperature measurement is estimated to be ±1.9°C.

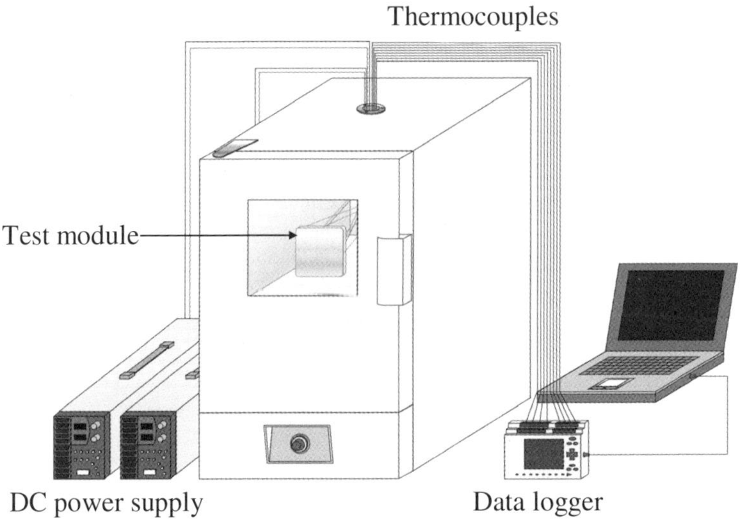

Fig. 10. Experimental apparatus using Paraffin.

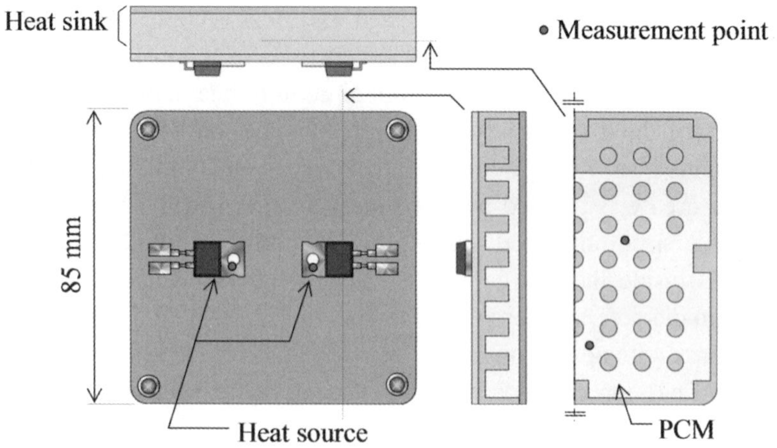

Fig. 11. Test module.

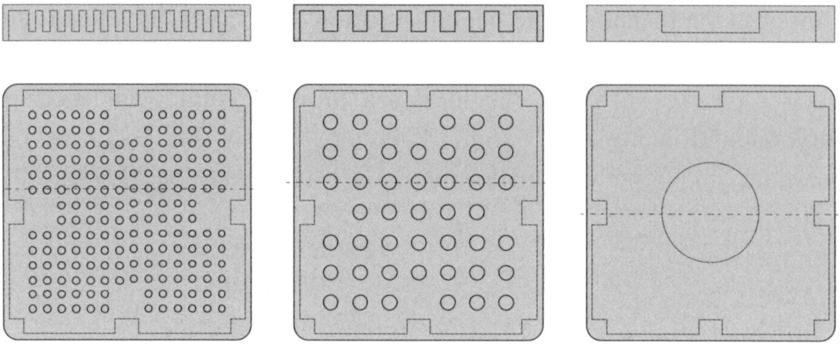

(a) d = 2.5 mm (b) d = 5 mm (c) d = 33.5 mm

Fig. 12 Heat sinks with different fin diameter.

Table 2 Properties of the PCM and aluminum (ADC50).

La[kJ/kg]	ρ[kg/m^3]	c_p[J/kgK]	λ[W/mK]	La[kJ/kg]	T_{melt}[degree C]
Paraffin	927	1970	0.23	197	40-71
ADC50	2700	963	96.0		

3.1.1. *Experimental results*

Figures 13 and 14 show the temperature measurement results of the PCM and the heat source for three fin diameters, 2.5, 5, and 33.5 mm. Heat dissipation is constant 30 W. Temperature of the heat source on which the heat sink without PCM was mounted, is also plotted in Figure 5. Figures 4 and 5 show that the temperature continues to increase with the elapse of time after power is applied until the PCM temperature reaches its melting point. Until PCM temperature reaches its melting point, dissipated heat is stored as the sensible heat of the substrate, the heat sink, and the PCM. The temperature rise is suppressed in the following 500 to 1000 seconds, during which the PCM temperature reaches around 60°C. This is due to the absorption effect of latent heat when the PCM melts. A phase of the PCM completely changes into liquid at 1200 seconds.

The temperature of heat source, the fin diameter of which is 33.5 mm, is about 4°C higher than other two cases during the PCM melts. This difference is caused by a decrease of heat conduction area by a increase of fin diameter. Heat source temperature histories are same after 1200 seconds.

Assuming that the heat source permit temperature rises up to 125°C, Figure 14 shows the case without PCM can operate for up to 1000 seconds, and the case with PCM for up to 2000 seconds. Thus, the thermal absorption effect of PCM doubles the operating time of the equipment. Therefore, the PCM has a cooling effect on transient heat dissipation and suppresses temperature rise. The heat sink weighs 125 g and the incorporated PCM weighs 37 g, increasing total weight by 30%. Therefore, a reduction in the weight of cooling system is feasible by utilizing the high thermal storage capacity of PCM.

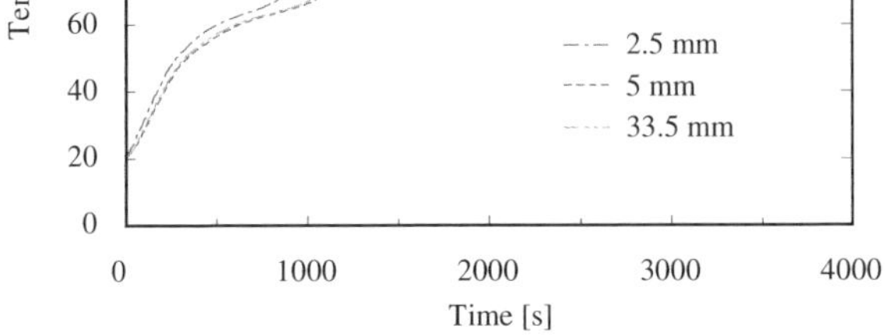

Fig. 13. Temperature of the PCM at heat dissipation 30 W with three fin diameters.

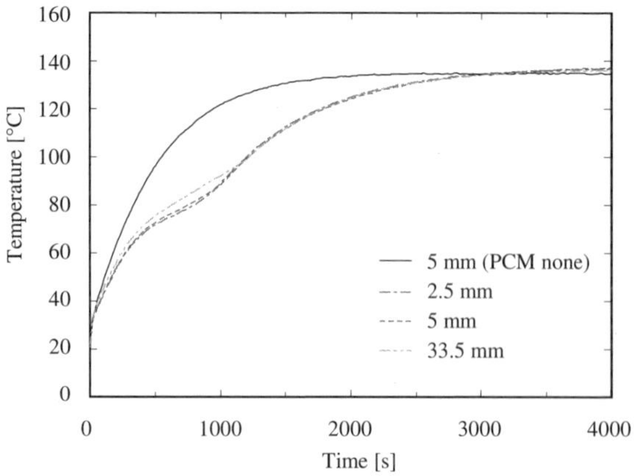

Fig. 14. Temperature of the heat source at heat dissipation 30 W with three fin diameters.

Figures 15 and 16 show the temperature measurement results of the PCM and the heat source for three heat dissipations, 15, 20, and 30 W. Fin diameter 5 mm are employed. Temperature rise gradients of the PCM and the heat source increase as heat dissipation increase. And Figure 6 shows time which is required for complete melting of PCM, decreases as heat dissipation increases.

Temperature of the heat sources for 15 and 20 W do not surpass the temperature limit 125°C even at steady state as shown in Figure 7. Thus, There are not operating time limit for 15 and 20 W.

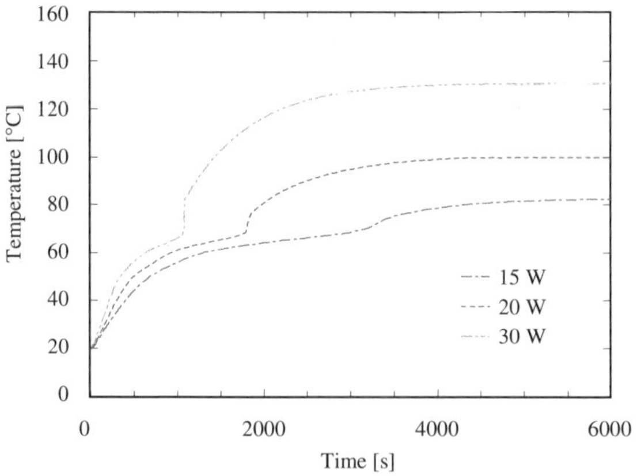

Fig. 15. Temperature of the PCM for three heat dissipations.

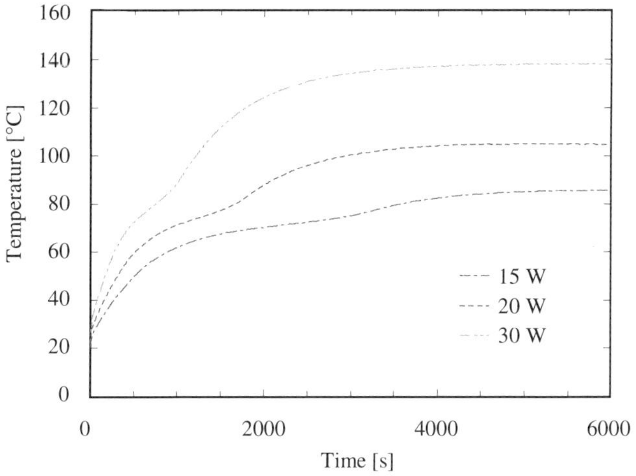

Fig. 16. Temperature of the heat source for three heat dissipations.

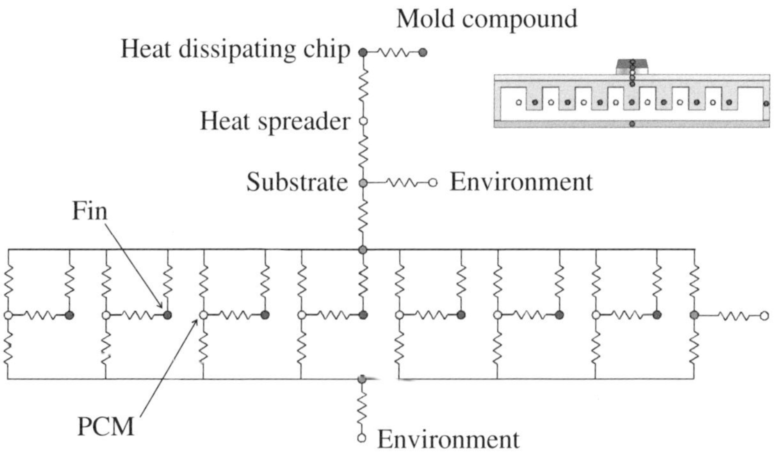

Fig. 17. Thermal network model (All capacitors are omitted in the figure.)

Thermal resistance between the heat dissipating chip and the heat spreader is set as 2.3°C/W from the commercial catalog.

3.2. *Modeling of Latent Heat*

Latent heat of phase change processes are taken into account as equivalent specific heat which is a function of the temperature of PCM. Equivalent specific heat is plotted in Figure 9. The varying process of specific heat of the PCM is determined based on experimental result with different scanning calorimetry (DSC). Thermal capacitance of nodes of PCM is calculated using c_s (1970 J/(kg·K)) which is specific heat at solid phase, until the temperature of the node reaches 40°C. After the temperature of the node surpasses 40°C, equivalent specific heat is used for calculation. Equivalent specific heat is follows

$$c_{equ} = c_s + \Delta c \tag{7}$$

An increment by latent heat Δc is defined as a linear function of the temperature. After the temperature surpasses 71°C, specific heat at liquid phase c_l is used for calculation. A heat quantity which integrates Δc from 40°C to 71°C is equal to the latent heat of the PCM (197 kJ/kg).

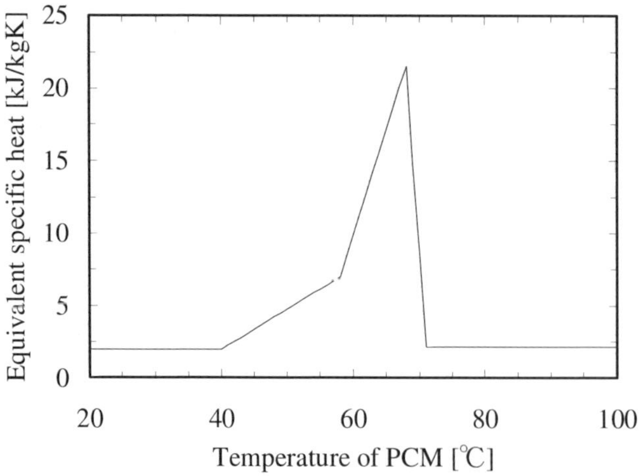

Fig. 18. Equivalent specific heat.

3.3. *Comparison Between Experimental and calculated results*

Figures 10 and 11 show a comparison between experimental and numerical temperature of the PCM and the heat source, respectively for three fin diameters at P = 30 W. Numerical results are good agreement with experimental result. Numerical result accurately predicts a melting process of the PCM which is important to design a cooling module using PCM. Numerical temperature histories of the heat source are almost matched with experimental result and accurately predict the temperature at steady state.

Figures 21 and 22 show a comparison between experimental and numerical temperature of the PCM and the heat source, respectively for three heat dissipations at d = 5 mm. Differences of melting time and steady state temperature of the PCM by different heat dissipations are accurately predicted in numerical analysis. The analysis result and the experimental results matched with different heat dissipation, proving that the thermal network method is an effective tool for thermal design. It appears possible to formulate guidelines to keep the temperature of the heat source within the allowable range by using a suitable type of PCM in the design process.

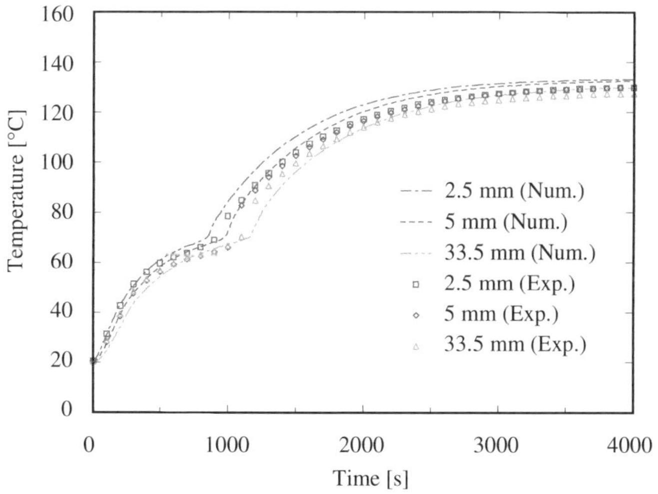

Fig. 19. Comparison of numerical and experimental temperature of the PCM for three fin diameters.

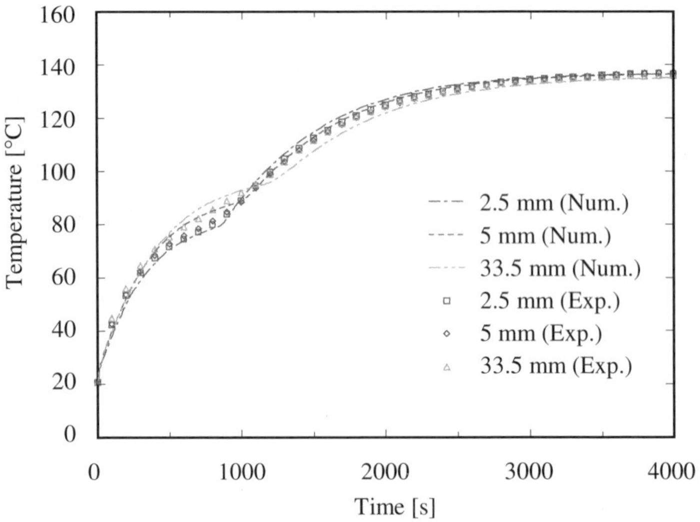

Fig. 20. Comparison of numerical and experimental temperature of the heat source for three fin diameters.

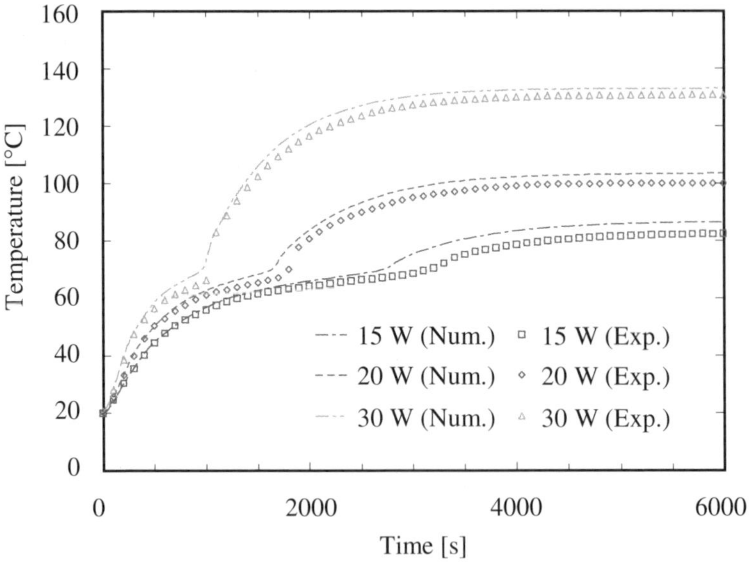

Fig. 21. Comparison of numerical and experimental temperature of the PCM for three heat dissipations.

Furthermore, varying power operation mode experiment and numerical analysis were performed to verify cooling effect of the PCM and accuracy of thermal network method at practical operation of electronic device. The heat dissipation was on with the power of 30 W for 1000 seconds and then the power was off for 1000 seconds and again the power was on with 30 W for 1000 seconds and off for 1000 seconds. And this procedure was repeated until 6000 seconds. The maximum heat source temperature was controlled under the limit temperature 125°C during all the operation time. When power supply is stopped, the temperature decreases to the melting point of the PCM and the temperature drop is decelerated due to emission of latent heat during solidification.

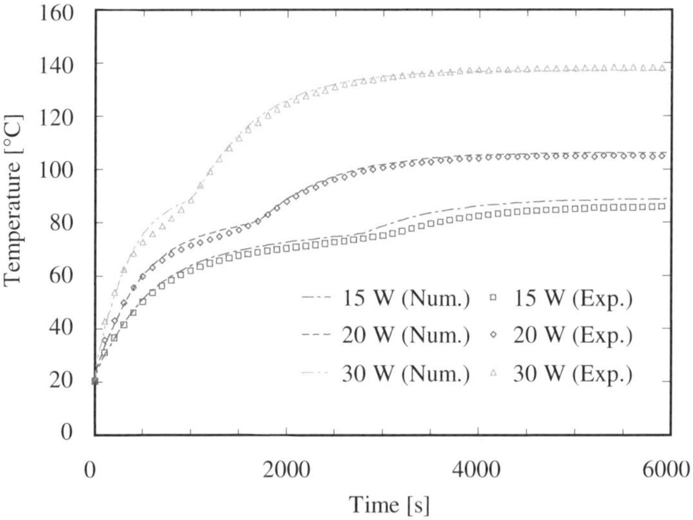

Fig. 22. Comparison of numerical and experimental temperature of the heat source for three heat dissipations.

Numerical temperature is in good agreement with experimental result and accurately predicts the maximum heat source temperature. Notice that numerical temperature decreasing history is accurate. Therefore, estimation of thermal resistances by natural convection and radiation are valid.

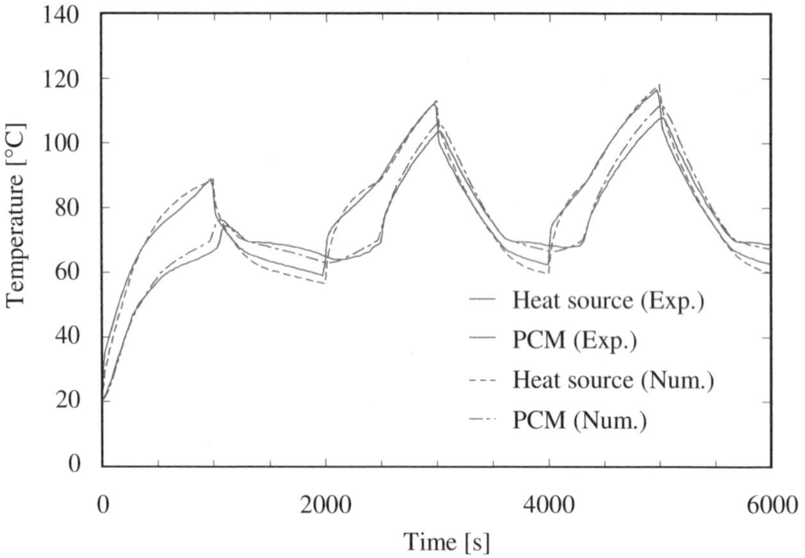

Fig. 23. Comparison of numerical and experimental temperature of the PCM and heat source for varying power operation at fin diameter 5 mm.

4. Conclusion

We built a module of an electronic device using paraffin wax as a PCM and measured its temperature experimentally. We also attempted to estimate the temperature of an electronic module incorporating a PCM using the transient thermal network method. As a result, we were able to verify that the temperature rise of the heat source could be controlled by latent heat effect of the PCM, extending the usable time of the equipment. In this analysis, the calculated results and the experimental were in good agreement. Thus, the thermal network method has been proved to be an effective thermal design tool, even when phase change process occurs.

Nomenclature

A = area, m^2
B = experimental parameter
C = thermal capacitance, J/K
c_p = specific heat, J/kg K

d = thickness
F = shape factor
g = gravity acceleration
Gr = Grashof number = $g \cdot \beta \cdot \Delta T \cdot d^3/\nu^2$
h = phase changing thickness, m
k = thermal conductivity, W/mK
L = reference length
La = latent heat for PCM. J/kgK
Nu = Nusselt number = $\alpha \cdot L/\lambda$
Q = heat flow, W
R = thermal resistance, K/W
T = temperature
ΔT = temperature rise
t = time, s
V =Volume, m^3

Greek

α = average thermal transfer coefficient
β = thermal expansion coefficient
ε = emissivity
λ = thermal conductivity
ν = kinematic viscosity
ρ = density
σ = Stefan-Boltzmann constant

Subscript

c = convection
r = radiation

References

[1] M. Hodes, R. D. Weinstein, S. J. Pence, J. M. Piccini, L. Manzione, and C. Chen, *ASME Journal of Electronic Packaging*, 124, 419–426 (2002).
[2] S. Krishnan and S. V. Garimella, *IEEE Transaction on Components and Packaging Technology*, 27, 1, 191–199 (2004).

[3] E. M. Alawadhi and C. H. Amon, *Inter Society Conference on Thermal Phenomena*, 466–475 (2002).

[4] D. Pal and Y. K. Joshi, *ASME Journal of Electronic Packaging,* 119, 40–50 (1997).

[5] R. A. Wirtz, N. Zheng, and D. Chandra, *Proceedings of IEEE Semiconductor Thermal Measurement and management Symposium*, 74–82 (1999).

[6] N. Zheng and R. A. Wirtz, *Proceedings of IMECE,* 1–8 (2000).

[7] R. D. Weinstein, T. C. Kopec, A. S. Fleischer, E. D'Addio, and C. A. Bessel, *ASME Journal of Heat Transfer,* 130 (2008).

[8] A. S. Fleischer, K. Chintakrinda, R. Weinstein, and C. A. Bessel, *Proceedings of ITHERM*, 561–566 (2008).

[9] C. A. Bauer and R. A. Wirtz, *Proceedings of IMECE*, 1–7 (2000).

[10] K. Lafdi, O. Mesalhy, and A. Elgafy, *ASME Journal of Electronic Packaging,* 130 (2008).

[11] M. Ishizuka, S. Nakagawa, and K. Koizumi, *ASME-MECE*, IMECE2006-13105 (2006).

[12] M. Ishizuka, K. Hisano T. Sasaki and Y. Fukuoka, *Proceedings of 1995 International Electronic Manufacturing Technology Symposium, Japan*, 463–467 (1995).

[13] M. Ishizuka, *IEEE Transactions on Components, Packaging Technologies,* 27, 2, 239–243 (2004).

[14] S. Takakuwa, M. Ishizuka, S. Nakagawa, *Proceedings of IPACK2009,* IPACK2009-89218 (2009).

CHAPTER 11

OPTIMIZATION UNDER UNCERTAINTY FOR ELECTRONICS COOLING DESIGN

Karthik K. Bodla,[1] Jayathi Y. Murthy,[2] and Suresh V. Garimella[1]

[1] *Cooling Technologies Research Center, an NSF IUCRC*
School of Mechanical Engineering and Birck Nanotechnology Center
Purdue University
West Lafayette, IN 47907-2088 USA
sureshg@purdue.edu

[2] *Department of Mechanical Engineering*
The University of Texas at Austin
Austin, TX 78712-0292 USA
jmurthy@me.utexas.edu

Optimization under uncertainty is a powerful methodology used in design and optimization to produce robust, reliable designs. Such an optimization methodology, employed when the input quantities of interest are uncertain, produces output uncertainties, helping the designer choose input parameters that would result in satisfactory thermal solutions. Apart from providing basic statistical information such as mean and standard deviation in the output quantities, auxiliary data from an uncertainty based optimization, such as local and global sensitivities, help the designer decide the input parameter(s) to which the output quantity of interest is most sensitive. This helps the design of experiments based on the most sensitive input parameter(s). A further crucial output of such a methodology is the solution to the inverse problem – finding the allowable uncertainty range in the input parameter(s), given an acceptable uncertainty range in the output quantity of interest.

A brief review of generalized polynomial chaos (gPC) based uncertainty propagation and subsequent optimization under uncertainty (OUU) is presented, with application to heat transfer in microelectronics cooling applications. The non-intrusive analysis methodology, implemented in DAKOTA, an open source design and analysis kit, is verified by comparison with computationally expensive intrusive alternatives employed in the literature. The usefulness of the method in arriving at strict and robust optimal solutions in a computationally inexpensive way is demonstrated with application to a pin-fin heat sink design problem. The results so obtained are compared with a traditional, deterministic approach in which the effect of uncertainties is only heuristically accounted for. A second example problem considered is the design and numerical optimization of a manifold microchannel heat sink for which large uncertainties in dimensional parameters may be present owing to manufacturing limitations.

1. Introduction

Uncertainties in engineering designs may be broadly classified as aleatory and epistemic. Aleatory or irreducible uncertainties are those that result from an inherent variability in inputs, such as those due to the presence of uncertain geometric parameters, while epistemic or reducible uncertainties result from a lack of knowledge, such as those due to unknown physical phenomena [1]. The challenge is to efficiently quantify these uncertainties and include them in optimization procedures to produce designs that meet the prescribed reliability and robustness levels. Electronics cooling designs, for instance, are prone to a number of uncertainties. A desktop computer heat sink sees different heat generation levels depending on the processor load. Similarly, the flow speed from the fan can be uncertain and variable. Thermal solutions such as microchannel heat sinks suffer from uncertainties in geometric parameters owing to inherent limitations of the fabrication techniques [2].

A conventional (deterministic) optimization approach to an engineering problem in the presence of such uncertainties ignores the uncertainties and treats the uncertain variables as certain, *i.e.*, it assumes them to be fixed at their means or bounds (if the extent and nature of the uncertainty is known) and proceeds with the optimization. Once the optimal design parameters are arrived at, a factor of safety may be then included, based on prior experience, to take the inherent variability and other uncertainties into consideration. This simple, easy-to-implement and computationally inexpensive approach is the most widely used method. However, in the presence of large uncertainties, or when there are strict constraints (such as expensive design parts, or designs where overall mass is a critical parameter), the method fails to produce a truly optimal design that is robust and reliable in addition to being optimal. The conventional approach does not quantify uncertainties or fold them into the optimization procedure; it accounts for them heuristically through a factor of safety applied after the fact [3].

To perform optimization under uncertainty (OUU), or probabilistic optimization, optimization techniques must be combined with uncertainty quantification (UQ) procedures. The OUU approach may be used to produce designs that meet specified robustness and reliability levels, *i.e.*, for "design for robustness" and/or "design for reliability" [4]. The goal of the former design problem is to produce a design that is less sensitive to variable inputs, such as a design with a low standard deviation in the output quantities. In contrast, in the latter problem, the objective is to produce designs with a lower probability of failure; for this problem, performance statistics at the tails of the distributions are of interest. An example of the latter problem is a design that reduces the probability of the output parameter falling below a threshold value. The former design problem is often the simpler one to address of the two, and does not always require UQ [4]. In the latter problem, UQ is performed and design parameters are estimated based on the tail (end) statistics of the output probability density function (PDF). This problem places a greater demand on UQ, and is often computationally more expensive compared to the former problem.

Optimization of electronics cooling equipment has been widely studied [5]. With the advent of electronics that dissipate large amounts of heat and are densely packed, there is an increasing need to optimize the existing cooling solutions and to design newer and more efficient ones such as those employing liquid jet impingement and phase change. However, to date, air cooling remains the most widely used approach, owing to its inherent simplicity, low cost and low maintenance [6], and also because air is the ultimate heat rejection medium. Extended surfaces such as pin fins and parallel-plate fins aid heat transfer from a hot surface to the surroundings, and are often employed in conjunction with air cooling [7]. Optimizing dimensional parameters such as fin-to-fin spacing and number of fins for a given set of input parameters is a commonly studied optimization problem in heat transfer. Bar-Cohen and Rohsenow [8] and Bejan [9] developed correlations for optimal spacing between parallel plates for natural convection heat transfer. Similar studies for forced convection were performed by Nakayama *et al.* [10] and Bejan and Sciubba [11]. Ledezma *et al.* [12] demonstrated the existence of an optimal spacing between pin fins in crossflow and impinging flow. Based on theoretical arguments relating to similarity of flow and heat transfer, the form of the correlation was borrowed from previous studies on cross-flow over staggered cylinders [9] and flow between parallel plates [11]. Some of the recent studies on optimization of pin-fin heat sinks include those by Kang [13] and Khan *et al.* [14].

Optimization studies of this kind neglect the uncertainties in the problem being investigated, such as uncertain fin dimensions and the variability in applied heat flux. The objective of the present study is to assess the effect of these uncertainties. Optimization under uncertainty is performed for the representative case of impinging flow on pin fins considered by Ledezma *et al.* [12]. Problem parameters such as base and fin dimensions and flow speeds are matched with [12]. The OUU framework developed here is verified against simple analytical and 2D problems such as flow in a channel with uncertain viscosity ([15, 16]). The deterministic computational model is first verified for the flow speeds and dimensions considered by Ledezma *et al.* [12]. The pin-fin geometry is then optimized by considering uncertainties in geometric and operating conditions. The cost of multiple flow and heat transfer simulations is alleviated by using response functions derived from generalized polynomial chaos (gPC) [17]. Optimization under uncertainty is performed, and robust designs that limit the standard deviation in Nusselt number are obtained. Further, a manifold microchannel heat sink such as the one considered by Kermani [18] is optimized employing the OUU methodology developed here. Uncertainties in geometric parameters are considered and a single objective optimization – minimization of overall thermal resistance – is performed at fixed pumping power.

2. Methodology

In this section, the uncertainty quantification (UQ) methodology and the subsequent nested approach to optimization under uncertainty (OUU) are briefly described. Readers are referred to comprehensive literature on the subject (*e.g.*, [17] and [1]) for a detailed discussion.

2.1. *Uncertainty Quantification*

Uncertainty quantification (UQ) refers to the process of determining the effect of input uncertainties on the output quantities of interest. These uncertainties, as previously mentioned, are generally categorized as either aleatory (*i.e.*, resulting from irreducible, inherent variability) or epistemic (resulting from a lack of knowledge, and thus potentially reducible) [1]. Aleatory uncertainties are typically addressed using probabilistic methods. Approaches to epistemic uncertainty vary. Both probabilistic (Bayesian) as well as non-probabilistic methods have been used to quantify these uncertainties. In the present work, the analysis is restricted to aleatory uncertainty, to which probabilistic methods such as polynomial chaos expansions (PCE) may be successfully applied.

The most common UQ methods are based on random sampling techniques. Based on the input probability distribution function (PDF), samples are drawn randomly employing algorithms such as Monte Carlo (MC) and Latin Hypercube Sampling (LHS). For each sample drawn, the function evaluation (simulation/analytical) is performed and when the entire variation of the input variable(s) is covered, the PDF of outputs may be computed along with various other response statistics [1]. Thousands of evaluations must be performed to cover the entire range of the input PDF; the number of evaluations becomes exceptionally large as the number of uncertain variables increases. Other methods, such as the sensitivity method based on moments of samples, exist but are less robust and depend on the model assumptions [1].

A less computationally intensive UQ method for modest numbers of input random variables is the polynomial chaos expansion (PCE) method. In this work, we employ the generalized Polynomial Chaos (gPC) approach, using the Wiener-Askey scheme originally proposed by Xiu and Karniadakis [17]. In the gPC approach, Hermite, Legendre, Laguerre, Jacobi and generalized Laguerre orthogonal polynomials are used to model the effect of uncertain input variables described by normal, uniform, exponential, beta and gamma PDFs, respectively,. These orthogonal polynomial selections are shown to be optimal for these distribution types since the inner product weighting function and its corresponding support range correspond to the PDF for these continuous distributions [17]. In theory, this selection of the optimal basis allows for exponential convergence rates, which are far better than the random sampling techniques mentioned previously [1]. Both intrusive and non-intrusive variants of gPC have been proposed [17].

The intrusive approach employs a Galerkin projection-based reformulation of the original governing equations to arrive at the stochastic equivalents. New solvers are required for solving this modified set of governing equations. As noted by Najm [15], a wide class of polynomial and non-polynomial functions may be readily transformed with this approach. Ghanem and Spanos [19] first outlined intrusive polynomial chaos UQ constructions in the context of the Galerkin finite element method.

The stochastic collocation method is a non-intrusive method based on gPC [17]. In this approach, the set of polynomials previously mentioned are used as an orthogonal basis to approximate the functional form between the stochastic output response and each of its random inputs. Deterministic simulations are performed at the collocation points in random space, and exploitation of the orthogonality properties of the polynomial basis functions allows the determination of the coefficients in the polynomial expansion. Existing solvers may thus be employed. Details may be found in [17] and in the comprehensive review by Eldred [1].

The chaos expansion for a response R takes the form:

$$R=\sum_{j=0}^{\infty}\alpha_j\psi_j(\xi) \tag{1}$$

Each of the terms $\psi_j(\xi)$ consists of multivariate polynomials which involve products of the corresponding one-dimensional polynomials in the random variable, ξ. In practice, the number of terms in Eq. (1) is truncated, and hence only a finite number of evaluations is needed to calculate the polynomial representation of the response function R. A number of options exist for choosing the specific evaluation points j, such as quadrature grids for each random variable (which translates to a tensor product grid in multiple dimensions, *i.e.*, when there is more than one random variable) and sparse gridding techniques [1]. One such sparse grid technique called the Smolyak sparse grid has been proven to be highly efficient, requiring fewer computations than a tensor product grid when there is a large number of uncertain parameters. We employ a Smolyak sparse grid in this work.

Once the response function is constructed based on deterministic runs at the collocation points ascertained by the Smolyak algorithm, it is then used as a surrogate model. PDFs of the response R may be computed by sampling the space of input random variables using random sampling algorithms such as MC and LHS. The output response statistics, for example, the mean, standard deviation, or indeed, the entire output PDF, may then be evaluated. Such UQ analyses also provide useful auxiliary information, such as that for sensitivity of outputs to inputs. Based on the reported Sobol's indices [20], information such as the sensitivity of a particular output metric to a particular uncertain input parameter may be obtained, which aids in the design of experiments. For example, based on this type of UQ analysis and the sensitivity information thus obtained, the less-sensitive parameters may be considered as deterministic (certain). Subsequent (refined) UQ may then be performed by resolving the more sensitive parameters more accurately, thereby reducing the computational effort significantly. In this work, we use the open-source UQ and optimization toolkit, DAKOTA [21], for performing the uncertainty quantification and the subsequent optimization.

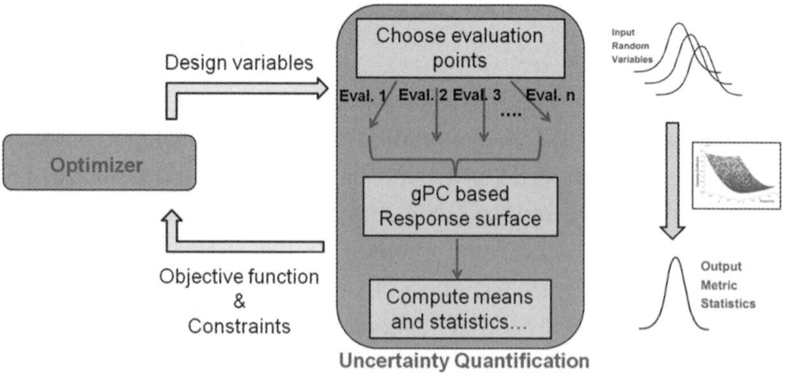

Fig. 1. Nested approach to optimization under uncertainty employed in this work.

2.2. *Optimization Under Uncertainty*

Optimization under uncertainty (OUU) refers to the optimization of a design by taking into consideration the uncertain input parameters and the corresponding output response statistics. Eldred *et al.* [4] provided an overview of the various OUU formulations available in DAKOTA. Of these, we use the most direct approach called the nested approach to OUU in the present work. The UQ loop is nested within the optimization loop. Starting with initial guess values for the design variables (*i.e.*, initial conditions for the optimization loop), a complete uncertainty quantification is performed using the gPC UQ methodology described in the preceding section. Based on the output response metrics from the UQ loop, the optimizer evaluates the objective function and the corresponding constraints (if it is a constrained optimization problem, such as a design for robustness and/or reliability) and checks for convergence. If the convergence criterion is not met, new values for the design variables are selected and the optimization proceeds. The nested OUU approach used in this work is shown schematically in Figure 1.

At convergence, the set of design variables satisfying a given set of constraints and also simultaneously minimizing (or maximizing) an objective function of interest are obtained. DAKOTA is equipped with a number of gradient- and non-gradient-based optimization algorithms.

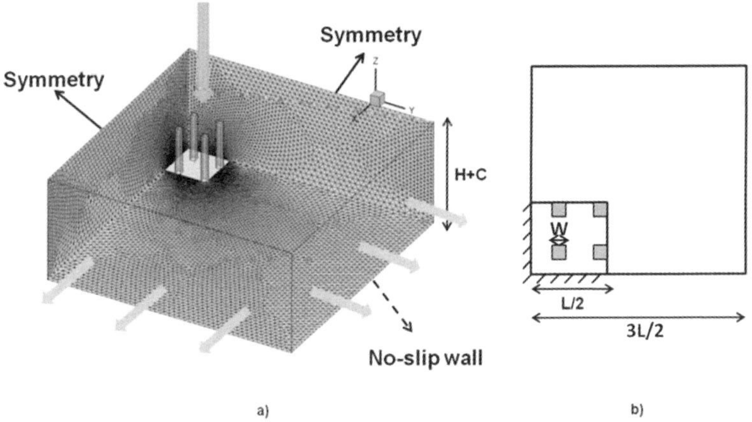

Fig. 2. (a) Isometric view with boundary conditions, and (b) top view, of pin-fin geometry considered.

We choose the gradient-based optimization algorithm known as the Fletcher-Reeves conjugate gradient method for unconstrained optimization and the method of feasible directions for constrained optimization, based on the CONMIN library [22]. Also, it may be noted that the method is not just limited to global optimization; DAKOTA is equipped with optimizers to deal with multiple local optima in addition to the global ones.

3. Numerical Modeling

This section describes the governing equations and boundary conditions and details of the numerical model.

3.1. *Governing Equations and Boundary Conditions*

The continuity, momentum and energy equations for the steady, constant-property flow of an incompressible Newtonian fluid are given by:

$$\frac{\partial}{\partial x_i}(\rho u_i)=0 \tag{2}$$

$$\frac{\partial}{\partial x_j}\left(\rho u_j u_i\right) = -\frac{\partial P}{\partial x_i} + \mu\left(\frac{\partial^2 u_i}{\partial x_j x_j}\right) \tag{3}$$

$$\frac{\partial}{\partial x_i}\left(\rho C u_i T\right) = \frac{\partial}{\partial x_i}\left(k\frac{\partial T}{\partial x_i}\right) \tag{4}$$

The computational domain for the pin-fin case is shown in Figure 2. Exploiting symmetry, only a $1/4^{th}$ section need be considered. The flow enters the computational domain at the top, impinges on the fin surfaces, cools the fins and leaves the domain at the two outlets along the sides on which the pressure is prescribed. The computational domain is suitably extended in the wake of the pin fins to account for flow reversal. Also, following Ledezma *et al.* [12], the fins are assumed to be isothermal, and hence only the fluid domain is considered for analysis. Based on their prior study [23] and other studies on air-cooled heat sinks used in practice (for example, Nakayama *et al.* [10] and Matsushima *et al.* [24]), Ledezma *et al.* [12] asserted that for an optimization analysis, the isothermal assumption yields results similar to those with a non-isothermal fins assumption. For a more thorough analysis, the solid domain must also be considered in the framework developed as part of the present work. A no-slip boundary condition is prescribed for all the solid-fluid interfaces. Figure 2 shows the various boundary conditions employed in the pin-fin heat transfer problem considered.

3.2. *Solution Procedure*

An effective optimization approach demands automation of the computational sequence shown in Figure 1. For this purpose, we use the in-built pre-processor in DAKOTA, called DPREPRO [21]. Once the computational model (meshing and subsequent analysis) is parametrized, actual values for the parameters may be obtained directly using DPREPRO, with little or no human intervention. The whole process is automated using a simple script written in Python. Also, for efficient computations, the inner loop UQ evaluations at the Smolyak collocation points are conducted in parallel, thereby speeding up the overall procedure. Further, for the deterministic simulations at the Smolyak collocation points, we use the parallelized CFD capabilities of FLUENT [25]. User-defined functions are written for generating the outputs in the format required by DAKOTA. After the first outer-loop iteration (which requires many inner-loop UQ evaluations), the results are passed back to the optimizer, which then decides the next set of design variables. The process is repeated until the convergence criterion and the constraints are satisfied.

The geometry is parametrized for meshing, and the journaling features of the meshing package CUBIT [26] are employed for generating the meshes at the Smolyak collocation points. The actual values for the parameters in the parametrized journal file are updated using DPREPRO for individual function evaluations. The governing equations are solved using the commercial CFD package FLUENT [25] as described earlier. The equations are first solved with a first-order upwind scheme for a few iterations. Using the flow and temperature fields so obtained as initial conditions, the equations are solved with a second-order upwind scheme until convergence. Pressure-velocity coupling is addressed via the SIMPLE algorithm, along with an algebraic multigrid algorithm (AMG) for solving the linearized system of governing equations. The governing equations are suitably under-relaxed to ensure proper convergence. Also, the default convergence criteria in FLUENT, based on the scaled residuals, are reduced by two orders for all the governing equations to enforce a stricter convergence criterion. Further details of the numerical method may be found in [27].

Grid-independence is first established by considering three levels of mesh refinement and analyzing the percentage error in the overall Nusselt number with respect to the finest mesh size. This study is conducted at the nominal design point. The number of elements for the three levels, for the case of 5 × 5 fins is approximately 238,000, 514,000 and 1,455,000, respectively. It is observed that the relative percentage difference in Nusselt number between the coarse mesh and the finest mesh is approximately 0.75%, while that between the intermediate and the finest mesh is 0.44%. The results from the intermediate mesh are therefore assumed sufficiently accurate for the purposes of this work, and unless otherwise stated, the results reported in this work are based on this mesh.

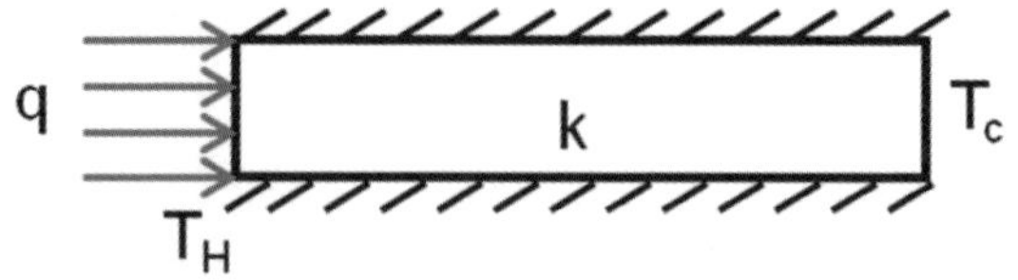

Fig. 3. Schematic diagram of the heater block design problem.

A coarse assessment of the computational time required for the OUU approach developed is also performed. Employing parallelized CFD solvers and mesh sizes of about 0.5 million, each inner loop evaluation required approximately 1.5 hours of real time employing 3 Intel E5410 CPUs. The computational time also depends on the level of the Smolyak grid, which determines the number of inner loop evaluations per outer loop evaluation. With 32 available CPUs, we could perform 10 simulations in parallel at any time, leading to a total time of approximately 10-12 hours per optimization run, corresponding to 70-80 total evaluations (inner loop × outer loop). A good way to reduce the computational time is to first optimize using the deterministic approach, locate the optimum input parameters, and then begin the probabilistic optimization with the these optimal inputs as the initial guessed values. Further reduction in computational time may be realized by assessing the required Smolyak grid level *a priori*, such as by just performing the uncertainty quantification analysis, and measuring the mean and standard deviation values of the outputs, predicted employing various sparse grid levels. Little change in the response function values with an increase in sparse grid level would then mean that lower sparse grid levels would suffice.

4. Results and Discussion

We now present the results corresponding to deterministic and probabilistic optimization for several test cases considered in this study.

Table 1. Optimization results for heater block design problem.

Parameter	Analytical Solution (Eq. (6))	Deterministic Optimization	Probabilistic Optimization
Q	17.0	17.0	Mean = 17.0 Std. deviation = 7.0
K	100.0	100.0	Mean = 100.0 Std. deviation = 30.0
T_c	325.0	325.0	Mean = 325.0 Std. deviation = 25.0
Initial value of A (mm^2)	--	20.0	20.0
Initial value of L (mm)	--	10.0	10.0
Final value of A (mm^2)	42.50	42.48	47.44
Final value of L (mm)	5.0	5.0	5.0
Volume ($A\times L$) (mm^3)	212.5	212.4	237.2

4.1. *Heater Block Design*

As a verification case for the OUU framework developed, the first problem considered in this study corresponds to the design of a simple heater block, for which an analytical solution exists.

The basic setup of the problem is shown in Figure 3. The uncertain parameters considered in this problem are the heat input q, the thermal conductivity of the heater block k, and the temperature at the cold end T_c. The block is assumed to be insulated at the sides and the objective is to design the block with minimum volume, $A\times L$. Here, A is the area of cross-section, and L is the length. Additional constraints in the problem are posed by restricting the mean value of the temperature at the hot end to a design temperature value of 345 K. The uncertain variables are considered to be Gaussian random variables, *i.e.*, normally distributed with mean and standard deviation values as shown in Table 1. The constrained minimization problem may be formally stated as:

$$\begin{aligned} &\textbf{Minimize} \quad A \times L \\ &\textbf{Such that} \quad \overline{T_h} \leq T_{design} \\ &\qquad 10.0 \leq A \leq 100.0 \\ &\qquad 5.0 \leq L \leq 20.0 \end{aligned} \tag{5}$$

For 1-D steady-state heat conduction, at fixed values of the uncertain variables, the temperature value at the hot end, T_h, may be calculated as follows [7] :

$$T_h = T_c + \frac{qL}{kA} \tag{6}$$

4.1.1. *Deterministic Optimization*

As a first step, the optimization is performed for the case of deterministic values for the uncertain variables. The uncertain parameters, q, k and T_c, are all fixed at their mean values. The initial values for the optimization loop are chosen to be 10 and 20 for L and A, respectively. The final optimized values for the design variables A and L are shown in Table 1. It may be noted that the values correspond very well with the analytical solution, obtained by setting $T_h = T_{design}$ and choosing A by fixing L or *vice-versa*. Also, the optimization process is assumed converged if the value of the objective function changes by less than 0.1% for three or more successive iterations.

4.1.2. *Probabilistic Optimization*

For probabilistic optimization, uncertainties are considered as part of the optimization procedure, as noted earlier. Beginning with the same initial guess values as for the case of the deterministic optimization, function evaluations for the inner UQ loop are performed. A Smolyak sparse grid of level 2 as described in [21] is used for this purpose, and based on evaluations at the collocation points, the polynomial response surface is obtained. This response surface is then used as a surrogate model and samples based on the LHS algorithm are drawn to cover the input variable range. The corresponding output values are estimated for each drawn sample and once a sufficient number of samples is drawn, output response metrics such as the mean output value may be obtained. The mean T_h value thus calculated is used by the optimizer to evaluate the inequality constraint in Eq. (6). The process is repeated until convergence; the convergence criterion used here is the same as in the case of the deterministic optimization. The corresponding optimized values for the design variables are shown in Table 1. It may be observed for the probabilistic case that the design variables assume more conservative values. In particular, for the present problem, an approximately 12% larger volume is recommended from the probabilistic case as compared to the deterministic one. The OUU approach thus allows the user to obtain a quantitative estimate of how conservative the design should be to account for uncertainties in the problem, and is more robust than a heuristic factor-of-safety approach.

4.2. *Channel Flow with Uncertain Viscosity*

A 2D channel flow with uncertain viscosity was considered by Le Maitre *et al.* [16]. This was one of the first studies that investigated the use of polynomial chaos UQ methods for incompressible laminar flow at moderate Reynolds numbers, as noted by Najm [15]. Le Maitre *et al.* [16] used an intrusive PCE method to modify the governing equations, Eqs. (2)–(4) to account for uncertain viscosity. Also, they employed a finite difference-based projection method for solving the governing equations. In the present work, we use the same problem (problem P1 from [16]) to verify our modeling framework. The problem statement is briefly summarized here and additional details are available in [16]. We note that the stochastic projection method of Le Maitre *et al.* is an intrusive PCE method, for which the governing equations must be modified as must the corresponding solver for performing the UQ [15]. However, we use a non-intrusive PCE method and use the original governing equations (Eqs. (2)–(4)) and the standard commercial solver, FLUENT.

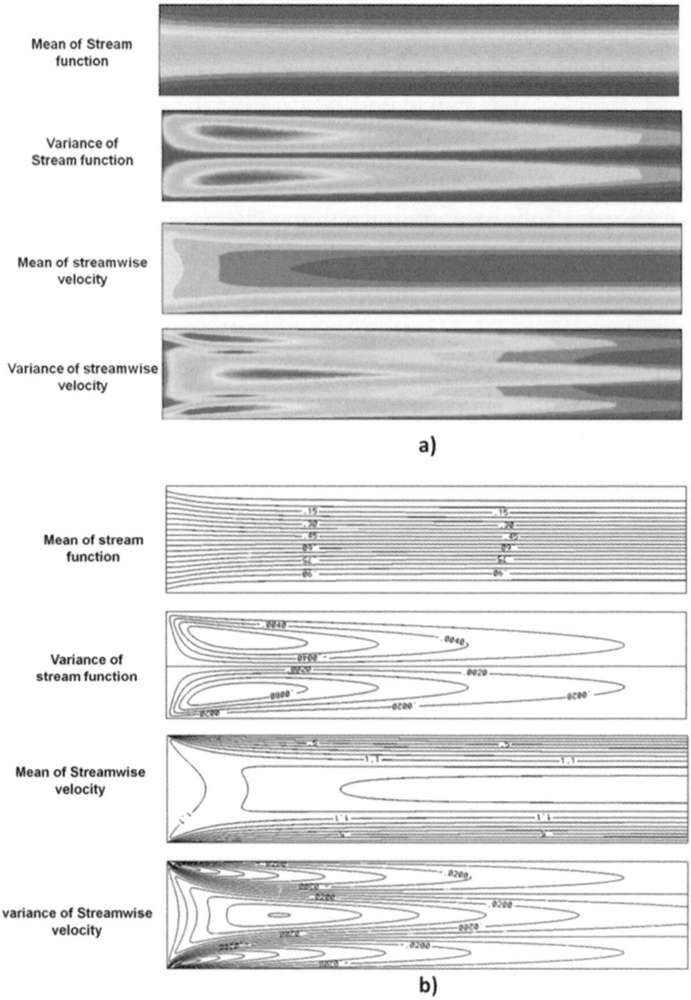

Fig. 4. Two-dimensional channel flow with uncertain viscosity for the case of Re = 81.24: (a) Current results, and (b) results from Le Maitre *et al.* [16].

The viscosity is assumed to be spatially uniform. At each location in the domain, it is considered to be a Gaussian random variable with a predefined mean and standard deviation. The boundary conditions however are deterministic, *i.e.*, a constant, uniform flow boundary condition is imposed at the inlet, while a no-slip condition is imposed on the two bounding walls. The flow is developing initially but attains fully developed flow conditions some distance downstream from the inlet. Figure 4 shows the contours of the mean and variance of stream function and streamwise velocity for a Reynolds number of 81.24 and a ratio of standard deviation to mean of the kinematic viscosity of 0.2, as used in [16]. It may be observed that thc results match qualitatively with those of Le Maitre *et al.* and this loosely verifies our framework. The development length, as inferred from the velocity data, is found to be about four times the channel width, which is in close agreement with theory ($x_{fd}/l \approx 0.05$ Re, [7]) and the results of Le Maitre *et al.*, which further verifies our results.

In the preceding two examples, we have demonstrated that UQ can be performed non-intrusively and in a computationally inexpensive manner. The approach adopted here requires fewer function evaluations than other UQ methods. The process is non-intrusive in that new stochastic solvers such as those used by Le Maitre *et al.* [16] are not needed. Instead, uncertainties are quantified through the output-input response function approach described.

4.3. *Pin-fin Optimization*

In this problem, we optimize the width of the pin fins cooled by impinging flow in the presence of uncertainties, and the corresponding results are compared with a conventional, deterministic approach.

At a constant input velocity of impingement, an optimum pin spacing exists [12]. As the spacing between the fins tends to zero (infinite fin limit), there is little penetration of flow between the fins, and the fins are under-utilized. Similarly, for a very large fin spacing (i.e., the zero fin limit), the reduction in surface area leads to reduced heat transfer. This leads to the existence of an optimum spacing (see, for example, [11], [23]). Ledezma *et al.* [12] performed numerical optimization of the fin-spacing for an impinging flow configuration. Their results were first validated against their experiments. The number of fins was then optimized as the design parameter for different input parameters such as fin height, width and Reynolds number. The optimal pin spacing was found to decrease as the Reynolds number increased, and the peak of the Nusselt number versus fin spacing curve flattened as the Reynolds number was decreased. For the range of Reynolds numbers considered, a fin count value of 4 or 5 along the side of the heat sink base was found to be optimal. However, their methodology did not employ a formal optimization procedure. Also, a discrete parameter, the number of fins, was chosen as the design variable, and simulations with different values of the fin count were conducted to arrive at an optimum.

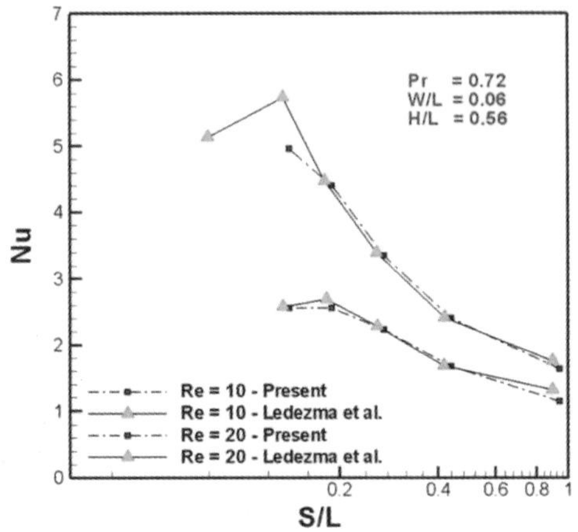

Fig. 5. Nusselt number versus non-dimensional fin-spacing. The results from Ledezma *et al.* [12] are shown for comparison.

In the present work, as we employ a formal optimization procedure; the number of fins is the discrete parameter and is fixed at $n = 5 \times 5$ fins spanning the entire base area, while the continuous parameter, the fin width W, is optimized. An optimization approach based on the Fletcher-Reeves conjugate gradient method [21] is employed for this purpose. Simulations are first performed for the set of conditions considered by Ledezma *et al.* [12], *viz.*, a non-dimensional fin width W/L of 0.06, a non-dimensional fin height H/L of 0.56, and a Prandtl number of 0.72. The simulations are performed for two different Reynolds numbers and the results, along with those from Ledezma *et al.*, are shown in Figure 5. Length parameters are non-dimensionalized by the length of the heat sink base, L, consistent with Ledezma *et al.* So also, the Reynolds number and the Nusselt number are defined based on the fin-width, W, as follows:

$$Re = \frac{\rho U_{in} W}{\mu}; \; Nu = \frac{qW}{L^2 (T_s - T_\infty) \times k_f} \tag{7}$$

In Eq. (7), U_{in}, T_s and T_∞ are the inlet velocity, specified fin surface temperature and ambient air stream temperature, respectively.

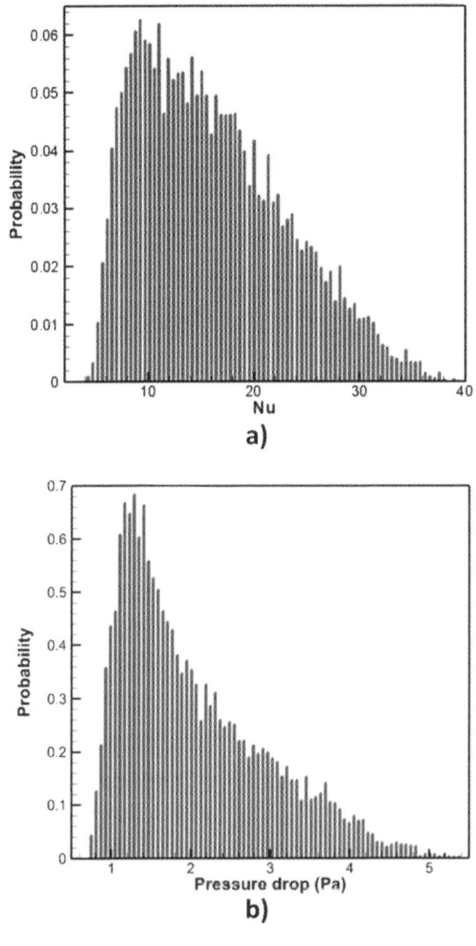

Fig. 6. PDFs of (a) Nusselt number, and (b) pressure drop for the uniformly distributed input parameters in Table 2.

The small differences between the present results and those from Ledezma *et al.* as observed in Figure 5 may be attributed to the coarser meshes considered by Ledezma *et al.*, as well as the uniform, structured nature of their meshes, with very few cells included in between the fins. In the present work, we have employed a boundary-layer mesh for resolving the fin surfaces, and ensured that there are at least 5-10 computational cells between successive fins in the cross-flow direction. In the following sections, we first describe the results of uncertainty quantification, and then the probabilistic and deterministic optimization results, corresponding to two cases of uncertainties commonly found in electronics cooling applications.

Table 2. Uncertainty quantification for pin-fin heat sinks.

Parameter	Deterministic approach	Probabilistic approach
	Inputs	
L (mm)	45.0	Uniform random, min. = 40.0 and max. = 50.0
t (mm)	3.5	Uniform random, min. = 2.0 and max. = 5.0
W (mm)	3.5	Uniform random, min. = 2.0 and max. = 5.0
H (mm)	20.0	Uniform random, min. = 10.0 and max. = 30.0
c (mm)	10.0	Uniform random, min. = 8.0 and max. = 12.0
n	5	5
	Outputs	
Nu	15.70	Mean = 16.124 Std. deviation = 6.99
ΔP (Pa)	1.79	Mean = 2.04 Std. deviation = 0.92

4.3.1. *Uncertainty Quantification*

The input parameters considered for UQ are: length of the base, L; thickness of the base, t; width of the fin, W; height of the fin, H; and clearance between the top of the fin and the inlet plane, c. Without loss of generality, these variables are all assumed to be uniformly distributed random variables. The corresponding variation in the input parameters is shown in Table 2. Again the number of fins is fixed at $n = 5 \times 5$.

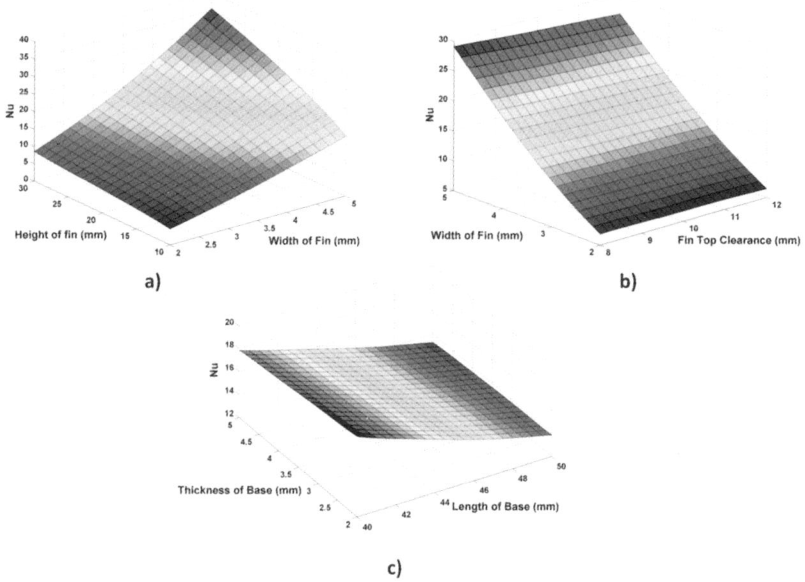

Fig. 7. Response surface plots of Nusselt number as a function of various input parameters.

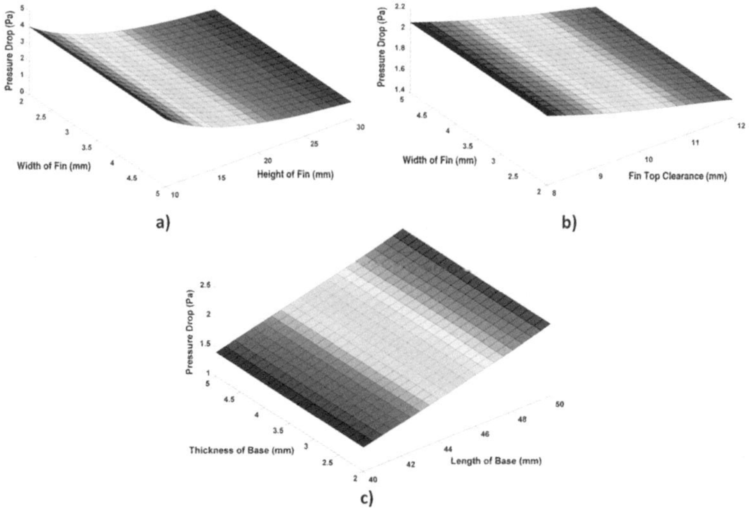

Fig. 8. Response surface plots of pressure drop as a function of various input parameters.

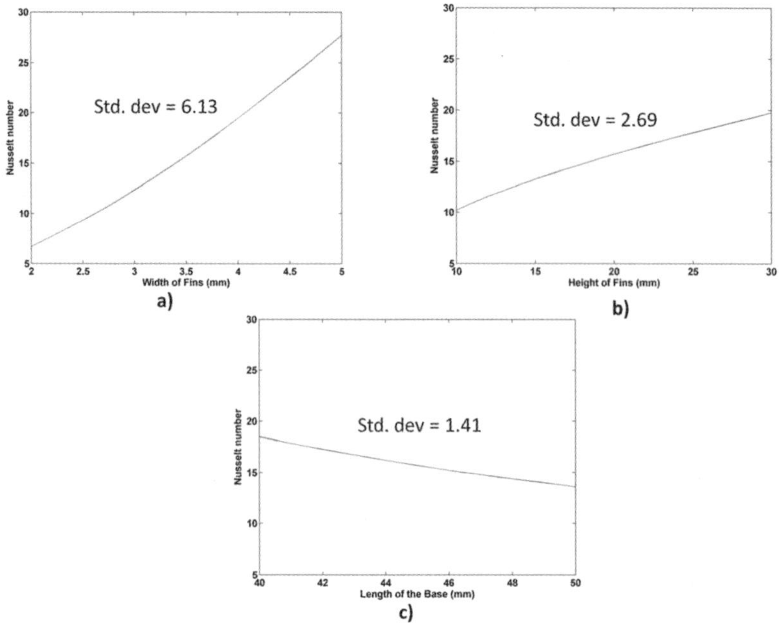

Fig. 9. Plots of Nusselt number shown for three most sensitive input parameters: a) fin width, b) fin height, and c) length of the base, for OUU of pin-fin heat sinks.

A level 2 Smolyak sparse grid (*i.e.*, response function containing terms up to fifth order polynomials) is used for uncertainty quantification which, for 5 uniformly distributed uncertain variables, translates to 71 evaluations for constructing the response surface. The inlet velocity for these runs was fixed at a value of 0.5 m/s. Following the procedure previously described, once the response surface is obtained, 10,000 samples are drawn randomly and output response statistics are calculated. The PDFs of Nusselt number and pressure drop are plotted in Figure 6, corresponding to the range of the uncertain inputs considered. Representative response surface plots for the outputs, Nusselt number and pressure drop, are shown in Figure 7 and Figure 8 respectively, as a function of various input variables. Also, the corresponding mean and standard deviation in the output quantities are reported in Table 2. As a comparison, simulations are also performed by fixing the uncertain variables at their means and the corresponding results are included in Table 2. It may be observed that the mean values of the output variables, as obtained from the UQ simulations, are different from the values obtained by assuming the uncertain variables to be deterministic and assigning them their respective mean values. Also, in order to assess the sensitivity of outputs to the various uncertain inputs, simulations are performed by fixing all input variables at their respective mean values except the input parameter relative to which the sensitivity is being assessed. By doing so, the input variables relative to which the outputs show the highest standard deviation are identified as the most sensitive variables.

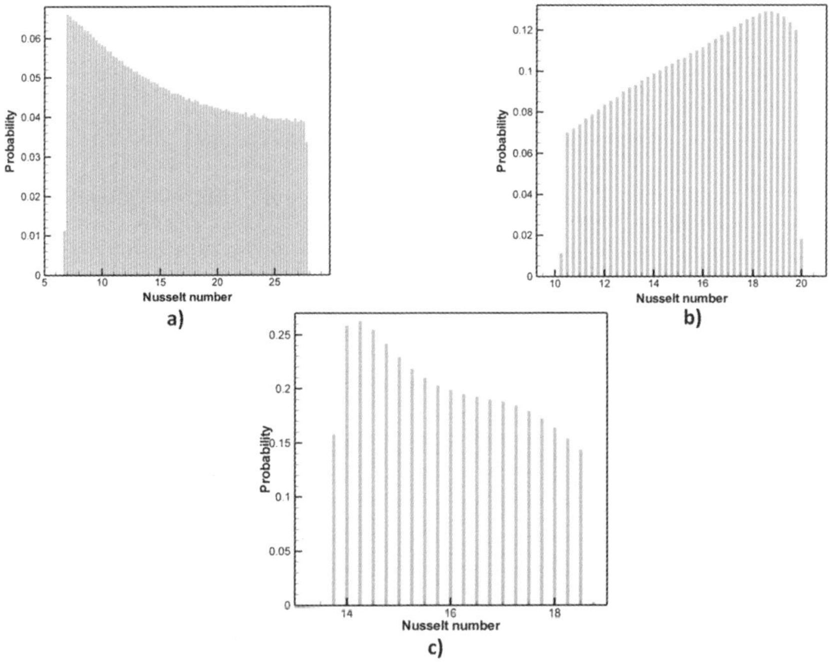

Fig. 10. Nusselt number PDFs shown for the three most sensitive input parameters: a) fin width, b) fin height, and c) length of the base, for OUU of pin-fin heat sinks.

Table 3 lists the standard deviation in *Nu* as the various input parameters, *viz.*, *L*, *t*, *c*, *W* and *H* are varied. Variations in the input variables similar to those in Table 2 are considered. Of all the input variables, it is clear that the Nusselt number is most sensitive to variations in the width *W* and height *H* of the fins. Sensitivity information such as this is useful in helping the designer decide which input parameters are most important. Such information also aids in the design of future experiments (DOE), as noted by Eldred *et al.* [4]. For further assessing the dependence of outputs on these sensitive input variables, the response surfaces are plotted in Figure 9. The shape of the response functions from Figure 9 suggests that a lower-order Smolyak grid is sufficient for this problem, and hence a Smolyak sparse grid of level 1 (*i.e.*, response surface containing terms up to second order polynomials) is employed in this work for the remainder of the results. However, simulations were also performed with a level 2 sparse grid to verify this choice, and similar results, with less than a 0.1% difference, were obtained. The corresponding PDFs of the Nusselt number for the case of single parameter variation are shown in Figure 10.

Table 3. Sensitivity analysis for pin-fin heat sinks.

Variable input parameter	**Range of variation in the input parameter**	**Std. deviation in *Nu***
L (mm)	Uniform random, min. = 40.0 and max. = 50.0	1.41
t (mm)	Uniform random, min. = 2.0 and max. = 5.0	0.21
C (mm)	Uniform random, min. = 8.0 and max. = 12.0	0.35
W (mm)	Uniform random, min. = 2.0 and max. = 5.0	6.13
H (mm)	Uniform random, min. = 10.0 and max. = 30.0	2.69

We may thus perform an initial approximate UQ analysis using, for example, low-accuracy polynomial expansions, and based on the sensitivity information, the most sensitive parameters may be identified. The uncertainty in only the most sensitive parameters may then be resolved better.

For probabilistic optimization the UQ loop is nested within an outer optimization loop, as noted earlier. To demonstrate the idea, we use an OUU approach for "design for robustness". That is, the optimization is performed to choose the design variables that not only maximize (or minimize) the cost function, but also restrict the standard deviation in the output quantities, which is posed as an additional constraint in the optimization procedure. Two sets of simulations are performed. In the first case, uncertainties in dimensional parameters arising out of manufacturing limitations are considered. An example for such a case is microchannels, where large deviations in the channel dimensions have been observed [2]. In the second case, uncertainties in thermal and flow conditions are considered.

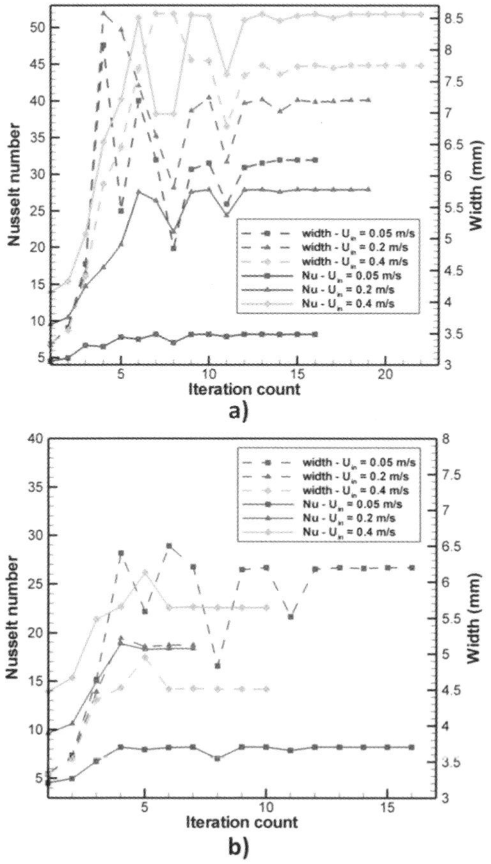

Fig. 11. Convergence history for OUU of pin-fin heat sinks with uncertain dimensions using (a) deterministic optimization for maximizing Nusselt number, and (b) probabilistic optimization for maximizing mean value of Nusselt number.

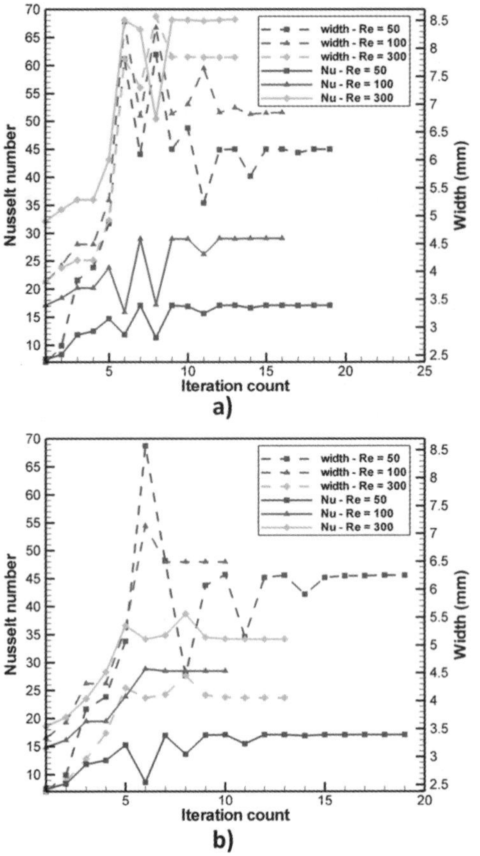

Fig. 12. Convergence history for OUU of pin-fin heat sinks with uncertain operating conditions using (a) deterministic optimization for maximizing Nusselt number, and (b) probabilistic optimization for maximizing mean of Nusselt number.

4.3.2. *OUU with Uncertain Dimensions*

As a first demonstration case, uncertainties in the thickness of the base t and the fin-top clearance c are considered for different inlet velocities. The length of the base L and height H are fixed at 47.6 mm and 26.66 mm, respectively, to match the values from [12]. Again, uniformly distributed uncertain variables, with minimum and maximum values as reported in Table 4, are considered for this analysis. For these sets of simulations, a sparse grid of level 1 is considered for the UQ and optimization is performed, starting with an initial guess value for the width, W.

Table 4. Input parameters for optimization of pin-fin heat sinks for the case of uncertain dimensions.

Parameter	Deterministic optimization	Probabilistic optimization
L (mm)	47.6	47.6
t (mm)	3.5	Uniformly distributed with minimum = 2.0 and maximum = 5.0
H (mm)	26.66	26.66
c (mm)	10.0	Uniformly distributed with minimum = 8.0 and maximum = 12.0
N	5	5
T_s-T_∞ (K)	20.0	20.0

For comparison, conventional (deterministic) optimization is also performed for fixed values of the uncertain variables, *i.e.*, with uncertain variables assuming their respective mean values. It may be observed from Table 5 that for the deterministic runs, as the velocity is increased, the optimum width of the fins shifts to higher values, consistent with the observation of Ledezma *et al.* [12]. This is because at lower velocities, the flow does not penetrate the pins adequately and the increase in fin width is detrimental to the overall heat transfer. On the contrary, at higher velocities, flow penetration is significant even at the larger fin widths and an increase in fin width enhances overall heat transfer by effectively increasing the surface area. For the probabilistic runs, a constraint bounding the standard deviation of the Nusselt number is also applied, and the optimization problem may be defined as:

$$\begin{aligned} &\textbf{Maximize} && \overline{Nu} \\ &\textbf{such that} && Std.Dev.(Nu) \leq 0.5 \\ & && 0.48\,mm \leq W \leq 8.75\,mm \end{aligned} \tag{8}$$

Table 5. Optimized width and corresponding Nusselt number for the pin-fin heat sink case with uncertain dimensions. The values of Nusselt number presented for the probabilistic case are mean values. Also, the bold values correspond to the inputs.

	Deterministic Optimization			**Probabilistic Optimization**		
U_{in} (m/s)	**0.05**	**0.2**	**0.4**	**0.05**	**0.2**	**0.4**
W (mm)	6.26	7.20	7.75	6.21	5.12	4.51
Nu	8.19	27.91	51.84	8.23	18.34	22.57

The limits on the design variable W are chosen so as to cover the entire range from very thin fins (infinite spacing limit) to very thick fins (zero spacing limit), with 8.75 mm being just less than the theoretical maximum width (theoretical maximum = $L/5$ = 9.52 mm).

The convergence history of the optimization procedure is plotted in Figure 11 along with results for the deterministic runs. The corresponding converged values for the width are shown in Table 5. It may be noted that the values of Nusselt number shown for the probabilistic case are mean values. By restricting the standard deviation in the output parameter, more conservative values result for the design variables. More importantly, the extent to which a design must be conservative in order to account for input uncertainties is quantified with this approach. Also, there is a significant reduction in the overall Nusselt number compared to the deterministic case, which again demonstrates the conservative nature of the design. The information from the convergence history in Figure 11 may also be used for obtaining a first estimate of the output values for a value of the design variable that is different from the final converged value.

4.3.3. *OUU with Uncertain Operating Conditions*

Heat sink designs are usually based on a thermal design power, TDP [28], which is a manufacturer-quoted maximum possible heat dissipation. However, in reality, depending on processor load, the heat sink may see different heat generation levels. Along similar lines, the flow speed from the fan can be uncertain and variable. Hence, to depict the operation of a real-world heat sink, a second demonstrative case with uncertainties in inlet velocity (Reynolds number) and input heat fluxes is considered.

As before, we seek to perform uncertainty propagation first, and then perform OUU to obtain the optimum fin width. The uncertain parameters are the Reynolds number and the pin fin surface temperature, which are assumed to be Gaussian random variables for this purpose, with mean and standard deviation values as listed in Table 6. As in the previous case of uncertain dimensions, L and H values are fixed at 47.6 mm and 26.66 mm, respectively. In addition, t and c are also held fixed at 2.0 mm and 10.0 mm, respectively. The standard deviation in the Reynolds number values considered is loosely based on the uncertainty analysis of Ledezma *et al.* [12], who noted a 5% uncertainty in Reynolds number in their experiments. A 3% standard deviation to mean ratio in the solid-fluid interface (fin and base) temperature is considered.

Table 6. Input parameters for optimization of pin-fin heat sinks for the case of uncertain operating conditions.

Parameter	**Deterministic approach**	**Probabilistic approach**
L(mm)	47.6	47.6
t(mm)	3.5	3.5
H(mm)	26.66	26.66
c(mm)	10.0	10.0
n	5	5
T_s (K)	320.0	Normally distributed, with mean = 320.0 and std. deviation = 10.0
Mean Re	50.0, 100.0, 300.0	Normally distributed, with mean =50.0, 100.0, 300.0
Std. deviation in Re	--	2.5, 5.0, 15.0

An additional constraint is imposed to restrict the standard deviation of the Nusselt number, as in the previous case. The corresponding optimization problem may be defined as below:

$$\begin{aligned} &\textbf{Maximize} && \overline{Nu} \\ &\textbf{such that} && Std.Dev.(Nu) \leq 1.0 \\ & && 0.48\,mm \leq W \leq 8.75\,mm \end{aligned} \tag{9}$$

The converged results for mean Nusselt number, along with Nusselt number values from the deterministic equivalent, obtained by assuming the mean values for the interface temperature and Reynolds number, are shown in Figure 12 and Table 7. Again, the conservative nature of the probabilistic optimization is clear. For the deterministic case, the value of the optimal fin width is observed to increase with *Re* as for the OUU with uncertain dimensions. Physical reasons for this behavior were previously explained. However, for the case of probabilistic optimization, due to the additional constraint restricting the standard deviation in *Nu*, no such clear trends are observed. From Table 7, it may be noted that as long as the constraint is not limiting, such as for *Re* = 50 and *Re* = 100, similar results are obtained with both deterministic and probabilistic approaches. At *Re* = 300, however, a very different optimum value for the width is returned compared to the deterministic case; this is because the constraint would be violated otherwise.

Table 7. Optimized width and corresponding Nusselt number for the pin-fin heat sink case with uncertain operating conditions. The values of Nusselt number presented for the probabilistic case are mean values. Also, the bold values correspond to the inputs.

	Deterministic Optimization			**Probabilistic Optimization**		
Re	**50.0**	**100.0**	**300.0**	**50.0**	**100.0**	**300.0**
W(mm)	6.19	6.85	7.84	6.25	6.49	4.05
Nu	17.10	29.07	68.16	17.09	28.45	34.18

Table 8. Optimization parameters and results for deterministic and probabilistic optimization of MMC heat sinks.

Parameter	Deterministic approach		Probabilistic approach
	Present	Ryu *et al.* [31]	
		Inputs	
q'' (W/cm^2)	100	100	100
Re	91	91	150
Pumping power (W)	2.5 to 2.7	2.56	2.5 to 2.9
Optimization constraint	--	--	Std. Dev. (R_{th}) <= 3×10^{-3}
Range of W_c (μm)	$10 \le W_c \le 100$	--	$10 \le \overline{W_c} \le 100$, $Std.Dev.(W_c) = 5$
Range of r	$0.5 \le r_c \le 4$	--	$0.5 \le \bar{r} \le 4$, $Std.Dev.(r) = 0.1$
		Outputs	
R_{th} (°C/W)	2.86×10^{-2}	3.10×10^{-2}	3.15×10^{-2}
W_c (μm)	16.85	16.5	30.94
$r(=L_{in}/L_{out})$	2.95	3.35	0.68

4.1. *Optimization of Manifold Microchannel Heat Sink*

The final problem considered concerns optimization of a manifold microchannel heat sink. A manifold microchannel (MMC) heat sink consists of a manifold system which distributes the coolant via multiple inlet-outlet pairs, thereby reducing the flow length of the coolant through the microchannels. A schematic illustration of the typical coolant flow path in a MMC heat sink is shown in Figure 13 [29]. This flow distribution has been shown to result in a lower overall pressure drop and thermal resistance compared to a conventional microchannel heat sink [30]. MMC heat sinks have been experimentally analyzed by Kermani [18], with application to solar cell cooling. Further, Ryu *et al.* [31] performed a numerical optimization of MMC heat sinks, employing a single unit cell, as shown in the inset of Figure 13 b). However, in this analysis, uncertainties such as those due to inherent limitations of the manufacturing technique were ignored. Recently, Sarangi *et al.* [29] performed a multi-objective optimization of MMC heat sinks, with and without inclusion of such uncertainties and quantified their effect on the optimal designs. They further identified the critical input parameters to which the outputs – heat transfer coefficient and pressure drop incurred – were most sensitive.

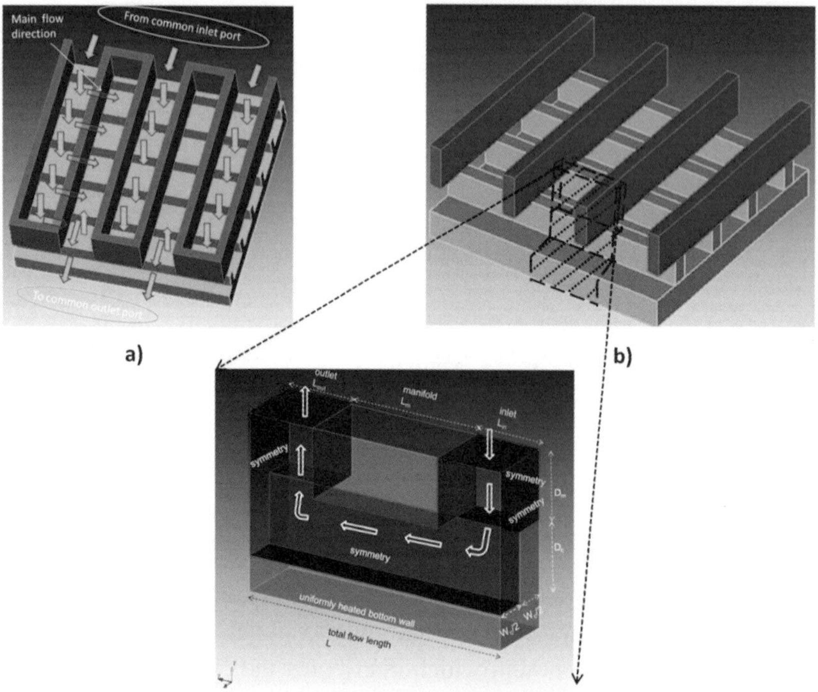

Fig. 13. Schematic diagram of a manifold microchannel heat sink: a) typical coolant flow path, and b) computational model with shaded region showing the 3D unit-cell employed for computations. Also shown is an inset of the unit-cell describing the various dimensional parameters of the solid and fluid domains and the boundary conditions employed [29].

In this study, we consider a geometry similar to that in Sarangi *et al.* [29] and perform optimization for two representative cases, a deterministic case and a probabilistic case. The unit-cell considered for analysis as well as the boundary conditions are shown in Figure 13 b). In the inset, the solid regions (the microchannel walls and the manifold) are both made of silicon. Water is the coolant and flows along the direction indicated in the fluid domain shown in blue. A constant heat flux boundary condition ($q^{''}$ = 100 W/cm^2, as in [31]) applied on the bottom wall serves as the heat input, simulating dissipation in an electronic chip. The conjugate heat transfer problem is solved considering both the solid and fluid regions, *i.e.*, the applied heat is first conducted through the solid domains including the manifold, and is then convected away by the coolant. A detailed description of the flow and heat transfer physics in this configuration is available in [29]. The optimization function for this case is the thermal resistance, defined as follows:

$$R_{th} = \frac{\overline{T_w} - T_{in}}{q^{''} \times A_{base}} \tag{10}$$

Here, T_{in} is the temperature of the coolant at the inlet, $\overline{T_w}$ is the average temperature of the heated base, and q'' is the applied heat flux at the base of area A_{base}. The thermal resistance is calculated over the entire chip area of 1 cm × 1 cm.

For the case of deterministic optimization, a case similar to that in Ryu *et al.* [31] is considered. That is, optimization at constant-pumping-power is performed to identify optimal values of microchannel width W_c (assumed equal to fin width, Figure 13) and manifold inlet-to-outlet width ratio r for minimizing the overall thermal resistance. This representative case corresponds to a Reynolds number (based on the hydraulic diameter of the microchannel) of 91 and pumping a power of 2.5 W, as in [31]. The optimization problem may be formally stated as below:

$$\begin{aligned} &\textbf{Minimize} && R_{th} \\ &\textbf{such that} && 2.5\,W \le Power \le 2.7\,W \\ &&& 10\mu m \le W_c \le 100\mu m \\ &&& 0.5 \le r \le 4 \end{aligned} \tag{11}$$

Further, the total flow length L, depth of the manifold D_m, depth of the microchannel D_c, and width of the outlet L_{out}, are fixed at 1000 μm, 500 μm, 150 μm and 114 μm, respectively. The results of the optimization are shown in Figure 14 a) and a comparison of the present results with those from Ryu *et al.* is presented in Table 8. The two sets of results are in good agreement, further validating our optimization procedure. The small differences observed may be attributed to the slightly different geometry (in the case of Ryu *et al.*, the fin width and the channel width are not equal, but the convergence history shows that they are close to each other during the entire optimization process); a further difference is that in the present computations, we impose a small range (≈0.2 W) around the value of constant power (2.56 W) suggested by Ryu *et al.*, for lowering computational effort. As a higher pumping power lowers the thermal resistance, our values more or less correspond to a constant power case of 2.7 W.

In the probabilistic case, a similar optimization is performed, with an added constraint to restrict the standard deviation of the thermal resistance to roughly 10% of the value obtained via the conventional deterministic approach. A Reynolds number of 150 and a pumping power of 2.75 W are employed. Also, uncertainties in geometric parameters W_c and r are considered. Further, by restricting the total flow length L (= 1000 μm) and outlet width L_{out} (= 114 μm) as in the aforementioned case, uncertainty in r translates to uncertainties in width of the inlet L_{in} and manifold length L_m, as noted in [29]. For this optimization, the depth of the microchannel D_c and the depth of the manifold D_m are also fixed at values of 200 μm and 500 μm respectively. As a representative case, W_c and r are assumed to be normally distributed with standard deviations of 5 μm and 0.1, respectively. The optimized mean values of the corresponding parameters are shown in Table 8. The probabilistic optimization problem for this case may be formally stated as follows:

$$\begin{aligned} &\textbf{Minimize} && R_{th} \\ &\textbf{such that} && 2.5\,W \le Power \le 2.9W, \\ & && Std.Dev(Rth) \le 0.003 \\ & && 10\mu m \le \overline{W_c} \le 100\mu m, \\ & && 0.5 \le \bar{r} \le 4 \end{aligned} \qquad (12)$$

Figure 14 b) shows the convergence history of the optimization procedure, while the optimal value of the thermal resistance for this case is shown in Table 8. Again, it may be noted that an increase in pumping power lowers the thermal resistance in general, and hence our values are expected to be closer to a case with constant pumping power of 2.9 W. This further demonstrates the utility of the OUU approach developed in arriving at optimal designs that, in addition to being optimal are also robust, or insensitive to variation in inputs.

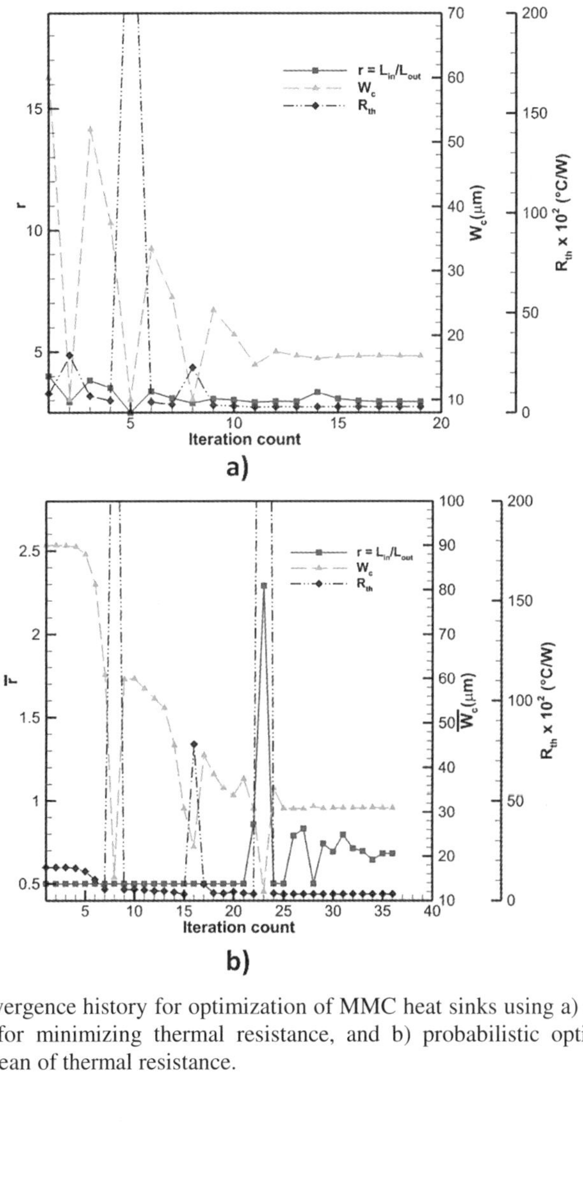

Fig. 14. Convergence history for optimization of MMC heat sinks using a) deterministic optimization for minimizing thermal resistance, and b) probabilistic optimization for minimizing mean of thermal resistance.

4. Conclusions

This work presents a flexible framework for performing optimization in the presence of uncertain inputs. The distinguishing feature of this framework is the use of a non-intrusive OUU approach. Using this, existing commercial solvers (or in-house codes) for solving fluid flow and heat transfer may be readily coupled with the open-source UQ and optimization toolkit, DAKOTA. The OUU framework developed here is first verified against a test case for which an exact solution exists. The framework is also tested by performing uncertainty quantification for a simple 2D channel flow with uncertain but spatially uniform viscosity. The results are verified against the results from Le Maitre *et al.* [16].

Heat sink design and optimization is of broad interest in electronics cooling. Traditionally, for such an optimization, the input variables are assumed to be deterministic, while in reality there could be uncertainties in both the dimensions and the operating conditions. In this work, a pin-fin heat sink under an impinging flow, similar to that considered by Ledezma *et al.* [12], and a manifold microchannel heat sink such as the one in [18], are considered for optimization. Two sets of uncertainties commonly encountered in engineering design problems, *viz.*, uncertainties in dimensions and in operating conditions, are explored, for the pin-fin heat sink case. The corresponding results are compared with their deterministic counterparts and the conservative nature of the probabilistic designs is quantified. Based on sensitivity information, the critical input parameters to which the output quantities are most sensitive are also identified. For the manifold microchannel heat sink, minimization at constant pumping power of overall thermal resistance is performed in the presence of geometric uncertainties. The approach developed is flexible, computationally inexpensive, and easy to implement, and may be readily used to perform effective OUU for a variety of electronics cooling designs, as demonstrated in this work.

Acknowledgements

The authors acknowledge the support for this work from industry members of the Cooling Technologies Research Center, an NSF Industry/University Cooperative Research Center.

Nomenclature

A	*Area, mm^2*
A_{base}	*Area of heat sink base, mm^2*
c	*Fin top clearance, mm*
C	*Specific heat, J/kgK*
D_c, D_m	*Depth of microchannel and manifold, μm*
H	*Fin height, mm*
k	*Thermal conductivity, W/mK*
L	*Length of the heat sink base, mm*
L_{in}, L_{out}	*Width of inlet and outlet, μm*
n	*Number of fins on a side*
Nu	*Nusselt number*
ΔP	*Pressure drop, Pa*
q	*Heat input, W*
r	*Inlet-to-outlet width ratio*
R	*Response function*
R_{th}	*Overall thermal resistance, °C/W*
Re	*Reynolds number*
s	*Fin-to-fin spacing, mm*
t	*Thickness of the heat sink base, mm*
T	*Temperature, K*
T_{in}	*Inlet temperature of coolant, K*
$\overline{T}_w$	*Average temperature of heat sink base, K*
u	*Velocity, m/s*
W	*Width of the fin, mm*
W_c	*Width of microchannels, μm*

Greek Symbols

α_i	*Coefficients in the response function*
ξ_i	*Random variable*
μ	*Fluid dynamic viscosity, N.s/m^2*
ρ	*Density, kg/m^3*
ψ_i	*Polynomials in the gPC response function*

Subscripts

f	*Fluid*
h,c	*Hot, cold ends*
in	*Inlet*
s	*Solid*
∞	*Ambient*

References

1. Eldred M. S., in *50th AIAA/ASME/ASCE/AHS/ASC Structures, Structural Dynamics and Materials Conference* (Am. Inst. Aeronaut. Astronaut, Palm Springs, CA, 2009).
2. Liu D., Garimella S. V., *J. Thermophys. Heat Trans.,* **18(1)**, 65 (2004).
3. Bodla K. K., Murthy J. Y., and Garimella S. V., *J. Heat Trans.,* **135(1)**, 011012 (2012).
4. Eldred M. S., Giunta A. A., Wojtkiewicz Jr S. F., and Trucano T. G., in *Proceedings of the 9th AIAA/ISSMO Symposium on Multidisciplinary Analysis and Optimization* (Am. Inst. Aeronaut. Astronaut, Buckheaed, Atlanta, GA, 2002) p. 4.
5. Ndao S., Peles Y., and Jensen M. K., *Int. J. Heat Mass Trans.,* **52(19-20)**, 4317 (2009).
6. Kitajo S., Takeda Y., Kurokawa Y., Ohta T., and Mizunashi H., in *Eighth Annual IEEE Semiconductor Thermal Measurement and Management Symposium*, (IEEE, New York, NY,1992). p. 119.
7. Incropera F. P., and De Witt D. P., *Fundamentals of Heat and Mass Transfer* (John Wiley & Sons, New Jersey, 2002).
8. Bar-Cohen A., and Rohsenow W. M., *Trans. ASME. J. Heat Trans.*, **106(1)**, 116 (1984).
9. Bejan A., *Convection Heat Transfer* (John Wiley & Sons, New Jersey, NJ, 2004).
10. Nakayama W., Matsushima H., and Goel P., *Cooling Tech. Electronic Equip.*, 195(1988).
11. Bejan A., and Sciubba E., *Int. J. Heat Mass Trans.*, **35(12)**, 3259 (1992).
12. Ledezma G., Morega A. M., and Bejan A., *J. Heat Trans.*, **118**, 570 (1996).
13. Kang H. S., *J. Heat Trans.*, **132(3)**, 034501 (2010).
14. Khan W. A., Culham J. R., and Yovanovich M. M., *IEEE Trans. Comp. Packaging Tech.*, **28(2)**, 247 (2005).
15. Najm H. N., *Ann. Rev. Fluid Mech.*, **41**, 35 (2009).
16. Le Maitre O. P., Knio O. M., Najm H. N., and Ghanem R. G., *J. Comp. Phys.*, **173(2)**, 481 (2001).
17. Xiu, D., and Karniadakis G. E., *SIAM J. Sci. Comp.,* **24(2)**, 619 (2002).

18. Kermani E., Dessiatoun S., Shooshtari A., and Ohadi M. M., in *Proceedings of the 59th Electronic Components and Technology Conference* (IEEE, San Diego, CA, 2009) p. 453.
19. Ghanem R. G., and Spanos P. D., *J. Eng. Mech.,* **117(10)**, 2351 (1991).
20. Sobol' I. M., and Kucherenko S., *Math. Comp. Simulation*, **79(10)**, 3009 (2009).
21. Eldred M. S., Giunta A. A., van Bloemen Waanders B. G., Wojtkiewicz S. F., Hart W. E., and Alleva M. P., *DAKOTA, A Multilevel Parallel Object-Oriented Framework for Design Optimization, Parameter Estimation, Uncertainty Quantification, and Sensitivity Analysis: Version 4.1 Reference Manual* (Sandia National Laboratories Albuquerque, NM, 2007).
22. Vanderplaats G. N., *CONMIN, A FORTRAN Program for Constrained Function Minimization: User's Manual*, (Ames Research Center and US Army Air Mobility R&D Laboratory, Moffett Field, CA, 1973).
23. Morega A. M., Bejan A., and Lee S. W., *Int. J. Heat Mass Trans.,* **38(3)**, 519 (1995).
24. Matsushima H., Yanagida T., and Kondo Y., *Heat Trans. - Japanese Research,* **21(5)**, 504 (1992).
25. User's Guide for FLUENT 6.0 (Fluent Inc., Lebanon, NH, 2002).
26. CUBIT 10.0 User's Manual, (Sandia National Laboratories, Albuquerque, NM, 2005).
27. Bodla K. K., Murthy J. Y., and Garimella S. V., *Numer. Heat Trans., Part A: Applications*, **58(7)**, 527 (2010).
28. Ganapathy D., and Warner E. J., in *11th Intersociety Conference on Thermal and Thermomechanical Phenomena in Electronic Systems* (IEEE, Orlando, FL 2008), p. 1242.
29. Sarangi S., Bodla K. K., Murthy J. Y., and Garimella S. V., in *Proceedings of International Mechanical Engineering Congress and Exposition*, (Am. Soc. Mech. Eng., Houston, TX, 2012).
30. Harpole G. M., and Eninger J. E., in *Proceedings of 7th Semiconductor Thermal Measurement and Management Symposium* (IEEE, Miltipas, CA, 1991) p. 59.
31. Ryu J., Choi D., and Kim S., *Int. J. Heat Mass Trans.,* **46**, 1553 (2003).

CHAPTER 12

HYDROPHILIC CNT-SINTERED COPPER COMPOSITE WICK FOR ENHANCED COOLING

Glen A. Powell, Anuradha Bulusu, Justin A. Weibel, Sungwon S. Kim, Suresh V. Garimella, Timothy S. Fisher*

Purdue University, School of Mechanical Engineering, and Birck Nanotechnology Center, West Lafayette, IN 47907
**tsfisher@purdue.edu*

The need to increase the performance and functionality of heat transfer devices in diverse areas ranging from power electronics and energy conversion to biotechnology and manufacturing has led to significant thermal transport challenges. Heat pipes have emerged as an attractive alternative for passive and compact cooling solutions due to the enhanced performance realized via liquid-vapor phase change. The choice of an appropriate wick structure is critical to the performance of a heat pipe as the capillary and boiling limits on performance are directly related to the wick structure. An ideal wick structure offers good capillary pumping and nucleate boiling characteristics while maintaining low thermal resistance. We present a heat pipe wick structure that incorporates nano and microscale elements to improve wicking and thin film evaporation. The sintered copper-carbon-nanotube composite wick structure is tested for thermal performance, and successfully supported high heat fluxes of the order of 500 W/cm^2 with low thermal resistances ranging between 0.2–0.4°C/W.

1. Introduction

Nanostructured materials and devices have received much recent attention for prospective application in new energy technologies. While much of the attention has focused on unique transport processes within the nanomaterial itself for energy conversion and storage applications such as photovoltaics, thermoelectrics, and batteries[1], a vast majority of global energy conversion processes in use today involve traditional single- or multi-phase fluid cycles for heat transfer processes. The present work describes the synthesis and experimentally determined multi-phase fluid performance of nanostructure-enhanced boiling surfaces. Further, the high fluxes dissipated through liquid-vapor phase change render the use of such structures suitable for high-flux electronics cooling applications.

Heat flux dissipation in modern electronics exceeds 100 W/cm^2 and continues to rise as the power density of chips increases, as emphasized in the 2007 International Technology Roadmap for Semiconductors[2]. The dissipation of even higher heat fluxes is required for laser diode arrays[3], avionics[4], and some high power electronics[5]. Many different solutions have been proposed to address such high heat fluxes[6-8]. Heat pipes (Figure 1) offer an effective passive approach for spreading heat or transporting heat away from a compact source so that it can be dissipated over a larger area by conventional means such as forced air convection. Heat pipes can achieve an effective thermal conductivity that is typically much greater than that of a bulk metal because the energy is transported by an exchange of latent heat. However, the maximum possible heat input for a heat pipe is governed by the ability of the wick to prevent dryout by pumping sufficient liquid (the capillary limit) and supporting a high critical heat flux (the boiling limit)[9]. The choice of wicking material governs the performance characteristics of a heat pipe. Several different types of structures commonly used as heat pipe wicks include grooved surfaces, stacks of metal screens, and sintered metal powders. Different configurations of heat pipes have been used to suit a wide variety of applications[10]. Heat pipes utilizing a copper shell with water as the working fluid are desirable for electronics cooling applications because of the high thermal conductivity of copper, excellent thermal properties of water, and relatively low boiling point at sub-atmospheric pressure. Hopkins *et al.*[11] studied wickless copper-water heat pipes with grooves providing the capillary action and achieved a heat flux dissipation of 90 W/cm^2.

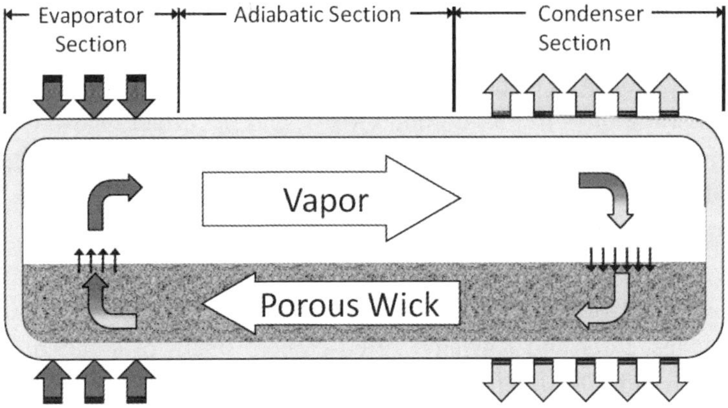

Fig. 1. Schematic diagram of heat pipe operation. Liquid evaporates from the porous wick when heated at the evaporator section, flows to the condenser section due to a local increase in vapor pressure, and is condensed back into liquid when cooled. Liquid is driven through the porous wick by capillary action back to the evaporator, completing the passive fluid flow loop in the device.

Li and coworkers[12,13] investigated capillary-fed boiling from sintered copper meshes and found that the thickness and porosity of the wick influenced the capillary-limited dryout heat flux; a maximum heat flux exceeding 200 W/cm^2 was reported.

In the current investigation, a sintered powder wick is chosen instead of screen or grooved structures since it can support a higher critical heat flux (CHF) and greater capillary pumping[11-14]. The maximum heat flux dissipated using copper-water heat pipes containing sintered metal wicks was reported to be in excess of 300 W/cm^2 by Gernert *et al.*[14], and the highest heat transfer coefficient was observed just prior to dryout.

Two distinct phase change regimes for heat transfer in heat pipes may be identified. At low heat fluxes, cooling occurs primarily due to evaporation from the liquid-vapor meniscus at the wick surface which remains saturated with liquid. For sintered wicks, Hanlon and Ma[15] showed that a large percentage of heat transfer takes place via thin-film evaporation from the liquid menisci formed in the wick pores in this regime. At higher heat fluxes, higher wick superheat temperatures cause incipience of vapor bubbles in the wick structure and the dominant mode of heat transfer becomes nucleate boiling at the wick-wall interface. Pioneering studies have been conducted by Bar-Cohen and co-workers on the incipience of nucleate boiling using dielectric fluids[16], effects of surface properties on pool boiling CHF[17], and the effect of using binary mixtures of dielectric liquids on pool boiling on silicon surfaces[18]. More recently, microporous coatings have been reported[19] to enhance pool boiling in dielectric liquids.

Novel wick structures have been designed with the goal of improving performance of both evaporation at low fluxes and boiling at high fluxes. Liter and Kaviany[20] used conical stacks of sintered copper spheres in order to separate the flow paths of returning liquid and rising vapor and reported a CHF three times greater than that of a random structure. Mughal and Plumb[21] investigated porous metal wicks with channels cut into the surface and found that the addition of channels increased the critical heat flux by 200%. They postulated that the addition of channels provided a path for vapor to escape the boiling surface.

Carbon nanotubes (CNTs) have been studied extensively because of their remarkable physical properties, including high strength, impressive thermal conductivity, and unique electrical properties[22]. In particular, the high thermal conductivity (comparable to that of diamond) of multiwalled arrays of CNTs grown by plasma-enhanced chemical vapor deposition (PECVD) suggests that CNTs could find use in thermal management[23]. While silicon provides an ideal growth substrate for the study of CNTs themselves, it is not ideal for exploiting the favorable thermal properties of CNTs in heat dissipation applications. Instead, CNT growth on copper, an excellent conductor of heat and electricity, may be more relevant for many engineering applications, particularly heat pipes. The growth of CNTs on bulk metal substrates (in particular Cu) has received limited attention compared to CNT growth on single-crystal Si wafers. However, copper has been explored as a growth substrate for use in electrical interconnects, thermal interface materials, and as boiling surfaces[24]. Growth of CNTs on a bulk copper substrate was utilized for investigation of flow boiling enhancement in a microchannel by Khanikar *et al.*[25]. This work built upon the study of Ujereh *et al.*[26], who studied CNT enhancement of nucleate pool boiling on Si and Cu substrates using a highly wetting dielectric fluid, FC-72. The CNT arrays, grown in this prior study by microwave plasma enhanced chemical vapor deposition (MPCVD), were able to mechanically withstand the boiling process and provide repeatable results. The mechanism for heat transfer enhancement was believed to be a combination of an increased heat transfer area and improved nucleation site density owing to the abundance of zero-cone-angle cavities provided by the interstices of the CNT array.

The present work reports the successful integration of carbon nanotubes with a sintered copper powder wick attached to a copper substrate for use in heat pipe applications, resulting in high CHF values and heat transfer coefficients. The fabrication methods and characterization of a hydrophilic CNT-sintered-copper composite wick is presented.

2. Experimental Methods

At first, a specialized three-layer wick structure was designed and fabricated for the purpose of achieving high heat transfer rates and extending CHF during boiling in the heat pipe evaporator section. The process flow for fabricating such a structure is outlined in Figure 2. The three layers of the structure consisted of: (i) a CNT array grown on a copper substrate and made hydrophilic by e-beam evaporation of Cu metal over the nanotubes, (ii) a layer of Cu particles sintered on to this CNT-coated substrate to form a porous wick, and (iii) a CNT array grown on the sintered wick and similarly functionalized with evaporation of Cu metal over the array. The processes used in the fabrication of these layers include: growth of CNTs on copper substrates by use of MPCVD, functionalization of those CNT arrays by Cu metallization using e-beam evaporation, sintering of Cu metal particles to form porous wicks in a tube furnace, and growth of CNT arrays on porous wicks. The bottom metallized CNT layer was included to enhance heat flow between the substrate and the sintered particles. The sintered powder wick layer was composed of 90-106 μm spherical Cu particles sintered at high temperature under a reducing forming gas atmosphere. The top metallized CNT layer of the structure was grown using the same

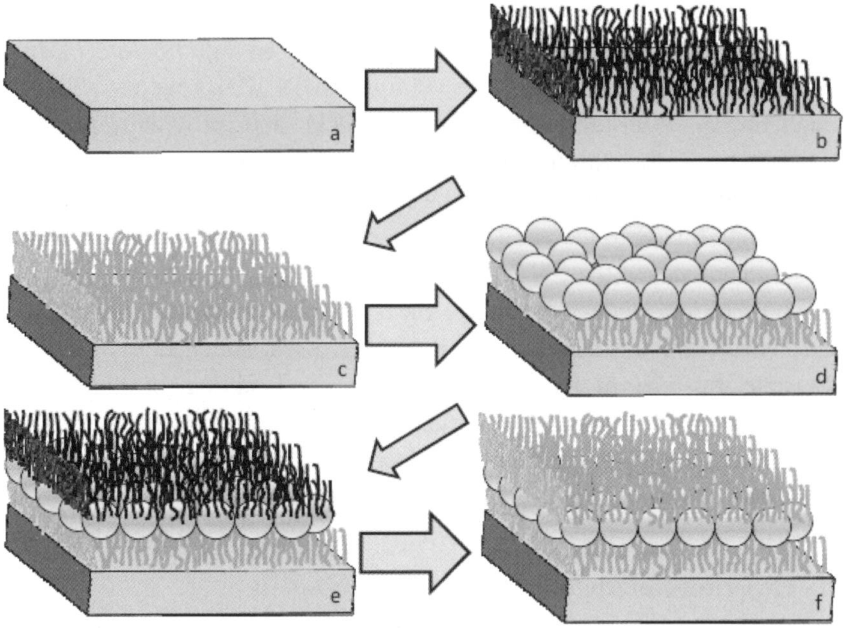

Fig. 2. Conceptual process flow for fabrication of a three layer micro-nano wick composite: (a) clean Cu Substrate, (b) growth of a CNT array, (c) metallization of CNT array by evaporation of Cu, (d) sintering of Cu particles onto metallized CNT array, (e) growth of CNT array on sintered Cu particles, and (f) metallization of top CNT array by evaporation of Cu.

synthesis parameters as the bottom layer of CNTs, but was grown on the sintered powder microstructure instead of a flat copper substrate.

The purpose of the top layer of metallized CNTs was to improve thin-film evaporation by creating nano-scale menisci in the interstices of the array, and to aid fluid feeding at the site of evaporation due to the wicking action of the nano-porous metallized CNTs. In addition to aiding in evaporative heat transfer, metallized CNT arrays on the top surface were expected to aid in boiling heat transfer as well by acting as fins for thin film evaporation and convective heat transfer within the vapor columns that form during nucleate boiling.

Carbon nanotubes were grown in a SEKI AX5200S MPCVD system. The copper substrates, 25.4 mm by 25.4 mm in area and 1 mm in thickness, were cleaned by degreasing with acetone and methanol. The growth of CNTs was achieved using a trilayer catalyst consisting of 100 nm of Ti, 10 nm of Al, and 3 nm of Fe deposited on a clean copper surface by means of e-beam evaporation. Samples were annealed in nitrogen before growth in the MPCVD. The operating conditions during growth were: 900°C with 50 standard cubic centimeters per minute (sccm) of H_2 and 10 sccm of CH_4 flowing through the chamber, with a 200 W microwave plasma at 10 torr for 10 minutes. The Ti layer acted as an adhesion layer for the catalyst stack and improved anchoring of the CNTs to the substrate. The Al acted as a surfactant to allow the Fe to break up into nanoparticles suitable for seeding growth[27]. In order to alleviate the inherent hydrophobicity of carbon nanotubes, they were coated with a layer of copper using a CHA e-beam evaporator. The average rate of deposition ranged from approximately 0.5 Å/sec to 2.5 Å/sec to achieve a nominal film thicknesses varying from 2 nm to 200 nm. An optimum thickness of 150 nm was used for the thermal test samples due to reasons to be discussed in the next section.

Next, a sintered powder wick was formed by placing loose copper powder over the metallized CNTs in a ceramic mold and heating to the required temperature under a controlled atmosphere in a tube furnace. After the copper powder (Alfa Aesar 42623, spherical, –100+325 mesh, 99.9% purity), which was sorted using sieves of –140+170 mesh (90-106 μm), was poured into the mold on top of the substrate, a straight-edge was used to level the powder in the mold. The ceramic molds were machined from a general-purpose alumina-silicate ceramic (McMaster-Carr 8479K) and fired prior to their first usage to increase strength and hardness. The mold and sample were then placed on a quartz boat in the tube furnace (Blue M model STF5531C), and exposed to 2.0 standard liters per minute (slpm) of forming gas — a mixture of 95% N_2 and 5% H_2. After purging with the forming gas for 30 minutes, the temperature was steadily ramped up for 30 minutes to 950°C, and held at this value for 60 minutes. The assembly was then allowed to cool, with forming gas continuing to flow, for approximately 150 minutes. The sample was removed after cooling to below 50°C, and was then placed in an airtight container to limit oxidation. Figure 3 shows the temperature profile during the sintering process along with

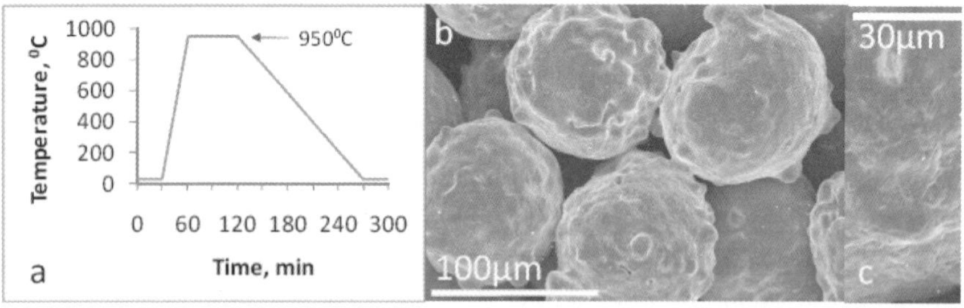

Fig. 3. Necking between 90-106 μm Cu particles that have been sintered under flowing forming gas: (a) sintering temperature profile used, (b) necking between groups of particles, and (c) necking between two particles.

field-emission scanning electron microscopy (Hitachi S-4800 FESEM) images that reveal necking between the spherical particles.

From these and other FESEM images, it was determined that the necking diameter between particles under such conditions was 15-20% of the particle diameter. Lastly, CNT growth and metallization was repeated on top of the sintered copper using the same synthesis procedure previously described for growth directly on the flat copper substrate.

3. Results and Discussion

CNT growth

Use of the trilayer catalyst yielded several different regimes for growth of CNTs on a flat copper substrate, as shown in Figure 4. The top pair of FESEM images, Figure 4(a) and Figure 4(b), show a short array of CNTs, less than 5 μm in thickness, grown in a uniform layer. The underlying Cu substrate can be seen, highlighted in Figure 4(b), demonstrating that the array is not dense. Figure 4(c) and Figure 4(d) show another common mode of CNT growth in clusters combined with large areas of exposed substrate where no growth has occurred. In contrast to the uniformly grown CNTs, Figure 4(c) shows CNTs which are tens of microns in length, and Figure 4(d) shows that the growth in the clusters is aligned.

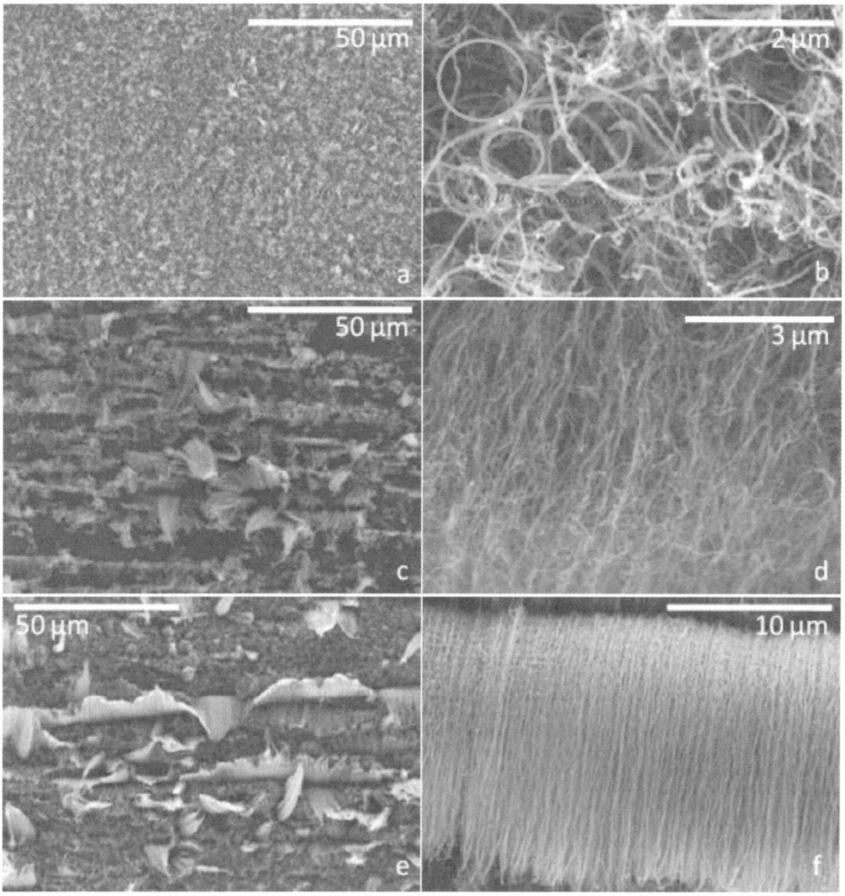

Fig. 4. Modes of growth of CNTs on Cu substrates with trilayer catalyst: (a) uniform film of CNTs and (b) underlying substrate revealed at higher magnification (indicated in circles), (c) long clusters of intermittent vertically aligned CNT growth with bare Cu in between and (d) a zoomed-in view of vertically oriented CNTs in such a cluster, (e) combination of vertically aligned CNT cluster and uniform film CNT growth shown beside, and (f) an image of aligned CNTs in a cluster from this growth mode.

These two modes are not mutually exclusive, as can be seen in Figure 4(e) and Figure 4(f). In these figures, a thin array of CNTs is a background for clusters of growth.

Clusters grown in this case, as shown in Figure 4(f), were also aligned. When utilizing the same nominal conditions stated above for growth along with the trilayer catalyst, all three of these growth morphologies (thin CNT film, CNT clusters, and CNT clusters among a thin CNT film) were encountered.

It is hypothesized that the thin arrays represent growth of CNTs from catalyst particles that remained attached to the substrate base and is interrupted by diffusion of Cu through the catalyst layer after only a few microns of growth. The longer clusters of growth represent a tip growth mode that is not interrupted by the diffusion of Cu, as the catalyst particles are elevated from the surface. Such base and tip growth modes have been described in the literature[28,29].

Sporadic growth using the trilayer catalyst was also observed to an even greater extent on the sintered Cu powder. In most cases, the initial growth attempt yielded poor CNT growth. As discussed previously, the probable cause for growth interruption is diffusion of the Cu through the Ti into the catalyst layer[30]. Therefore, it was found that for samples where little or no growth occurred, simply re-depositing the trilayer catalyst onto the surface and performing a second growth attempt yielded dense CNT arrays similar to those seen in growth on Si and other passive substrates.

Characterization of CNT arrays

CNT arrays were characterized using the FESEM to estimate the length, diameter, and density of the CNTs. Density is estimated for arrays of CNTs by counting the number of CNTs in a sample area. These numbers should be considered approximate due to the difficulty in counting obscured CNTs from a two-dimensional image and due to the inherent variability in density across the surface. Length estimates were obtained from observing images of CNTs near the sample edge tilted in the SEM to show a cross section. For CNTs on sintered wicks, the estimates of length were made by comparing the average diameter of the sintered particles before and after growth, and by comparing the average space between side-by-side particles before and after growth.

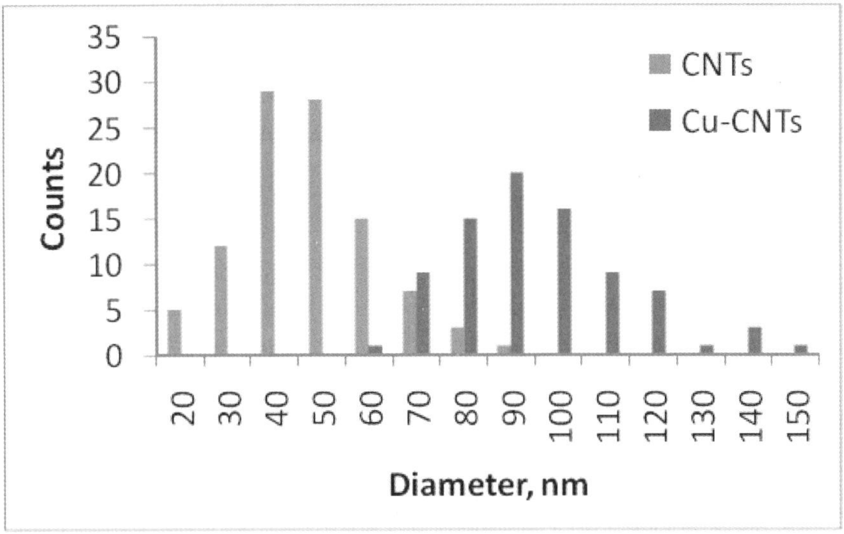

Fig. 5. Histogram of CNT and Cu-CNT (CNTs metallized with Cu via e-beam evaporation) diameters. CNTs were grown using a re-deposited trilayer catalyst and Cu-CNTs were metallized with 150 nm nominal Cu thickness.

The length of CNT arrays grown on Cu substrates was between 20 and 100 μm, with a median observed length of 40 μm. For CNT arrays grown on sintered wicks, the array height was much smaller, between 5 and 10 μm. This finding may be a result of the difference in catalyst composition due to the increased surface area of sintered powder as compared to a flat plate of copper, or due to differences in local plasma conditions induced by the comparatively larger surface roughness of the sintered particles. Diameter measurements were taken using SEM images at 50k magnification of CNTs and Cu-CNTs (CNTs metallized with Cu via e-beam evaporation) grown by re-deposition of trilayer catalyst after the initial growth attempt. By measuring the diameter of multiple CNTs throughout the sample, a histogram was constructed. All Cu-CNTs measured for the purpose of diameter estimation were metallized with an area-averaged nominal thickness of 150 nm of Cu (i.e., the thickness of the layer normal to the substrate that would be deposited on a flat surface). Due to an increase in the total surface area compared to a flat surface, the conformal coating thickness over the individual nanotubes was expected to be less than this value. Figure 5 shows this histogram for the CNTs and Cu-CNTs. The median-diameter CNT was 50 nm, while this value for the median Cu-CNT was 90 nm. Therefore, it may be surmised that a coating thickness of approximately 20 nm was added to CNTs upon the deposition of 150 nm of Cu.

The presence of amorphous carbon and defects on the CNT outer walls seemed to balance the high mobility of copper, preventing the agglomeration of copper into clusters, and leading to a uniform coating. The rate of deposition was observed to have an important effect on the quality of the metal coating. Rates of deposition greater than 4 Å/sec would either completely bury the nanotubes underneath the metal or would cause regions of the substrate to be devoid of nanotubes. In such cases, it is speculated that the higher deposition rate caused the metal atoms to impinge on the nanotubes with very high velocities causing the nanotubes to ablate from the substrate.

The amount of metal deposited also had an impact on the uniformity of the coating. For tubes that were approximately 10 μm in length, a nominal deposition thickness of at least 200 nm to 250 nm of metal would ensure uniform coverage that would coat the nanotubes lying below the surface. For lower nominal thicknesses of deposited copper (such as 150 nm or less), it was observed that the metal did not penetrate into the inner layers of the CNTs. This caused the nanotubes to clump, leading to a shift in the area covered by the nanotubes. The depth of penetration of Cu into the array was investigated because it was desirable for the nanotubes to be hydrophilic from root to tip for achieving the greatest heat transfer enhancement. Because CNTs at the edge of the sample were not thought to be representative of Cu penetration in the interior region, viewing the side of the sample was deemed a poor assessment of Cu penetration. Instead, the Cu-CNT array was viewed after it had been wetted by water and dried. The capillary forces during this wetting process caused collapse of the Cu-CNTs into bundles that revealed the copper penetration depth. In Figure 6(a), an array of Cu-CNTs viewed post-wetting shows that the bases of the CNTs have not been metallized. From this view, an estimate of the Cu penetration depth into the array is made by measuring how far the Cu extends down the length of the CNTs. From Figure 6(b) the penetration

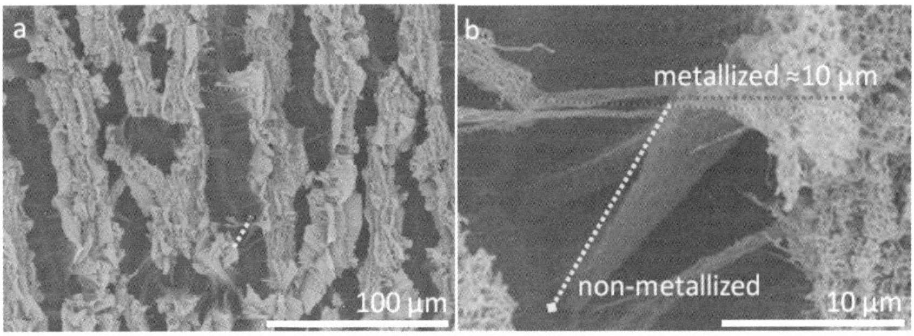

Fig. 6. SEMs of a Cu-metallized CNT array after wetting with water which show (a) formation of cellular structures following wetting, and (b) the penetration of the metallization into the array. The Cu coating reaches a depth of about 10 µm down from the CNT tips.

depth of the Cu metallization was estimated to be 10 µm. It is therefore concluded that typical Cu-CNT arrays grown on Cu substrates (with nanotube lengths of around 40 µm) were not fully coated with Cu to the base of the CNTs. However, the Cu-CNT arrays grown on sintered particles (with nanotube lengths of approximately 10 µm) were coated over most or all of their length.

Wetting of Cu-CNT arrays on Cu substrates

It is well known that dense arrays of CNTs exhibit hydrophobic behavior[31,32], and consequently, CNTs have been previously considered incompatible with water for multi-phase heat transfer enhancement. Achieving hydrophilic behavior would allow CNT arrays to be utilized with water for such enhancement. Following deposition of a Cu coating of 150 nm nominal thickness, the improvement in wettability of the CNTs with water was significant. Figure 7 shows images of water droplets on four different surfaces taken using a goniometer (CAM 100, KSV Instruments Ltd.) and backlit LED light source. The liquid-solid contact angle observed for a Cu substrate degreased to remove oil from the surface (Figure 7(a)) was 100 deg. A Cu substrate that had been etched with piranha solution (a 3:1 combination of sulfuric acid and hydrogen peroxide) to remove the native oxide layer from the surface following degreasing (Figure 7(b)) had a lowered contact angle of 49 deg.

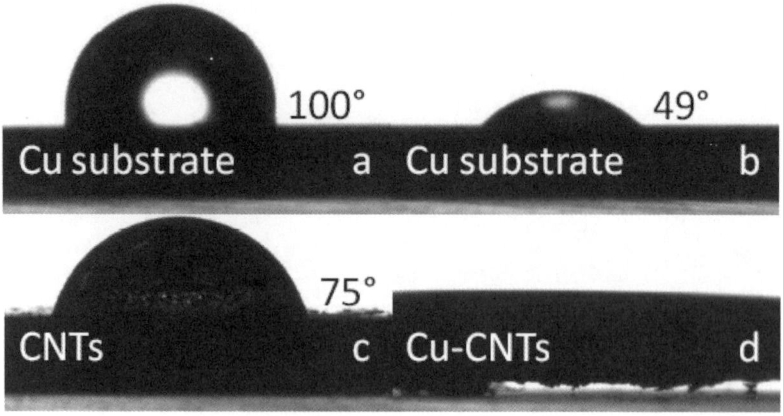

Fig. 7. Sessile drop tests were performed with water and images are shown immediately after placing a droplet on the surface for (a) a flat Cu substrate after degreasing, (b) a flat Cu substrate after etching with piranha solution, (c) a CNT array exhibiting CNT clusters among a thin CNT film, and (d) a CNT array of similar resulting growth morphology following metallization with 150 nm Cu. The water droplet placed on the Cu-CNT surface quickly spread, preventing estimation of the static contact angle.

For a CNT array (Figure 7(c)), an initial dynamic advancing contact angle of 75 deg was observed immediately after placing a water droplet on the surface. Over time, the droplet shown in Figure 7(c) slowly spread out as it was absorbed into the array by capillary action. Figure 7(d) shows a water droplet immediately after placement onto a metallized CNT array. The droplet spread through the array quickly enough that an estimation of initial dynamic advancing contact angle could not be made. The metallized CNT array clearly exhibited hydrophilic characteristics compared to the CNT array as-grown.

Raman spectroscopy characterization

Raman spectra of copper-coated CNTs were obtained using a 10 mW, 632.8 nm He-Ne laser. Figure 8 compares the Raman spectra of CNTs coated with nominal copper coating thicknesses of 150 nm and 250 nm. The spectra revealed a number of peaks corresponding to Cu_2O (92, 142, shoulder at 217, and 650 cm^{-1}) and Cu-OH/Cu_2O (540-580 and

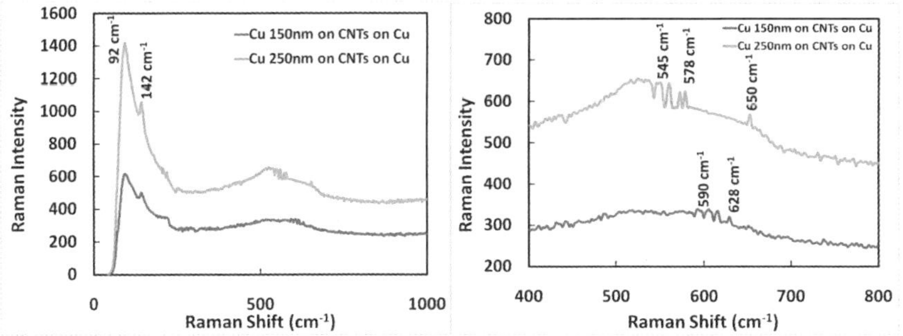

Fig. 8. (a) Broad and (b) magnified view of Raman spectra for copper-coated CNTs showing Cu-OH/Cu_2O peaks at 540-580 and 590-630 cm^{-1}.

590-630 cm^{-1})[33,34]. The evaporated copper conformally coating the nanotubes forms Cu-OH bonds when exposed to atmospheric water vapor. This presence of polar OH- groups on the surface of the copper-coated nanotubes is believed to render the nanotubes hydrophilic.

Sintering of copper powder

As previously discussed, the sintered powder wick layer was formed by placing 90-106 μm copper particles into a ceramic mold at high temperature in a reducing forming gas atmosphere. The typical appearance of a fabricated sintered copper wick is shown in Figure 9(a). Sintered copper powder attached to the copper substrate emerged from

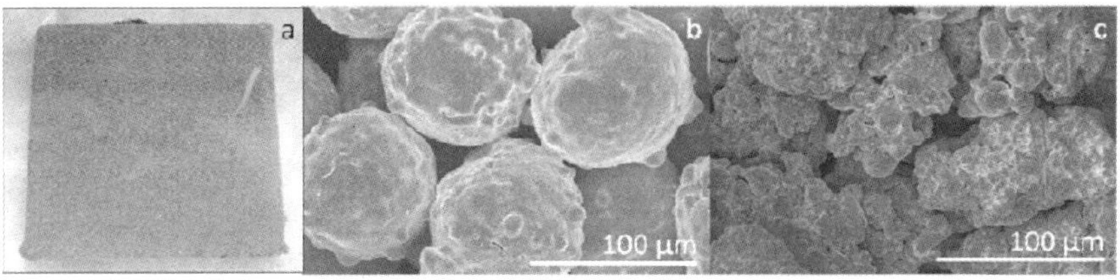

Fig. 9. A typical sintered sample: (a) post-sintering sample appearance, (b) air-atomized spherical sintered particles, and (c) water-atomized amorphous sintered particles.

the furnace with a salmon color typical of native copper, indicating that this process removed the oxide layer from the particles during sintering. Samples left exposed to atmosphere oxidized in a matter of hours, accompanied by a color change of the surface and associated hydrophobic behavior. This motivated storage of all samples in sealed containers. Two types of copper particles were used to fabricate sintered wicks. Purchased particles fabricated by air-atomization versus water-atomization form spherical (Figure 9(b)) versus amorphous (Figure 9(c)) particle shapes, respectively. Spherical particles were the primary focus of the current investigation. During sintering, particles formed interconnecting necks. Formation of these necks reduced the pore size and distance between the particles. This decrease in porosity caused the powder to contract within the ceramic mold. In the case of particles sintered onto metallized CNTs, the individual bonds between a Cu-CNT and particle are expected to be weaker than those between the particle and its adjacent particles. Therefore, as the powder contracted within the mold, bonds between the particles and Cu-CNTs broke as the particles moved relative to the substrate. While the movement was small, on the order of tenths of millimeters, the distance was much larger than the length of CNTs attached to the particles at their base.

Fig. 10. Shrinkage of the sintered powder layer relative to the substrate edge following sintering onto a layer of metallized CNTs: (a) photo showing the typical degree of shrinkage, measured to be as much as 0.76 mm, and (b) an SEM of the sample corner exhibiting shrinkage.

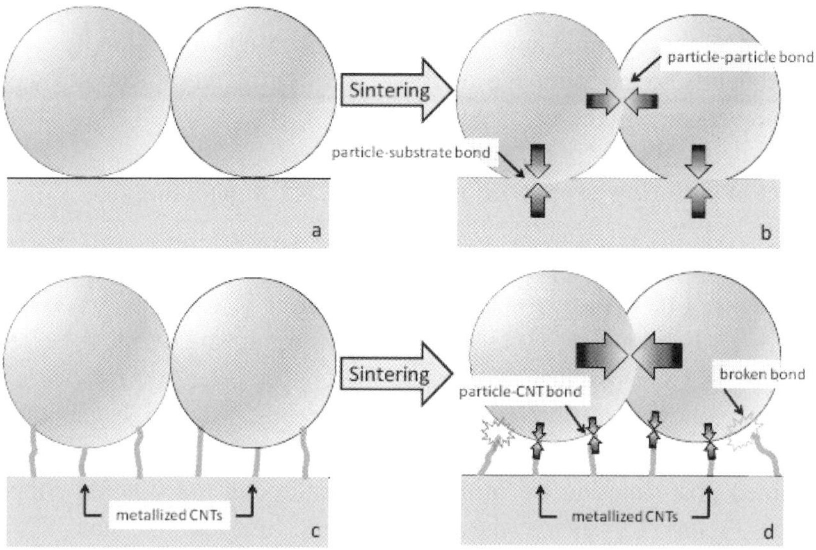

Fig. 11. Schematic illustration of the effect of sintering of particles to bare copper substrates versus sintering of particles to metallized CNTs. Shown in the top illustrations are (a) particles placed on a bare copper substrate and (b) bonds formed between particles and substrate post sintering. The lower illustrations show (c) particles placed on metallized CNTs and (d) lateral shrinking during sintering, causing breakage of particle-CNT bonds.

In the case of copper powder sintered onto metallized CNTs, therefore, the entire layer of particles often detached completely from the surface. Figure 10 shows the copper powder sintered on the Cu-CNT surface prior to detachment while Figures 11(c) and 11(d) schematically illustrate the breaking of the bonds between the metallized CNTs and the Cu particles during sintering. Alternatively, in the case of sintering copper powder directly onto a copper substrate, as illustrated in Figure 11(a) and 11 (b), bonds of relatively equal strength formed between the particles and the substrate and between adjacent particles. Therefore, the particles became well anchored, and little or no shrinking from the edges of the substrate was observed. To alleviate the structural incompatibility, the bottom CNT layer was omitted, and a two-layer configuration where the copper powder was directly sintered on to the copper substrate was adopted. Both the three-layer and two-layer configuration samples were subsequently tested, although the two-layer configuration samples were more mechanically robust.

Effect of sintering process on metallized CNT morphology

During sintering of copper powder to the metallized underlayer of CNTs grown using the trilayer catalyst, the underlayer of CNTs is exposed to sintering conditions of 950°C for 1 hour in forming gas. While the effect of forming gas is expected to reduce the copper layer and remove any surface oxides, the effect of the sintering conditions on the nanotubes was of interest. SEM images of the metallized CNT underlayer were obtained post-sintering by cutting a trench through the sintered copper layer. Figure 12 shows the post-sintering images of such CNTs metallized with 150 nm of copper. It is evident from the figure that the highly mobile copper coating on the CNTs seems to have undergone Ostwald ripening, resulting in the formation of nanoscale spheres on the tips of the CNTs. Ostwald ripening was also found to occur for metallized CNTs not attached to sintered particles, verifying that the ripening was a result of the mobile copper sourced from the CNT coating, and not from the sintered copper particles. Water droplets placed on the sample surface after observation of Ostwald ripening indicated that the

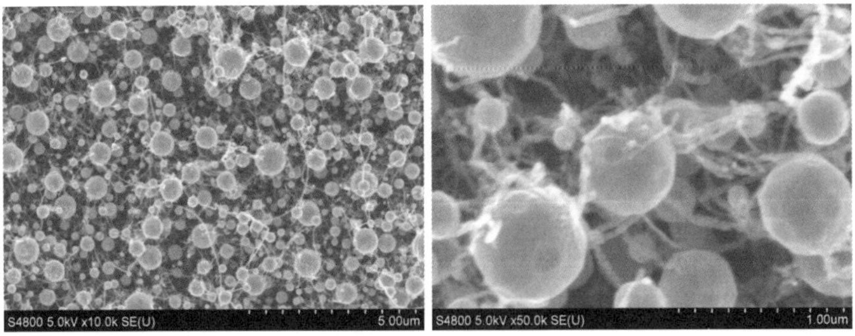

Fig. 12. SEM images of copper spheres formed on the tips of metallized nanotubes (with 150 nm nominal copper coating thickness) as a result of Ostwald ripening following exposure to sintered conditions (950°C for 1 hour).

bulk sintered copper wick remained hydrophilic. It is expected that the nanoscale spheres that form as a result of the ripening and annealing may act as re-entrant cavities for the vapor embryos in the superheated region and enhance nucleate boiling performance.

Characterization of CNT array wicking

Having already observed the dramatic difference in wetting performance of CNTs as compared to Cu-CNTs from the sessile drop tests, the ability of these samples to wick water against gravity was also investigated. Two CNT samples (bare and copper-metallized) grown on flat substrates were tested by placing the lower edge of the sample in a water bath and allowing liquid to rise up through the array to a static height. The Cu-CNT array was metallized with a nominal thickness of 150 nm Cu. The rise height of the samples is noted here only for the purpose of comparison between the wicks, as the static height may not correspond to the capillary pressure generated due to concurrent evaporation from the surface. As shown in Figure 13, the rise height of the Cu-CNTs was approximately twice that of the CNTs without metallization. Additionally, the observed rate of liquid rise was faster for the metallized CNTs. Qualitatively, the Cu-CNTs are more suited for wicking water than bare CNTs. It may be concluded that addition of Cu-CNTs to the sintered powder wick structure may aid in transporting liquid to the site of evaporation/boiling in a heat pipe.

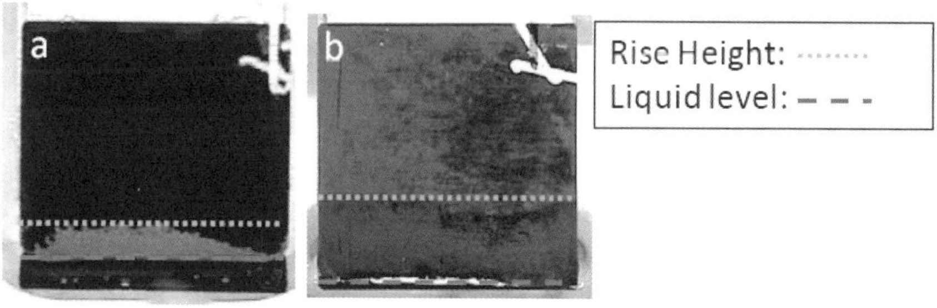

Fig. 13. Capillary rise of water up the surface of a CNT array grown by trilayer catalyst re-deposition on a flat substrate (a) with and (b) without 150 nm of copper metallization.

Thermal performance of CNT/sintered copper powder wicks

The primary motivation for developing wicks modified with Cu-CNTs was to assess the evaporation and boiling performance enhancement provided. A test facility with a closed chamber as described in [35] was used to create a saturated environment (similar to that in a heat pipe) in which to evaluate the capillary-fed boiling performance of the samples. In this chamber (shown in Figure 14), the bottom edge of the sample is immersed in the fluid bath, allowing for wicking of water up to the heat input location. A heat flux was applied to a 5 mm × 5 mm area by means of a copper heater block soldered to the back of the Cu substrate. Temperature measurements were made in the heater block, at the sample surface, and in the vapor in order to measure the input heat flux and estimate the temperature drop across the sample. Based on these measurements the thermal resistance is defined as the temperature drop across the sample divided by the total heat input.

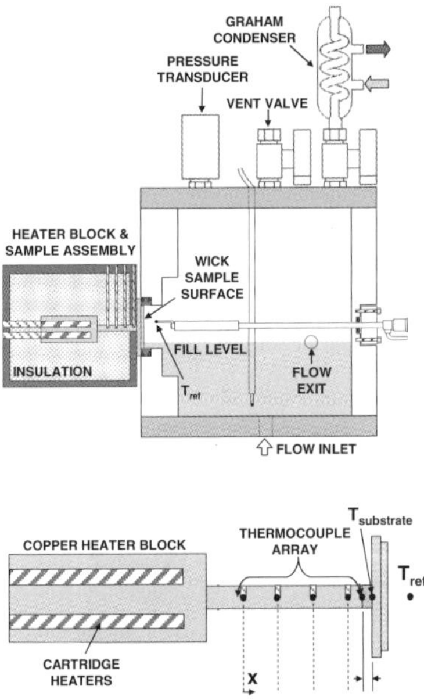

Fig. 14. Schematic diagram of the saturated test chamber and temperature measurement locations of the capillary-fed boiling thermal test facility used to evaluate the composite CNT-sintered powder wick performance.

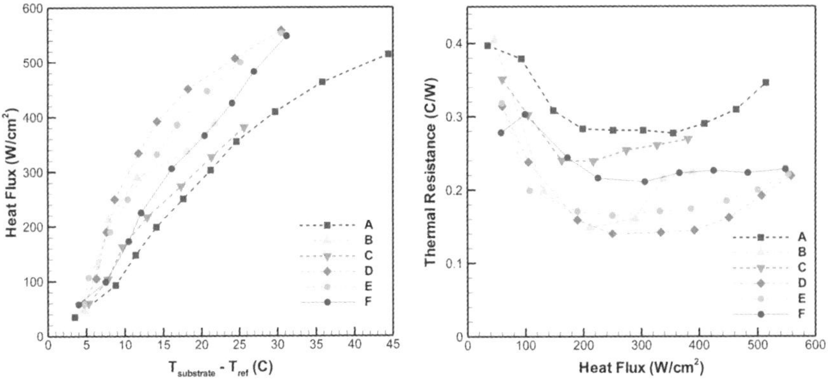

Fig. 15. Measured boiling curves and wick thermal resistance as a function of heat flux for 2-layer Cu-CNT sintered powder composite wicks.

The thermal test results are shown in Figure 15. In terms of thermal performance, the CNT-coated sintered powder samples described in prior work[36] achieved maximum heat fluxes of approximately 400 to 550 W/cm^2, with minimum thermal resistances in the range of 0.2 and 0.4°C/W. In general, the thermal resistance decreased with increasing heat flux as boiling began over an increasing area of the wick structure. The two best performing samples, D and E, had a thermal resistance below 0.2°C/W in the boiling regime and supported a heat flux in excess of 500 W/cm^2 with less than 30°C of substrate superheat. As a baseline comparison, a sintered copper powder wick without any CNTs was fabricated and tested (sample F in Figure 15). This shows that over a wide range of heat fluxes, the two-layer CNT-sintered powder composite test samples maintained a lower substrate temperature than the baseline sample.

Samples B and C, on the other hand, reached dryout at much lower heat fluxes of less than 400 W/cm^2 and samples A, B, and C all exhibited a thermal resistance larger than the baseline sintered powder wick. One possible reason for this premature dryout, increased thermal resistance, and inconsistency in the results was the difficulty in maintaining the samples in a hydrophilic condition due to rapid oxidation of the copper coating on the nanotubes. Testing was performed chronologically from samples A to F and for the later tests (specifically D and E), a more rigorous effort was made to limit atmospheric exposure time and oxidation between fabrication and testing. Considering this was the only difference between the fabrication and testing processes for all samples, it is expected that reduced oxidation of the surfaces preserved their hydrophilic nature and led to the improved performance.

Based on the thermal performance enhancement observed for several of the samples, there is significant promise offered by the use of sintered powder wicks modified with Cu-CNTs. Further development and optimization of the wick structures is required to better understand the performance trends and achieve consistent thermal performance.

4. Conclusions

The successful development and fabrication of a multi-layer micro/nano-wick composite consisting of sintered copper powder and copper-coated CNTs is presented. Inherently hydrophobic CNTs were coated with copper, making them hydrophilic and suitable for integration with sintered copper powder wicks used in two-phase water heat pipes. Thermal performance results indicated that heat fluxes greater than 500 W/cm^2 could be supported. For several composite samples, the thermal resistance was reduced compared to a baseline sintered powder wick, verifying the efficacy of this approach. The results demonstrate that it is possible to intelligently design wick structures by individually addressing the various factors that contribute to overall thermal performance. Metallization rendered carbon nanotubes hydrophilic which in turn allowed them to assist in thin film evaporation and boiling heat transfer. By incorporating nanoscale design elements to obtain improved heat dissipation, this micro-nano composite wick structure is a promising candidate for light-weight, small-volume, and high-performance heat pipe cooling solutions of the future.

Acknowledgments

This material is based upon work supported by the Defense Advanced Research Projects Agency (DARPA) and Space and Naval Warfare Systems Center (SPAWAR/SYSCEN) San Diego, California, under Contract No. N66001-08-C-2011.

References

1. Baxter, J., Bian Z., Chen, G., Danielson, D., Dresselhaus, M.S., Fedorov, A.G., Fisher, T.S., Jones, C.W., Maginn, E., Kortshagen, U., Manthiram, A., Nozik, A., Rolison, D., Sands, T., Shi, L., Sholl, D, and Wu, Y., 2009, "Nanoscale design to enable the revolution in renewable energy," Energy and Environmental Science, Vol. 2, pp. 559–588.
2. International Technology Roadmap for Semiconductors (ITRS), 2007, ITRS 2007 Edition, http://www.itrs.net/Links/2007ITRS/Home2007.htm.
3. Missaggia, L.J., Walpole, J.N., Liau, Z.L., and Phillips, R.J., 1989, "Microchannel heat sinks for two-dimensional high-power-density diode laser arrays," Quantum Electronics, IEEE Journal of, Vol. 25, pp. 1988–1992.
4. Mudawar, I., 2001, "Assessment of high-heat-flux thermal management schemes," Components and Packaging Technologies, IEEE Transactions on, Vol. 24, pp. 122–141.
5. Mahajan R., Chiu C., and Chrysler G., 2006, "Cooling a microprocessor chip. Heat spreading, choice of thermal interface materials, and proper heat-sink design can enhance cooling of microprocessor packages and systems", Proceedings of the IEEE, Vol. 94, 8, pp. 1476–1486.
6. Sauciuc I., Prasher R., and Chang J. Y., 2005, "Thermal performance and key challenges for future CPU cooling technologies," ASME 2005 Pacific Rim Technical Conference and Exhibition on Integration and Packaging of MEMS, NEMS, and Electronic Systems collocated with the ASME 2005 Heat Transfer Summer Conference (InterPACK2005), July 17–22, San Francisco, California, USA, pp. 353–364.
7. Bar-Cohen, A., Arik, M., and Ohadi, M., 2006, "Direct liquid cooling of high flux micro and nano electronic components," Proceedings of the IEEE, Vol. 94, pp. 1549–1570.
8. Garimella, S.V., Singhal, V., and Liu, D., 2006, "On-chip thermal management with microchannel heat sinks and integrated micropumps," Proceedings of the IEEE, Vol. 94, pp. 1534–1548.
9. Dunn, P.D., and Reay, D.A., 1973, "The heat pipe," Physics in Technology, Vol. 4, pp. 187–201.
10. Garimella, S.V., and Sobhan, C.B., 2001, "Recent advances in the modeling and applications of nonconventional heat pipes," Advances in Heat Transfer, Vol. 35, pp. 249–308.
11. Hopkins, R., Faghri, A., and Khrustalev, D., 1999, "Flat miniature heat pipes with micro capillary grooves," Journal of Heat Transfer, Vol. 121, pp. 102–109.
12. Li, C., Peterson, G.P., and Wang, Y., 2006, "Evaporation/boiling in thin capillary wicks (l)--wick thickness effects," Journal of Heat Transfer, Vol. 128, pp. 1312–1319.

13. Li, C., and Peterson, G.P., 2006, "Evaporation/boiling in thin capillary wicks (II)--effects of volumetric porosity and mesh size," Journal of Heat Transfer, Vol. 128, pp. 1320–1328.
14. Gernert, N.J., Toth, J., and Hartenstine, J., 2004, "100 W/cm2 and higher heat flux dissipation using heat pipes," Proceedings of the 13th International Heat Pipe Conference, Shanghai, China, September 21–25.
15. Hanlon, M.A., and Ma, H.B., 2003, "Evaporation heat transfer in sintered porous media," Journal of Heat Transfer, Vol. 125, pp. 644–655.
16. You, S.M., Simon, T.W., Bar-Cohen, A., and Tong, W., 1990, "Experimental investigation of nucleate boiling incipience with a highly-wetting dielectric fluid (R-113)," International Journal of Heat and Mass Transfer, Vol. 33, pp. 105–117.
17. Arik, M., and Bar-Cohen, A., 2003, "Effusivity-based correlation of surface property effects in pool boiling CHF of dielectric liquids," International Journal of Heat and Mass Transfer, Vol. 46, pp. 3755–3764.
18. Arik, M., and Bar-Cohen, A., 2010, "Pool boiling of perfluorocarbon mixtures on silicon surfaces," International Journal of Heat and Mass Transfer, Vol.53, pp. 5596–5604.
19. Arik, M., Bar-Cohen, A., and You, S.M., 2007, "Enhancement of pool boiling critical heat flux in dielectric liquids by microporous coatings," International Journal of Heat and Mass Transfer, Vol.50, pp. 997–1009.
20. Liter, S.G., and Kaviany, M., 2001, "Pool-boiling CHF enhancement by modulated porous-layer coating: theory and experiment," International Journal of Heat and Mass Transfer, Vol. 44, pp. 4287–4311.
21. Mughal, M.P., and Plumb, O.A., 1996, "An experimental study of boiling on a wicked surface," International Journal of Heat and Mass Transfer, Vol. 39, pp. 771–777.
22. Dresselhaus, M.S., Dresselhaus, G., and Jorio, A., 2004, "Unusual properties and structure of carbon nanotubes," Annual Review of Materials Research, Vol. 34, pp. 247–278.
23. Yang, D.J., Zhang, Q., Chen, G., Yoon, S.F., Ahn, J., Wang, S.G., Zhou, Q., Wang, Q., and Li, J.Q., 2002, "Thermal conductivity of multiwalled carbon nanotubes," Physics Review B, Vol. 66, pp. 165440.
24. Lin, N., Wang, H., Dixit, P., Xu, T., Zhang, S., and Miao, J., 2009, "Investigation of carbon nanotube growth on multimetal layers for advanced interconnect applications in microelectronic devices," Journal of the Electrochemical Society, Vol. 156, pp. K23-K27.
25. Khanikar, V., Mudawar, I., and Fisher, T.S., 2009, "Flow boiling in a micro-channel coated with carbon nanotubes," IEEE Transactions on Components and Packaging Technologies, Vol. 32, pp. 639–649.
26. Ujereh, S., Fisher, T.S., Mudawar, I., 2007, "Effects of carbon nanotube arrays on nucleate pool boiling," International Journal of Heat and Mass Transfer, Vol. 50, pp. 4023–4038.

27. Maschmann, M.R., Amama, P.B., Goyal, A., Iqbal, Z., Gat, R., and Fisher, T.S., 2006, "Parametric study of synthesis conditions in plasma-enhanced CVD of high-quality single-walled carbon nanotubes," Carbon, Vol. 44, pp. 10–18.
28. Kong, J., Soh H.T., Cassell A.M., Quate C.F. and Dai H., 1998, "Synthesis of individual single-walled carbon nanotubes on patterned silicon wafers," Nature, Vol. 395, pp. 878–881.
29. Sinnott S.B., Andrews R., Qian D., Rao A.M., Mao Z., Dickey E.C. and Derbyshire F., 1999, "Model of carbon nanotube growth through chemical vapor deposition," Chemical Physics Letters, Vol. 315, pp. 25–30.
30. Jeng J.S., and Chen J.S., 2002, "Interdiffusions and reactions in Cu/TiN/Ti/Thermal SiO2 and Cu/TiN/Ti/Hydrogen silsesquioxane multilayer structures", Journal of the Electrochemical Society, Vol. 149, pp. G455-G460.
31. Barber, A.H., Cohen, S.R., and Wagner, H.D., 2004, "Static and dynamic wetting measurements of single carbon nanotubes," Physical Review Letters, Vol. 92, pp. 186103.
32. Feng, L., Li, S.H., Li, Y.S., Li, H.J., Zhang, L.J., Zhai, J., Song, Y.L., Liu, B.Q., Jiang, L., and Zhu, D.B., 2002, "Super-hydrophobic surfaces: from natural to artificial," Advanced Materials, Vol. 14, pp. 1857–1860.
33. Gan, Z. H., Yu, G. Q., Tay, B. K., Tan, C. M., Zhao, Z. W., and Fu Y. Q., 2004, "Preparation and characterization of copper oxide thin films deposited by filtered cathodic vacuum arc," Journal of Physics D: Applied Physics, Vol. 37, pp. 81–85.
34. Mayer, S. T. and Muller, R. H., 1992, "An in situ Raman Spectroscopy study of the anodic oxidation of copper in alkaline media," Journal of the Electrochemical Society, Vol. 139, pp. 426–434.
35. Weibel, J.A., Garimella, S.V., and North, M.T., 2010, "Characterization of evaporation and boiling from sintered powder wicks fed by capillary action," International Journal of Heat and Mass Transfer, Vol. 53, pp.4204–4215.
36. Powell, G.A., 2009, "Controlled synthesis of CNT-based nanostructures for enhanced boiling and wicking," M.S. thesis, Purdue University.

CHAPTER 13

A CABINET LEVEL THERMAL TEST VEHICLE TO EVALUATE HYBRID DOUBLE_SIDED COOLING SCHEMES

Qihong Nie

Software, Electronics and Mechanical System Corporate Research Lab
3M Company, St Paul, USA
qnie@mmm.com

Yogendra Joshi

G. W. Woodruff School of Mechanical Engineering
Georgia Institute of Technology, Atlanta, USA
yogendra.joshi@me.gatech.edu

Packaging of power semiconductor devices presents some of the greatest thermal design challenges due to the resulting high heat fluxes. Advanced cooling techniques are desired to help meet these demands for current and future devices. A hybrid double-sided approach combining micro-channel liquid cooling, thermoelectric cooling, and forced air convection is investigated via a test vehicle for the thermal management of electronic cabinets. A reduction of 74% in the chip junction temperature rise was achieved by using double-sided cooling, compared to single-sided air convection. Further reduction can be achieved by utilizing thermoelectric cooling (TEC). Additional reductions of 22.4% and 6.5% were achieved by utilizing TEC in single-sided air cooling and double-sided cooling, respectively. The effect of water flow rates through the air-to-liquid heat exchanger and the microchannel heat sink on the chip junction temperature rise was insignificant, compared to the effect of TEC, and cooling configuration.

1. Introduction

The number of transistors integrated per chip has grown dramatically according to Moore's law.[1] As a result, the device level heat dissipation and total heat generation rates have increased dramatically for power electronic cabinets. For example, the total heat generated from large power semiconductors such as insulated gate bipolar transistors (IGBT), diodes, or thyristors can be as high as several kilowatts.[2] Therefore, the total heat dissipation of a power electronic cabinet may dissipate up to 12 kW.[3] For electronic devices such as integrated power converters based on IGBT operating in temperature above 100°C, thermal concerns are becoming critical. High operating temperature generally compromises performance of devices and may in some cases have an impact on reliability. Additionally, temperature cycles resulting from turning on and off large amounts of power can cause fatigue of the die and other elements of the chip package. How to efficiently manage such large amount of heat dissipation in power electronic cabinets is a unique challenge for thermal engineers.

Air cooling has been intensively used for the thermal management of electronic cabinets with low heat dissipation (typically below 15 kW), due to its easy equipment maintenance, low operation cost, and acceptable cooling efficiency.[4-9] Forced air cooling is widely used in most data center processing cabinets today.[8,9] As heat loads continue to rise, so does the challenge of cooling with air due to the limits of heat sink/air moving device performance, and rack level acoustic noise limitations. Liquids, primarily because of their higher density and specific heat, are much more effective in the removal of heat than air, making liquid cooling a desirable choice for increased heat loads.[10-12] Besides air and liquid cooling options, hybrid cooling combining multiple approaches is also increasingly utilized in a variety of applications, such as when electronics need to be isolated from harsh ambient environments to operate functionally and reliably. For such applications, the heat generated from the electronics is typically removed by air circulation inside the rack, and is transferred to the ambient through a liquid-to-air heat exchanger.[13-15]

Packaging of power semiconductor devices presents some of the highest heat fluxes. For example, metal oxide semiconductor-controlled thyristors generate heat fluxes from 100 to 200 W/cm^2 in 1995,[16] while today's power module can dissipate up to 500 W/cm^2.[17] Heat fluxes in some optoelectronics applications and military high power electronics exceed 1,000 W/cm^2.[18] Double-sided cooling can help to meet these demands for current and future semiconductor devices.[19-22] This concept was originally proposed by Gillot for power electronic modules and studied in details by Pang, as shown in Figure 1.[19,20] All components of the module are connected in parallel and are sandwiched between two direct bond cooper (DBC) substrates. Cooling devices can be attached to both sides of the module so that heat can be dissipated through both sides. This concept has been increasingly studied by simulation and experiments at device level.[23,24] In this work, this concept is first introduced at the system level by combining microchannel liquid cooling, thermoelectric cooling, and forced air convection.

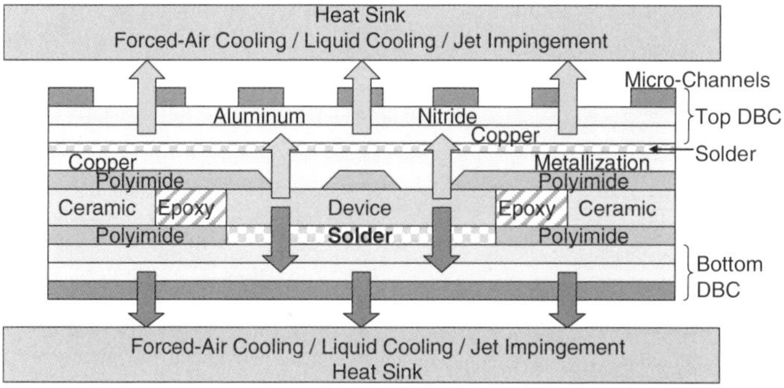

Fig. 1. Double-sided cooling of power electronic module[20].

2. Test Vehicle with Hybrid Double-Sided Cooling Technique

A test vehicle with hybrid and double-sided cooling is constructed, as shown in Figure 2. Its configuration is based on a prototype power conversion module (PCM) cabinet called PCM-1,[6] as shown in Figure 3. PCM-1 cabinets are used to distribute power in various ship zones. This test cabinet consists of four enclosures and one bottom bay, with each enclosure containing three packages, as shown in Figure 4(b). The air is isolated inside the cabinet and circulated by the blower and fans, as depicted in Figure 4(a). The detailed configuration of the package is illustrated in Figure 4(c).

Fig. 2. Physical test vehicle.

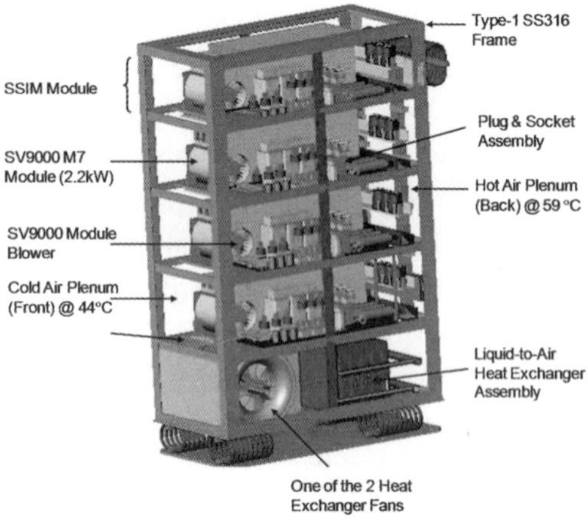

Fig. 3. PCM-1 cabinet[6].

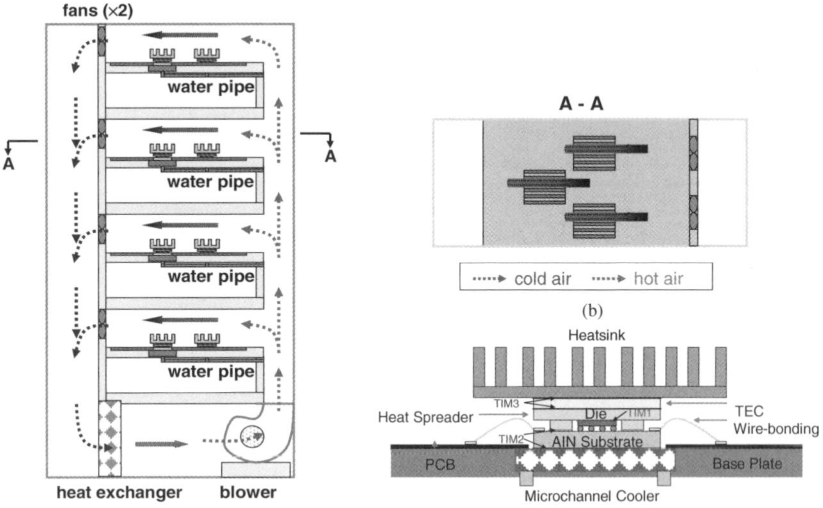

(a) (c)

Fig. 4. Schematic of (a) cabinet, (b) enclosure, and (c) package.

A thermoelectric module (TEC) with heat sink is attached to the top of the package to provide a heat flow path from the top. To achieve cooling from the bottom of the package, a microchannel cooler is directly attached to the substrate of the package. It is noted that wire-bonding technology, instead of the traditional flip chip technology was utilized, so that the substrate and the microchannel cooler can be contacted directly to reduce the contact thermal resistance, without introducing design complexity and electrical isolation issues. An aluminum nitride substrate is used in this configuration to enhance the cooling from the bottom of the package. With this configuration, a double-sided and hybrid forced air convection, thermoelectric cooling, and microchannel cooling approach is achieved.

3. Experimental Setup for Thermal Management Study of Cabinet

A schematic of the experimental setup is shown in Figure 5. A temperature control unit (TCU) stabilizes the temperature of the chilled water to within ± 0.56°C (± 1°F) from the chiller and drives the chilled water through flow meters to the heat exchanger and test chips.[27] The chilled water collects the heat from the microchannel cooler at each package and the heat exchanger at the bottom bay and goes back to the chiller through TCU, whose schematic is shown in Figure 6. Part of the hot water coming out of the test vehicle will return to the chiller through the TCU, and the rest will be recirculated within the TCU through the circulating pump inside the TCU. The immersion heater inside the TCU will be automatically switched on and heat the water if the recirculating water temperature is below the set point. A bypass loop is used to adjust the flow rate and pressure to the test vehicle. The cool sensor and heat sensor are used to monitor the return water temperature and supply water temperature so that the immersion heater can be switched on and off through the proportional integral derivative (PID) fuzzy control logic. The system safety is maintained through a safety thermometer and pressure relief valve. The maximum operating temperature and pressure of the TCU are designed to be 121°C (250°F) and 1.03 MPa (150 PSI), respectively. Two pressure gauges are used to monitor the water pressure at the inlet and outlet of the test vehicle, respectively. Another two pressure gauges are used to monitor the pressure drop across the microchannel cooler attached to the left chip package in the third enclosure from the bottom. The water flow rate through each microchannel cooler is adjusted by a flow meter. Eight DC power supplies are used to power the chips and TEC modules, and the temperature signals are collected through a data acquisition system and processed through a PC.

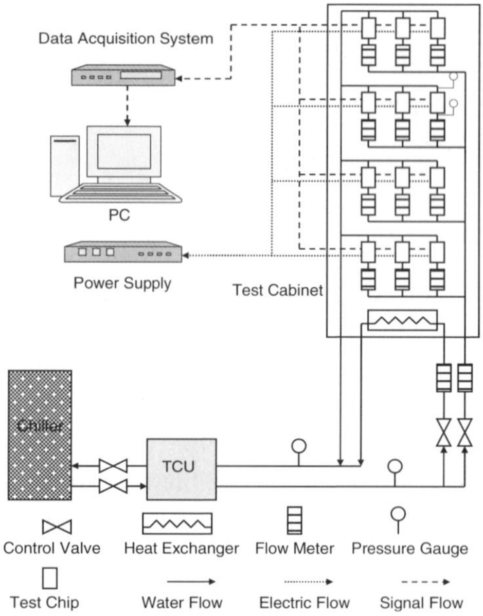

Fig. 5. Schematic of test flow loop.

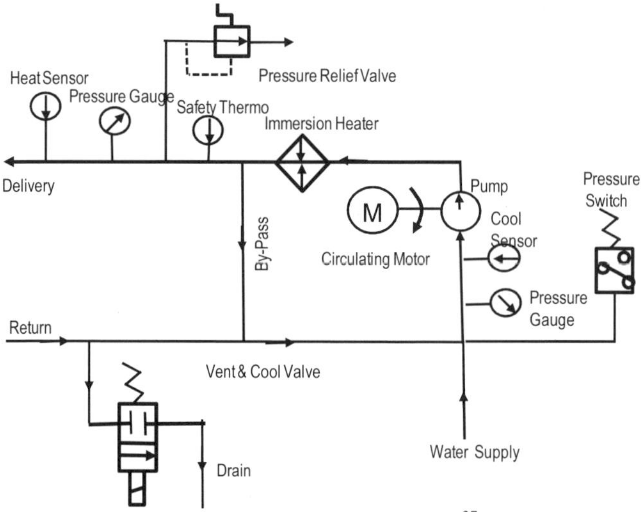

Fig. 6. Flow diagram of the TCU[27].

4. Calibration and Characterization

A schematic of the experimental setup is shown in Figure 5. A temperature control unit (TCU) stabilizes the temperature of the chilled water to within ± 0.56°C (± 1°F) from the chiller and drives the chilled water through flow meters to the heat exchanger and test chips.[27] The chilled water collects the heat from the microchannel cooler at each package and the heat exchanger at the bottom bay and goes back to the chiller through TCU, whose schematic is shown in Figure 6. Part of the hot water coming out of the test vehicle will return to the chiller through the TCU, and the rest will be recirculated within the TCU through the circulating pump inside the TCU. The immersion heater inside the TCU will be automatically switched on and heat the water if the recirculating water temperature is below the set point. A bypass loop is used to adjust the flow rate and pressure to the test vehicle. The cool sensor and heat sensor are used to monitor the return water temperature and supply water temperature so that the immersion heater can be switched on and off through the proportional integral derivative (PID) fuzzy control logic.

4.1. *Calibration of Thermal Module*

Thermal test dies (Delphi PST4-02) were used in the test vehicle to characterize the thermal performance of the package. The thermal test die is a flip chip type test die, whose layout is shown in Figure 7. Resistive heating in the thermal test die is accomplished by driving a current through a doped silicon well between a pair of bus bars, labeled R_s and R_f. The two-point measurement method was used to measure the heater resistance of the thermal test die in this work. The 4 R labeled pads accommodate four-point Kelvin connections. These measurements are expected to be more accurate than 2-point measurement when measuring small resistances in the mili-ohm or micro-ohm range. This measurement method may also need to be considered when the lengths of the wires between the device and multi-meter are long, due to the lead resistance associated with the wires.[25] Since the resistance of each thermal test die in this study is around 20 ohms, measured individually at the die location, the improvement in accuracy by the 4-point measurements is expected to be negligible.

At the top and bottom of the die are a pair of pads, labeled *D* in the diagram, which connect a serial five-diode temperature sensor network. A second temperature monitoring circuit uses a bridge network by connecting the "V" at the top of the die and the "G" at the bottom of the die with one sense pin "S" at the top of the die and the other sense pin "S" at the bottom of the die. The five-diode string from the center is duplicated in all four corners. The corner diode strings are connected in series such that each corner can be monitored individually, while driven by a single current source.

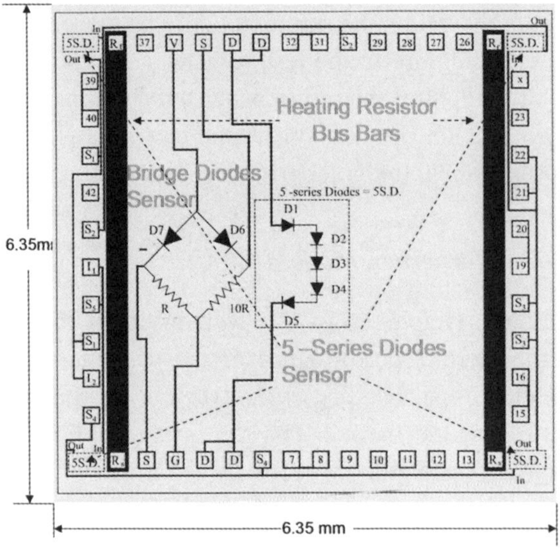

Fig. 7. Layout of thermal test die.

The electrical resistance of the heating circuit within thermal test die is designed at 20Ω. But this value may vary from die to die due to the manufacturing uncertainty and contamination. A multi-meter was used to measure the resistance R_c of each heating circuit, whose values are shown in Table 1. It is noted that the measurement uncertainty of the multi-meter is about 0.7% plus 0.01 ohm. Since long electrical wires are

Table 1. Electrical resistance of heating circuit

Chip #	1-1 (Middle)	1-2 (Right)	1-3 (Left)	2-1 (Middle)	2-2 (Right)	2-3 (Left)
R_t (Ω)	19.59	18.52	22.12	20.99	22.33	21.08
R_c (Ω)	19.81	18.82	22.12	21.72	23.1	21.2
Chip #	3-1 (Middle)	3-2 (Right)	3-3 (Left)	4-1 (Middle)	4-2 (Right)	4-3 (Left)
R_t (Ω)	18.32	20.68	22.41	22.45	22.37	21.08
R_c (Ω)	18.45	21.27	22.93	23.01	23.04	22.56

used to connect the thermal test die to the DC power supply, their external resistance also needs to be measured so that the correct voltage and current settings of the power supply can be selected. The total resistance R_t including the resistance for the thermal test die and the external resistances associated with the electric wires and solder connection are also shown in Table 1. An individual DC power supply is assigned to each test die so that the same power input to each test die can be obtained.

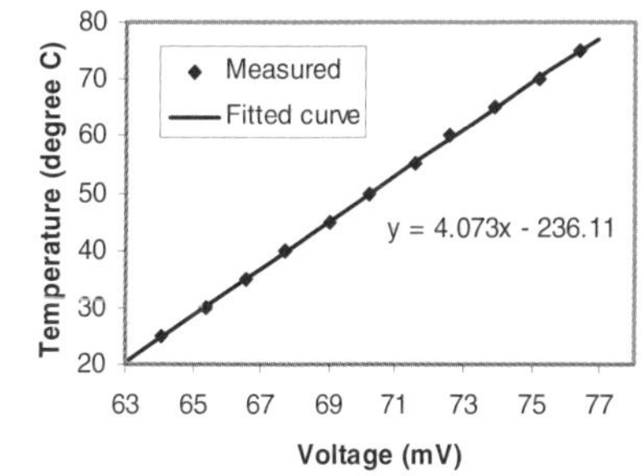

(a)

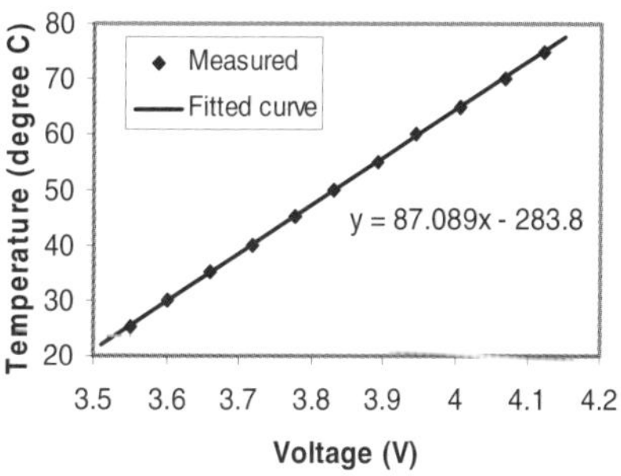

(b)

Fig. 8. Calibration curve of temperature diodes: (a) Bridge, (b) 5-Series.

For the 5-series diodes, designed temperature response across each is 2 mV/°C. For the bridge temperature diodes, the designed temperature response is 0.2 mV/°C. The thermal test die was placed inside an oven with embedded temperature indicator. The oven temperature can be accurately adjusted and read out through the indicator or calibrated thermocouple. The voltages of the temperature diodes are measured with a multi-meter. The correlations between the voltage and temperature measured by the oven temperature indicator are shown in Figures 8(a) and 8(b) for representative bridge temperature diodes and 5-series temperature diodes, respectively.

4.2. *Calibration of Thermocouples*

In order to ensure the accuracy of thermocouple measurements, an individual calibration was performed on each of the thermocouples used in experiments. Thermocouples were separately placed in a small tube of water within a thermocouple calibrator. Each thermocouple was calibrated against the water bath temperature at 20°C, 30°C , 45°C, 60°C, 75°C, and 90°C.[25] It was found that the maximum difference between any measured temperature and the one reported by the calibrator was approximately 0.35°C, which is within the uncertainty of measurements. This indicates that none of the thermocouples were systematically under or over-predicting the temperatures.

4.3. *Characterization of Microchannel Cooler*

The physical schematic of the micro-cooler (Curamik PolarXstream Cooler) is illustrated in Figure 9(a), with dimensions of 48 mm (W) × 48 mm (D) × 4 mm (H). It contains ten layers of microstructure as shown in Figure 9(b), which are stacked together alternatively in reversed direction. To decrease the computational modeling effort, a compact model is developed by focusing on only one channel area of the micro-cooler, as shown in Figure 10.[27] This model is based on the assumptions that the fluid is uniformly distributed to each channel, and heat flux across the bottom surface of the micro-cooler is also uniformly distributed. These assumptions are acceptable based on the measurement data provided by the manufacturer and the fact that highly thermally conductive pure copper is used for the entire structure.

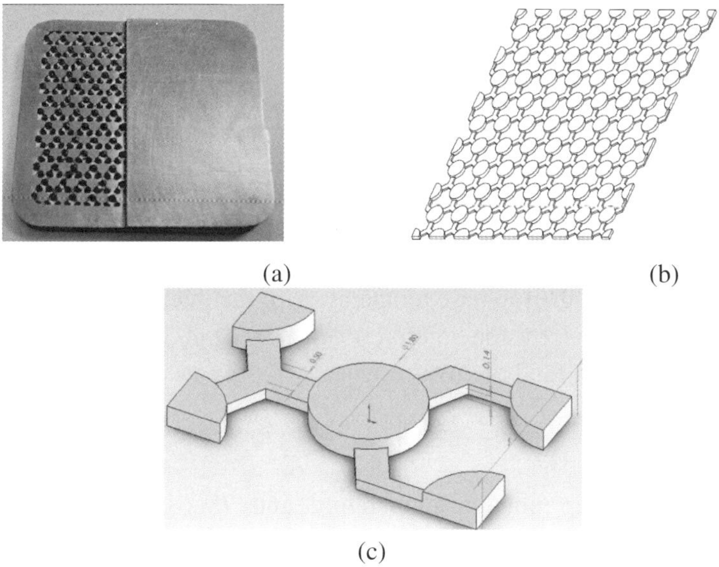

Fig. 9. Microchannel cooler (a) physical view (b) single layer of microstructure (c) unit cell of micro-structure.

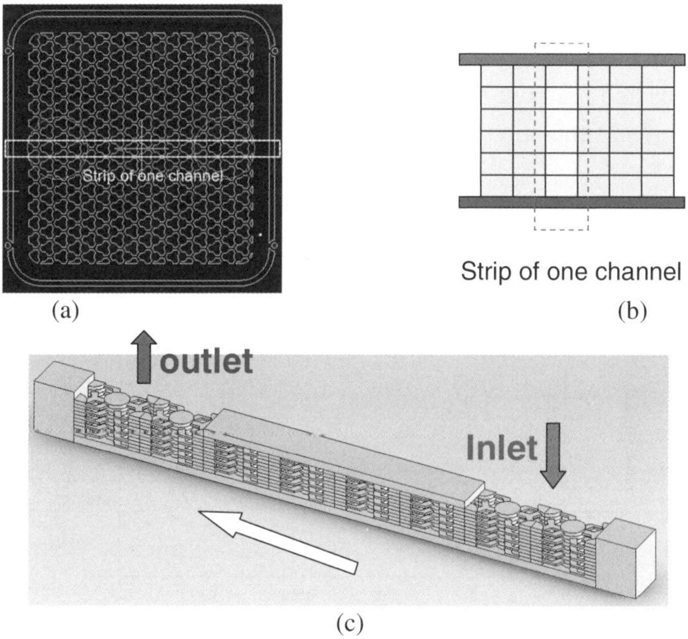

Fig. 10. (a) top view, (b) side view, (c) 3D compact model[27].

The surfaces of the micro-cooler are assumed under natural convection, except the center region (10 mm × 3 mm) as the heat source with a uniform heat flux of 2×10^6 W/m^2. To characterize the thermal resistance of this micro-cooler, the thermal resistance is defined by

$$R_{cooler} = \frac{T_b - T_w}{Q} \tag{1}$$

where $T_w = (T_{w,o} - T_{w,i})/2$ is the average water temperature across the micro-cooler, T_b is the top base temperature of the micro-cooler, and Q is the heat dissipation by the micro-cooler. Since it is practically very difficult to get T_b, a total thermal resistance of the package is defined as

$$R_t = \frac{T_j - T_w}{Q} = \frac{(T_j - T_b) + (T_b - T_w)}{Q} = R_{cooler} + R_{chip} \tag{2}$$

where T_j is the chip junction temperature, and R_{chip} is the chip thermal resistance. The thermal resistance R_{cooler} of the micro-cooler can be indirectly obtained once the total thermal resistance R_t is obtained

$$R_{cooler} = R_t - R_{chip} \tag{3}$$

The chip thermal resistance can be obtained by detailed numerical simulation conducted on the package. Figure 11 depicts the thermal resistances and pressure drops obtained by the experimental measurement and numerical simulation.[26]

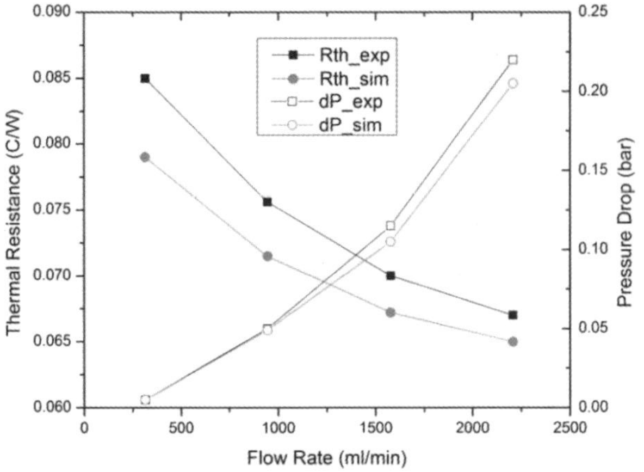

Fig. 11. Thermal resistance and pressure drop of microchannel cooler[27].

4.4. *Characterization of Liquid-to-Air Heat Exchanger*

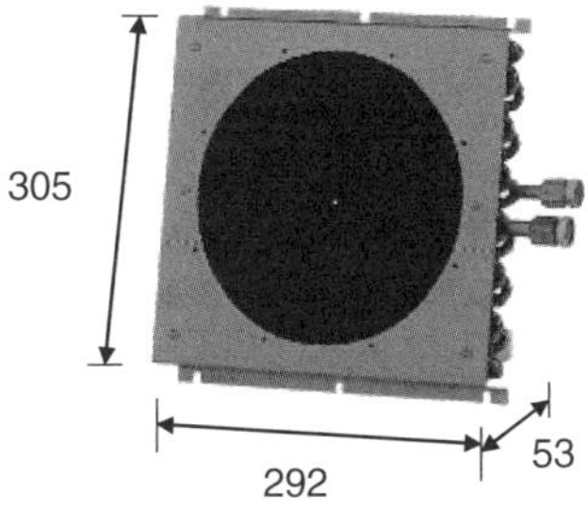

Fig. 12. Schematic of liquid-to-air heat exchanger.[28] The dimensions are in mm.

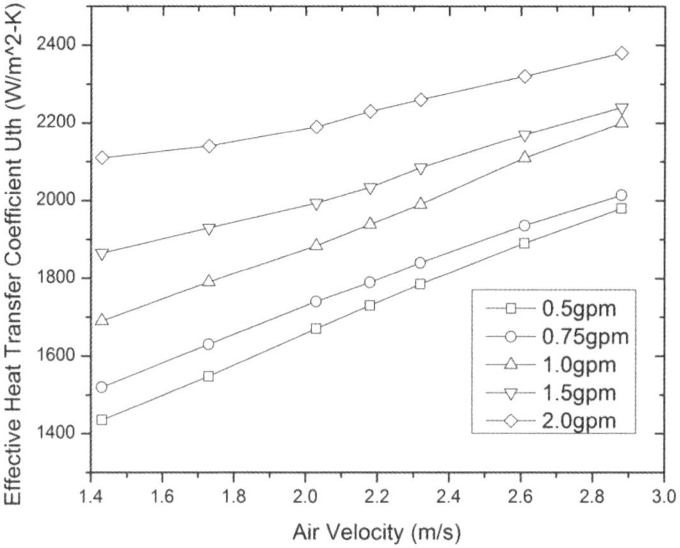

Fig. 13. Effective heat transfer coefficient vs. air velocity and water flow rate.

The schematic of the liquid-to-air heat exchanger (Lytron C6310) in the bottom bay of the cabinet is illustrated in Figure 12.[28] It is necessary to characterize the overall heat transfer coefficient of the heat exchanger to select the operating parameters. The standard counter-flow log mean temperature difference (LMTD) is widely used to determine the temperature driving force for heat transfer in flow systems:[29]

$$LMTD = \frac{Q}{F_{HX} U A_{HX}} = \frac{[(T_{a,i} - T_{w,o}) - (T_{a,o} - T_{w,i})]}{\ln[((T_{a,i} - T_{w,o}) / (T_{a,o} - T_{w,i})]} \tag{4}$$

where $T_{w,i}$, $T_{w,o}$ are the water inlet and outlet temperatures, respectively, and $T_{a,i}$, $T_{a,o}$ are the air inlet and outlet temperatures respectively. U_{HX} is the overall heat transfer coefficient of the heat exchanger. The correlation of overall heat transfer coefficients and air flow rates under different water flow rates is depicted in Figure 13. It can be seen that U_{HX} increases with air flow rate across the heat sink for a fixed water flow rate. The same trend holds when the water flow rate increases under a fixed air flow rate.

5. Experimental Studies with the Cabinet Test Vehicle

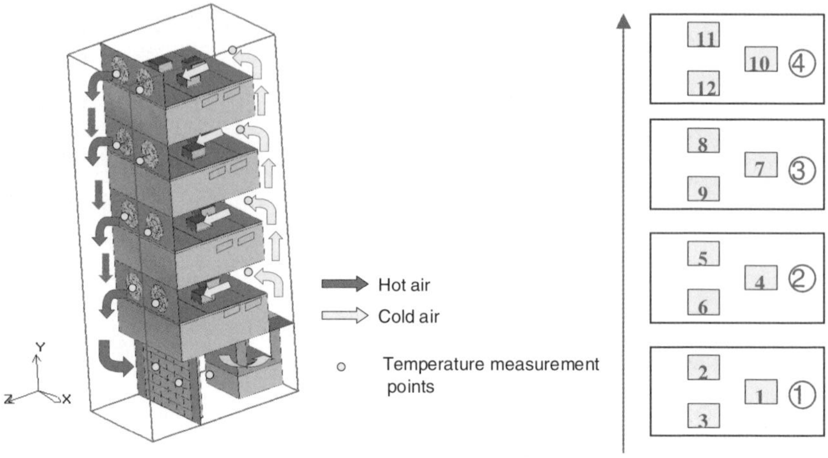

(a) (b)

Fig. 14. Temperature measurement points (a) air, (b) chip.

Experiments were conducted to evaluate the thermal performance of test vehicle with hybrid cooling using the approach described above. The air temperature measurement points and chip indices are shown in Figure 14. The number inside the circle refers to enclosure numbering and the number inside the rectangular refers to chip index. There are two thermocouples at the front of the heat exchanger (about 10 mm away from the front surface of the heat exchanger) to measure the inlet air temperature across the heat exchanger. Similarly, two thermocouples are put at the back of the heat exchanger to measure the backside air temperature. The averaged values from each pair of thermocouples are taken as the measurement values. Similarly, the average of the two thermocouples at the outlet of each enclosure is taken as the outlet air temperature of the enclosure.

5.1. *Measurement Uncertainty*

Error sources for the temperature measurement include the calibration uncertainty due to the thermocouple and uncertainties due to curve fitting. The latter is estimated to be within ±0.2°C. The uncertainty of T-type air temperature measurement is estimated to be within ±0.25°C from the calibration curve of manufacturer. A slightly higher uncertainty of K-type thermocouple measurement is estimated, which is within ±0.5°C. Combining these effects gives an uncertainty of ±0.45°C for T-type thermocouple measurement and ±0.75°C for K-type thermocouple measurement. The power dissipation is determined from the product of the voltage and current measured at the heating circuit inside the thermal test dies. In this experiment, a constant current is provided for each thermal test die, and the readings (products of the voltages and currents) from DC power supplies are taken as the measurements of power dissipations of test dies. Constant current source is better than constant voltage source in this case, due to voltage drop across external wires and solders. By measuring the voltage across the thermal test die, the power to the thermal test die can be obtained. The current measurement has ±0.2 % uncertainty, as indicated by the product manual and validated by calibration. For the voltage measurement, an uncertainty of ±0.3% results for the 30 V case. These uncertainties cause a ±0.6% uncertainty in power input measurement. Combining the uncertainties in temperature diodes and heat dissipation gives the uncertainties of total temperature measurement of chip at around 2%. The inlet chilled water temperature oscillates within ±0.56°C.

5.2. *Temperature Distribution Across the Thermal Test Die*

The temperature rise (over ambient of 21.5°C) of the thermal test die on the left of the second enclosure, monitored by six sets of diodes for a representative test case are shown in Figure 15. The heat load of each chip is 20 W, and water flow rate through the heat exchanger is 5.678 l/min (1.5 gpm). Almost a uniform temperature distribution is achieved, with the highest temperature occurring at the center of the test die as expected. The variation is about 1.3%, which indicates that the temperature at a single point is sufficient to characterize the chip junction temperature. Unless stated otherwise, the chip junction temperature mentioned in the following is the monitored value by the five-series of diodes at the center of the package (marked by the square sign in Figure 15).

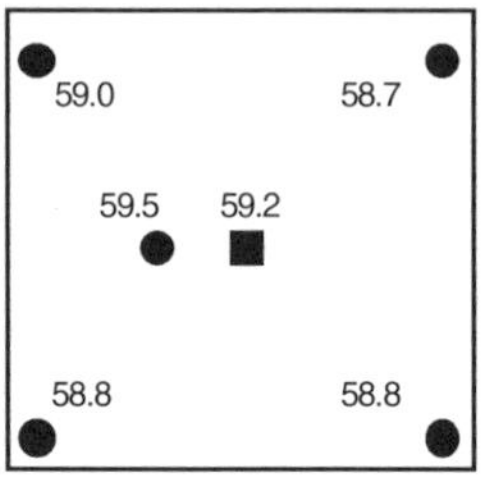

Fig. 15. Chip temperature distributions.

5.3. *Single-Sided Forced Air Convection Cooling (SAC)*

The control valve to the micro-channel coolers is switched off so that there is no water flow. The heat generated within the package under these conditions will primarily be dissipated through the top and out of the test vehicle walls. Approximately 40 minutes are needed for the system to reach steady state for each test case.

5.3.1. *Effect of Flow Rate Through Heat Exchanger*

The effect of water flow rates through the heat exchanger at the bottom bay on the chip junction temperature was investigated under different heat loads. Three different flow rates (3.028 l/min, 5.678 l/min, and 9.464 l/min) are considered. There is no power to the TEC modules for these experiments, so only forced air convection is considered. The correlation between the chip junction temperatures rise over cabinet internal ambient (13.5°C) and flow rates is illustrated in Figure 16 under chip load of 10 W. It can been seen that the chip temperature rise decreases with the increase of water flow rate, with a maximum reduction of 3.5% when the water flow rate increases from 3.028 l/min to 9.464 l/min. It is estimated that significant heat transfer (up to 430 W) occurs into the cabinet from the ambient by knowing room ambient temperature (e.g. 21.5C), average internal air temperature, and cabinet frame panel thickness and thermal properties. The change of water flow rate does not change the air temperature inside the system significantly due to this heat transfer. The effect of the water flow rate on the chip junction temperature is therefore not significant. More significant reduction of chip temperature via increasing the water flow rate is expected by using increased cabinet insulation, which would correspond to a situation not encountered usually in practice.

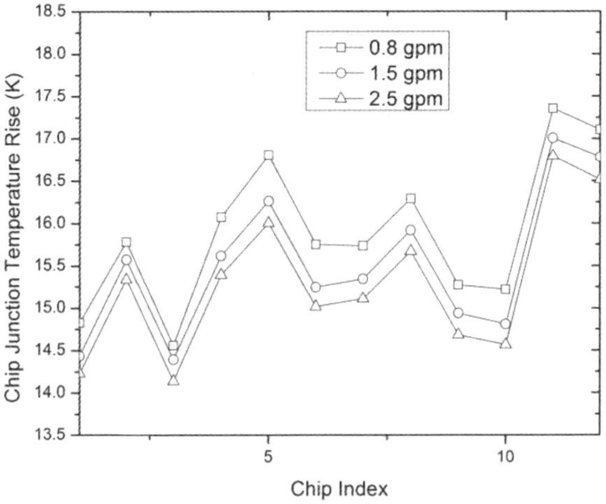

Fig. 16. Chip junction temperature rise vs. flow rate.

5.3.2. *Effect of TEC*

It is interesting to investigate how much benefit the TEC modules can bring to the forced air convection cooling. The TEC current is fixed at 0.8 A. The water flow rate to the heat exchanger is 3.028 l/min (0.8 gpm), and the heat load is 20 W per chip. No water flows through the micro-cooler so that SAC is considered. Figure 17 depicts the chip junction temperature rise. A reduction of up to 12 K or 20.4% is achieved with TEC for the current test case.

The effect of the TEC current on the heat dissipation from the package was also investigated. The same test case as above is repeated for three different currents (0.3 A, 0.8 A, 2.0 A). Figure 18 shows the chip junction temperature rise. Lowest chip junction temperatures are achieved for 0.8 A, with the second lowest at 2.0 A, and only 3°C is reduced under 0.3 A. A current of 0.2 A is insufficient to dissipate 20 W from the chip, and a current of 2.0 A is beyond the optimal operation point of the TEC.

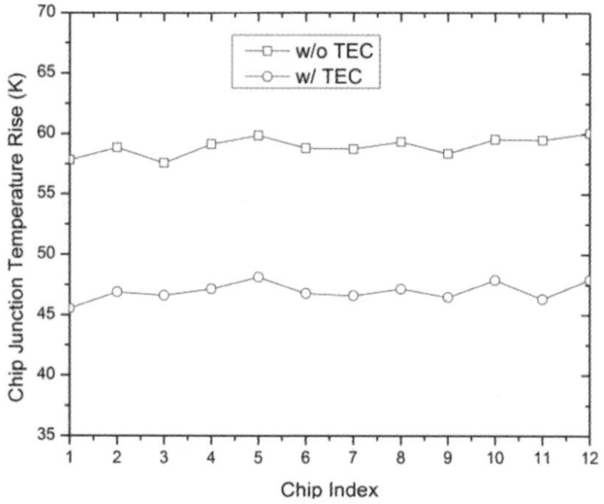

Fig. 17. Chip junction temperature rise vs. TEC.

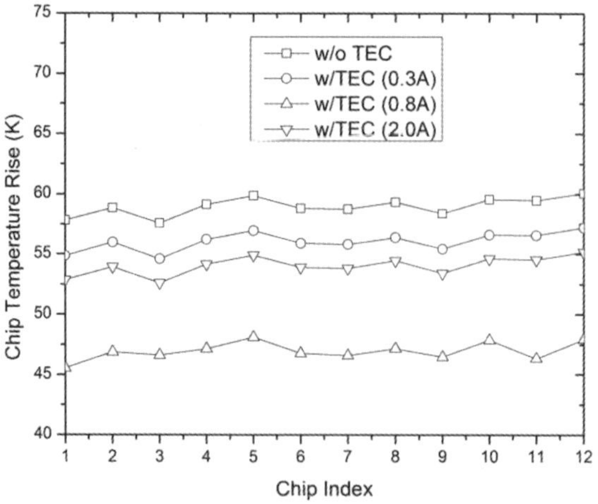

Fig. 18. Chip junction temperature rise vs. TEC current.

5.4. *Single-Sided Water Cooling (SWC)*

The fans and blower are powered off and the control valve to the heat exchanger is switched off. The control valve to the micro-coolers is also switched off, so that the generated heat will primarily be dissipated through the bottom of the package. Since the water inlet temperature to each micro-cooler is the same, all of them are expected to have the same performance. Thus only the chip package on the left of the third enclosure is tested. The chip junction temperature rise under various flow rates, shown in Figure 19, decreases weakly with increase of the water flow rate. The reason for this is that the thermal resistance of the micro-cooler is less than 8% of the total thermal resistance of the package.

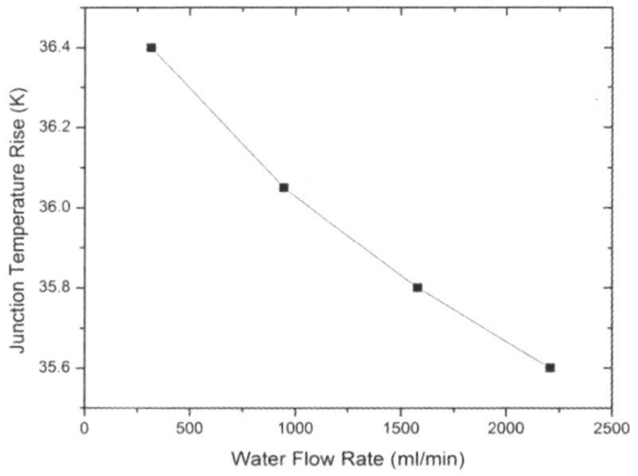

Fig. 19. Chip junction temperature rise vs. water flow rate.

5.5. *Double-Sided Cooling (DSC)*

Both the control valves to the heat exchanger and micro-coolers are switched on, so that double-sided cooling takes effect. The results by DSC are compared to those by SAC and SWC under different scenarios.

5.5.1. *Effect of Double-Sided Cooling*

To investigate the benefit of DSC, a test case with a heat load of 20 W per chip is considered. The water flow rate through the heat exchanger is 3.028 l/min (0.8 gpm), and the water flow rate through micro-cooler is

2.208 l/min (35 gph). No TEC module is powered on for this case. The chip junction temperature rise achieved by DSC is shown in Figure 20. The results by SAC and SWC are also shown for comparison. It can be seen that a reduction up to 72% in chip junction temperature is achieved with SWC, compared to SAC. The chip junction temperature is further reduced with DSC to 74%, compared to SAC, and 6.2% compared to SWC. Although the current benefit of DSC over SWC is not significant, e.g. only about 1 K reduction for the chip junction temperature, DSC may still be a good option when the cold water temperature is close to ambient temperature or the test vehicle is well insulated from the ambient.

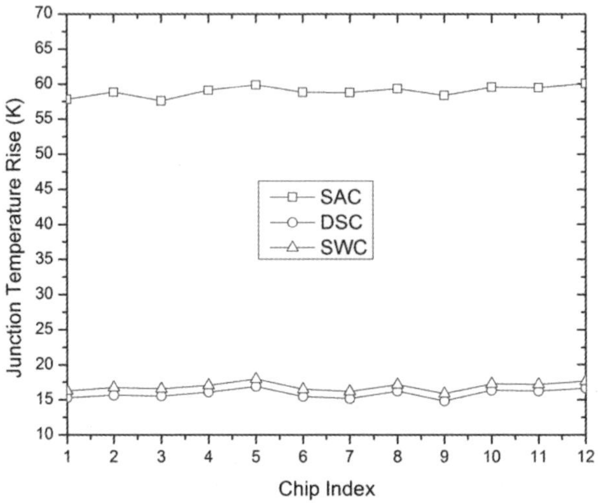

Fig. 20. Chip junction temperature rise vs. cooling methods.

5.5.2. *Effect of TEC*

The effect of TEC on double-sided cooling is also investigated. The same test as above is conducted, except that the TEC modules are powered with an electric current of 0.8 A. Figure 21 shows the chip junction temperature rises with DSC when the TEC modules are switched off and on, respectively. A reduction of 0.9 K is achieved when the TEC modules are switched on. A further reduction is possible for an optimal operating current to the TEC modules.

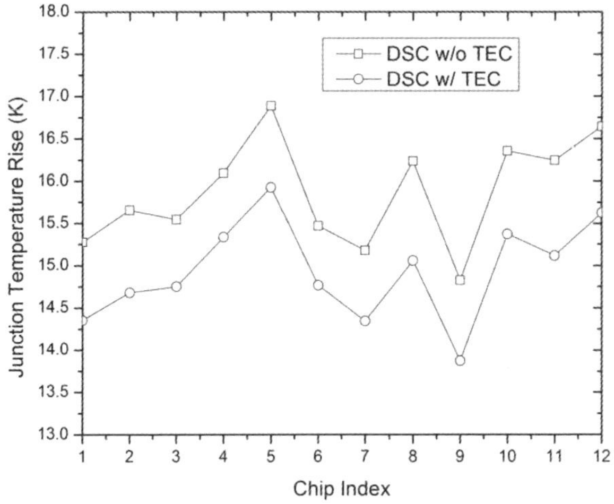

Fig. 21. Chip junction temperature rise of DSC vs. TEC.

5.6. *Transient Measurement*

A transient test was conducted for the third enclosure counted from the bottom. The ambient air enters the enclosure, and flows across the three packages and exits the enclosure through the two exhaust fans. The chip packages are initially at ambient temperature (21.5°C), and then the fans and chips are powered on (20 W/chip). For the demonstration, only SAC with TEC was investigated during this test. The current to the TEC modules is set at 0.8 A. Figure 22 depicts the chip junction temperature variation over time. The system reaches steady state after 150 seconds.

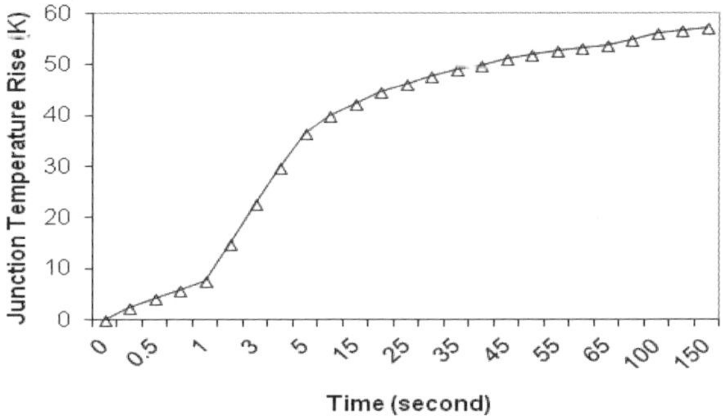

Fig. 22. Chip junction temperature rise over time by SAC.

6. Conclusion

A hybrid double-sided approach combining micro-channel liquid cooling, thermoelectric cooling, and forced air convection is investigated via a test vehicle for the thermal management of power electronic cabinets. Significant reductions of junction temperatures are achieved for chip packages by utilizing double –sided cooling technique. Further reduction can be achieved by adding TEC cooling devices to the existing cooling solutions. The impact of various parameters such as water flow rate and TEC current on chip junction temperatures is investigated.

The effect of the water flow rates through the heat exchanger on the chip junction temperature is not significant. Only a maximum reduction of 3.5% is achieved when the water flow rate increases from 3.028 l/min to 9.464 l/min. This is because up to 430 W is transferred from the ambient through the cabinet walls. It was also observed that the microchannel water flow rate does not affect the chip junction temperature significantly. This is because the thermal resistance of the micro-cooler is less than 8% of the total thermal resistance of the package.

The chip junction temperatures were decreased by 74% and 6.5% by double-sided cooling, compared to single-sided air convection and single-sided water cooling, respectively. The advantage of double-sided cooling over single-sided cooling will be larger when the chilled water temperature is higher, e.g. close to room temperature, which is more favorable when considering energy efficiency of cooling. The chip junction temperatures decreased up to 20.4% when the TEC modules are switched on for single-sided air cooling case. Smaller reduction (e.g. 0.9°C) was observed for double-sided cooling cases. The TEC performance mainly depends on the current to the TEC, and the cold and hot side thermal resistances. For a certain current to the TEC and cold side thermal resistance, smaller hot side thermal resistance will significantly improve the thermal performance of TEC. For double-sided cooling, the thermal resistance of the hot side (i.e. that of the heat sink and TIM between the heat sink and TEC) is about 1.2 K/W under the air flow velocity of 2.0 m/s. Decreasing this thermal resistance will make double-sided cooling with TEC superior to single-sided cooling.

Acknowledgments

This work was supported by the Office of Naval Research under contract N00014-04-1-0335, monitored by Dr. Mark Spector. I would like to thank those who provided valuable comments on the early drafts.

References

1. ITRS, International Technology Roadmap for Semiconductors 2007, 1 (2007).
2. D. B. Tuckerman and R. F. W. Pease, *IEEE Electron Device Letter*, Vol. 2, (1981), p. 126.
3. The Uptime Institute Inc, 2005-2010 Heat Density Trends in Data Processing, Computer Systems, and Telecommunications Equipment, V2.1 (2010).
4. ASHRAE, American Society of Heating, Refrigeration and Air Conditioning Engineers, 1 (2005).
5. M. Ohadai and J. Qi, Semiconductor Thermal Measurement and Management Symposiums. 1 (2004).
6. L. Burton, Multiscale Thermal Modeling Methodology for High Power Electronic Cabinet, (Georgia Institute of Technology, Atlanta, 2007), p. 18.
7. C. D. Patel, C. E. Bash, C. Belady, L. Stahl, and D. Sullivan, *IPAC'01.* (2001).
8. T. Ogushi and G. Yamanaka, *Proc. of 68th JSME.* (1990).
9. A. Latrobe, M. Cadre, and J. P. L. Jannou, *IEEE INTELEC'86* , 1 (1986).
10. T. Kowalski and A. Radmehr, IEEE 16th Semiconductor Thermal Measurement and Management Symposium, (2000).
11. N. Rolander, An Approach for The Robust Design of Air Cooled Data Center Cabinets, (Georgia Institute of Technology, Atlanta, 2005), p. 50.
12. J. Rambo and Y. Joshi, *ASME Summer Heat Transfer Conf. & InterPACK'05,* (ASME, Scan Francisco, 2005).
13. R. Schmidt, *Electronics Cooling*, Vol. 11, 3 (2005).
14. S. R. LaPlante, N. Aubry, D. Porter, *et al.*, *Electronics Cooling,* Vol. 12, 4 (2006).
15. Z. Zuo, L. R. Hoover, and A. L. Phillips, *IEEE Transaction on Component and Packaging Technology*. Vol. 25, 4 (2002), p. 629.
16. R. E. Simons and R. C. Chu, *IEEE 16th SEMI-THERMTM Symposium*. (IEEE, San Jose, 2000).
17. T. Kobayashi, M. Nakamura, T. Ogushi, A. Iwamaru, and M. Fujii, *Heat Transfer—Asian Research*, Vol. 43, (2001), p. 267.
18. J. Wei, J. of Fujitsu Science Technology, Vol. 43, (2007), p. 122.
19. C. Gillot, C. Schaeffer, C. Massit, and L. Meysenc, *IEEE Trans. on Components and Packaging Technologies,* Vol. 24, (2001), p. 698.

20. Y. Pang, Assessment of Thermal Behavior and Development of Thermal Design Guidelines for Integrated Power Electronics Modules, (Virginia Tech, Blacksburg, 2005), p. 64.
21. I. Mudawar, IEEE Trans. on Components and Packaging Technologies, Vol. 24, (2001), p. 122.
22. T. Steiner and R. Sittig, The 12th International Symposium on Power Semiconductor Devices and ICs, (IEEE, Pula, Croatia, 2000).
23. J. Yin, J. D. V. Wyk, W. G. Odendaal, and Z. Liang,, *The 39th IEEE/IAS Annual Meeting*, (Seattle, USA, 2004).
24. B. C. Charboneau, F. Wang, J. D. v. Wyk, D. Boroyevich, Z. Liang, and E. P. Scott, *The 40th IEEE/IAS Annual Meeting*, (Hongkong, 2005).
25. T. R. Kuphaldt, All About Circuits: Volume I - DC, (2003).
26. G. Nelson, Development of An Experimentally-Validated Compact Model of A Server Rack, (Georgia Institute of Technology, Atlanta, 2007), p. 43.
27. Q. Nie, Experimentally Validated Multiscale Thermal Modeling of Electronic Cabinet, (Georgia Institute of Technology, Atlanta, 2008), p. 135.
28. LYTRON, Copper 6000 Series Heat Exchangers, (LYTRON Inc., 2007).
29. F. Incropera, D. Dewitt, T Bergman, *Fundamentals of Heat and Mass Transfer,* (John Wiley & Sons Inc, 2006), p. 711.

CHAPTER 14

ENERGY EFFICIENCY AND RELIABILITY RISK MITIGATION OF DATA CENTERS THROUGH PROGNOSTICS AND HEALTH MANAGEMENT

Jun Dai, Michael Ohadi*, and Michael Pecht

Center for Advanced Life Cycle Engineering (CALCE)
University of Maryland, College Park, MD, USA 20742
**ohadi@umd.edu*

Cooling systems (primarily air conditioners) account for a large share of energy consumption in data centers. In 2009, about 40% of the energy consumed by data centers was for cooling. One method that reduces energy consumption of data centers is known as free air cooling, where ambient air is used to cool equipment directly, thereby reducing the energy consumed for cooling and conditioning. However, the implementation of free air cooling changes operating environment, including temperature and humidity, and may have a significant impact on performance and reliability. Maintaining the high availability of data centers requires reliability methods that provide early warning of impending failures, identify failure locations, and isolate failure causes, while taking into account the life cycle conditions during system service. This paper presents an analysis of the potential risks arising from free air cooling based on the expected operating conditions. A prognostics-based approach to assess and mitigate those risks is then presented. A case study will be used to show the implementation process.

1. Introduction

Data centers are the buildings, facilities, and rooms that contain data servers and telecommunication, cooling, and power equipment [1]. The continuous growth of the communication and data storage industries has resulted in a rapid increase in the energy consumption of data centers, to the extent that data centers now represent an important share of national

* Corresponding author.

energy consumption and carbon footprint. In recent years, there has been a substantial increase in the power densities of data centers, with the power density doubling nearly every 2.5 years as an expected corollary of Moore's Law. The global energy consumption of data centers reached 237 terawatt hours (TWh)† in 2010, accounting for about 1.3% of the world's electricity use [2]. In the U.S. alone, data centers consumed about 76 TWh of electricity in 2010, accounting for about 2% of total U.S. electricity consumption [2]. Furthermore, more than 40% of energy consumption is spent on cooling in a typical data center [3], as shown in Figure 1.

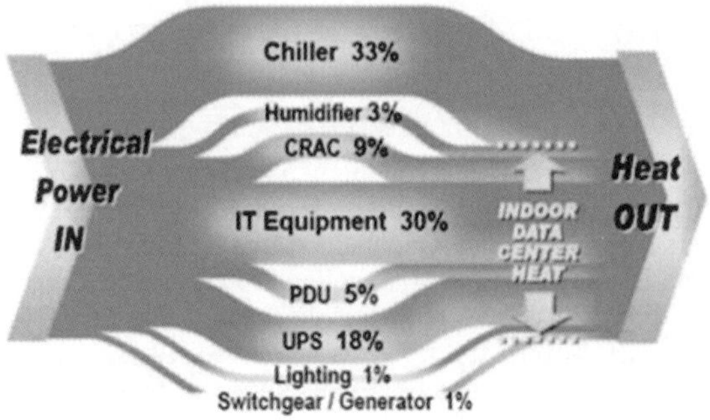

Fig. 1. Power consumption distribution in data centers [3].

A 2012 biannual survey by Emerson Network Power polled members of the Data Center Users' Group, an association of data centers, IT, and facility managers, about the greatest issues facing data centers. Energy efficiency was cited as the number-one concern, followed by availability and infrastructure monitoring as the second and third concerns, respectively [4]. The cost of the electricity necessary to remove the heat generated from server racks has continued to rise to the point where the four-year energy costs of operating many data centers exceed their purchase price. Free air cooling, which uses outside ambient air as the primary cooling medium, saves energy for the cooling of servers and other electronics. Numerous successful examples of free air cooling have

† This value is the midpoint of the low bound and high bound of the energy consumption estimation [5].

demonstrated substantial energy savings. As a demonstration project, Intel utilized free air cooling for an IT data center for ten months, saving about 2.87 million USD in the operation of a 10-MW data center (a 67% savings in total energy). Microsoft and Google used free air cooling to replace traditional chillers in some of their new data centers in Europe in 2009 [6],[7], and Facebook also implemented free air cooling in its data centers, resulting in significant energy savings [8].

A systematic examination of the impact of free air cooling on the performance and reliability of telecommunication equipment is needed. The implementation of free air cooling changes the operating environment, such as the temperature and humidity, and may have a significant impact on reliability. Maintaining the high availability of data centers requires reliability methods that provide information about impending failures, identify failure locations, and isolate failure causes, while taking into account the life cycle conditions during system service.

The Implementation of Free Air Cooling. Free air cooling typically involves the use of airside economizers when the ambient air conditions are within the required operating condition ranges. Figure 2 shows an example of an airside economizer [9]. An airside economizer consists of sensors, ducts, dampers, and containers. An airside economizer supplies the appropriate volume of air within the right temperature and humidity range (set by operators) to satisfy cooling demands. This set temperature range can be based on the recommended operating ranges given by published standards, such as the American Society of Heating, Refrigerating, and Air-Conditioning Engineers (ASHRAE) 2011 Thermal Guidelines [10], Telcordia GR-63-CORE [11], and Telcordia GR-3028-CORE [12]. If the outside ambient air conditions are within this range, or if they can be brought within the range by mixing cold outside air with warm return air, then outside air can be used for data centers for cooling via an airside economizer fan. If the conditions achievable by economization and mixing of outside air are outside the pre-set ranges, a cooling and heating control system will be used to adjust the supply air conditions to within the pre-set ranges. The energy saving potential of free air cooling varies with the supply air temperature and the setting of the room temperature.

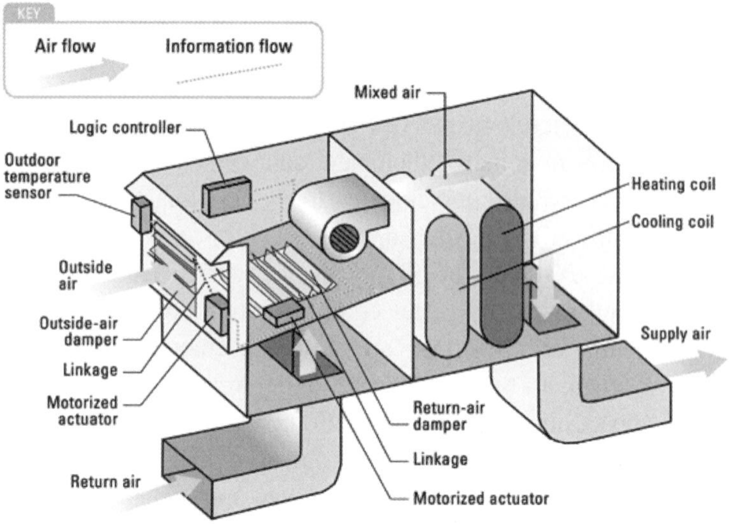

Fig. 2. Schematic of airside economizer with airflow path [9].

Previous Research. Research on free air cooling in the past has primarily focused on maximizing energy savings. Bulut and Aktacir [13] studied free air cooling in a case study in Istanbul, Turkey. They found that the seasonal transition months (April, May, September, and October) usually had the highest energy saving potential. Lazaro *et al.* [14] tested the phase change material (PCM) air real-scale heat exchangers used in free air cooling and then developed a design based on the experimental results. Dovrtel *et al.* [15] incorporated weather forecasts into the free air cooling control system and then optimized the regime of operating conditions to improve energy efficiency. However, these studies did not address possible risks from free air cooling. Dai *et al.* [16], [17] implemented prognostics and health management (PHM) techniques to mitigate the reliability risks of telecom equipment under free air cooling conditions, but they only detected anomalies and did not predict failure.

Some companies have researched the reliability of equipment under free air cooling and obtained preliminary results. For example, Intel and Dell claimed that there was only a small difference between the equipment failures under free air cooling and those under traditional air-conditioned cooling [5], [18]. However, the Intel project duration of ten months was short compared to the typical five- to ten-year lifetimes of telecom equipment in data centers. Dell used constant high temperature and humidity (40°C and 85% RH), but the operating conditions under free air cooling are more likely to be a thermal cycling environment. As a result, previous research lacks conclusive evidence whether or not free air cooling impacts the reliability of telecom equipment in data centers.

In this chapter, we analyze the potential risks arising from free air cooling based on the expected operating conditions of data centers. The paper then describes the development of a prognostics-based approach to detect anomalies and predict the equipment failures of data center equipment already in operation. A case study will be used to show the implementation process.

2. Potential Reliability Risks Associated with Free Air Cooling

Generally, there are two types of operating condition requirements specified for data centers: 1) the recommended conditions, which serve as guidance for data center operators to maintain high reliability while practicing energy-efficient measures; and 2) the allowable conditions, in which IT manufacturers test their equipment to verify that the given equipment functions within the specified environmental operating conditions [19]. The 2011 version of the ASHRAE thermal guidelines for data centers [10] sets the recommended operating range at 18–27°C (64.4–80.6°F), and 5.5°C dew point (DP) to 60% relative humidity (RH) and 15°C DP. The allowable condition ranges depend on the specified data center classes[‡], as shown in Figure 3. Details of the corresponding environmental specifications are shown in Tables 1 and 2. In most conventional data centers, the operating conditions are maintained within 20–25°C [20] and 35% RH–55% RH [21]. However, to maximize the energy savings, the operating conditions of free air cooling may deviate considerably from the ASHRAE-recommended conditions. These changes in operating conditions pose reliability risks to data center equipment. Specifically, the increased operating temperature,

[‡] Additional details on ASHRAE data center classifications can be found in [10].

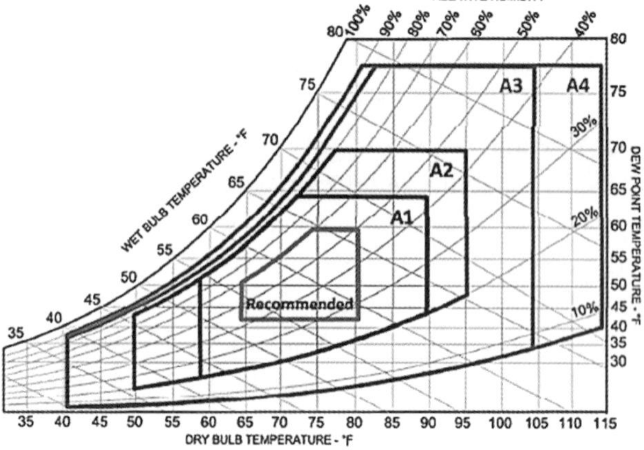

Fig. 3. ASHRAE psychrometric chart for data centers [10].

Table 1. Environmental specifications for 2011 ASHRAE thermal guideline classes [10].

Class	Product Operation			Product Power Off		
	Temperature	Humidity range	Maximum dew point	Temperature	Humidity range	Maximum dew point
			Recommended			
A1 to A4	18°C to 27°C	5.5°C DP to 60% RH and 15°C DP				
			Allowable			
A1	15°C to 32°C	20% to 80% RH	17°C	5°C to 45°C	8% to 80% RH	27°C
A2	10°C to 35°C	20% to 80% RH	21°C	5°C to 45°C	8% to 80% RH	27°C
A3	5°C to 40°C	-12 °C DP & 8% RH to 85% RH	24°C	5°C to 45°C	8% to 85% RH	27°C
A4	5°C to 45°C	-12°C DP & 8% RH to 90% RH	24°C	5°C to 45°C	8% to 90% RH	27°C
B	5°C to 35°C	8% to 80% RH	28°C	5°C to 45°C	8% to 80% RH	29°C
C	5°C to 40°C	8% to 80% RH	28°C	5°C to 45°C	8% to 80% RH	29°C

Table 2. Class specifications of data centers based on the 2011 ASHRAE thermal guidelines [10]

Class	Definition
A1	Typically a data center with tightly controlled environmental paramcters (dew point, temperature, and relative humidity) and mission-critical operations; types of products typically designed for this environment are enterprise servers and storage products.
A2	Typically an information technology space or office or lab environment with some control of environmental parameters (dew point, temperature, and relative humidity); types of products typically designed for this environment are volume servers, storage products, personal computers, and workstations.
A3/A4	Typically an information technology space or office or lab environment with some control of environmental parameters (dew point, temperature, and relative humidity); types of products typically designed for this environment are volume servers, storage products, personal computers, and workstations.
B	Typically an office, home, or transportable environment with minimal control of environmental parameters (temperature only); types of products typically designed for this environment are personal computers, workstations, laptops, and printers.
C	Typically a point-of-sale or light industrial or factory environment with weather protection and sufficient winter heating and ventilation; types of products typically designed for this environment are point-of-sale equipment, ruggedized controllers, or computers and personal digital assistants (PDAs).

uncontrolled humidity, and lack of contamination control may cause some adverse effects on the data center equipment, as described below.

Increases in operating temperatures beyond the recommended values can lead to a decrease in lifetime and performance of the components used in a data center. For example, the recommended temperature for batteries in data centers is usually 25°C. Increased temperatures may accelerate the corrosion of bipolar plates, which reduces battery reliability and battery lifetime. The battery lifetime is expected to drop by 50% when the operating temperature increases 50% percent from 25°C [22]. On the other hand, decreased temperature below the recommended values may reduce battery capacity, which is undesirable as well.

The increased humidity range under free air cooling conditions also poses risks to telecom equipment depending on the extent to which an increase exceeds the recommended limits in the ASHRAE thermal guidelines [10] or Telcordia standards [11], [12]. In the Intel case, the humidity ranged from 4% to more than 90% [5]. This uncontrolled humidity may make some failure mechanisms more active. For example, conductive anodic filament (CAF) formation and creep corrosion are accelerated by high humidity, while electrostatic discharge (ESD) is more common in low humidity. This may impact telecom equipment reliability and lifetime.

Free air cooling may also cause chemical contamination in a data center, leading to adverse chemical effects (e.g., obstruction of the cooling airflow and copper-induced creep corrosion), mechanical effects (e.g., electrical short-circuiting and increased friction), and electrical effects (e.g., circuit impedance changes) on data center equipment. There are two kinds of contaminants: particulate contaminants, such as smoke and dust; and gaseous contaminants, such as SO_2. Particulate contaminants can be minimized by filters, but gaseous contaminants are difficult to control. It has been reported that sulfur-bearing gases increase the failure rate of data center equipment [23], which will become worse when data centers are located in an industrial area or an area with other sources of pollution.

3. Risk Assessment and Mitigation by Prognostics and Health Management

Predictive methods to assess and mitigate the risk associated with the use of free air cooling are useful for data centers, particularly for data centers already in operation, since it is not practical to re-qualify equipment for use under free air cooling conditions. There are at least three reasons why re-qualifying equipment is impractical. First, it is not practical to interrupt equipment services for test purposes, especially in mission-critical data centers with 24/7 service requirements. Second, even if the test were attempted, the tested equipment would lose some of its useful life in the test process. This would offset the energy savings of free air cooling. Third, it is nearly impossible to gather an appropriate sample size of equipment already in operation for the tests. Prognostics and health management (PHM) can overcome these difficulties and be used as a retrofitting technique to assess and mitigate risks in order to enable the implementation of free air cooling in data centers that were not initially designed for this cooling method [17].

3.1. PHM Approaches

Prognostics and health management assesses the reliability of a product (or system) under actual operating conditions [24]. It provides advance warning of failures, minimizes unscheduled maintenance, reduces equipment downtime, and improves product design. PHM includes the physics-of-failure (PoF) approach, the data-driven approach, and an approach that combines both (the fusion approach) [25].

The PoF approach utilizes information of a product's life-cycle loading and analyzes failure mechanisms to assess product reliability [24]. This approach monitors the product's loading conditions and then calculates the cumulative damage due to the identified failure mechanisms of the product under the operating conditions. Early warning of failures can be provided based on the calculated cumulative damage of the product. A flowchart of the PoF-based PHM approach is shown in Figure 4. This approach includes failure modes, mechanisms, and effects analysis (FMMEA); the details can be found in [26].

The first step in this approach is to identify the key components and functions based on an analysis of the product material properties and geometries. Then, the potential failure modes are identified based on the expected product lifecycle operating loading. The next step is to identify the potential failure causes, potential failure mechanisms, and the failure model. According to historical data, previous experiences, maintenance records, and other available information, the failure mechanisms can be prioritized based on their occurrence rates and consequences. All of the analysis processes need to be documented. Then, the monitoring parameters and locations of the high priority failure mechanisms are chosen, and the fuse and canary devices[§] are used to monitor these parameters and locations, which can be used to estimate the product's remaining useful life (RUL). In addition, data reduction and load feature extraction from the operating loading conditions are used to calculate the stress/strain and damage and then estimate the product RUL.

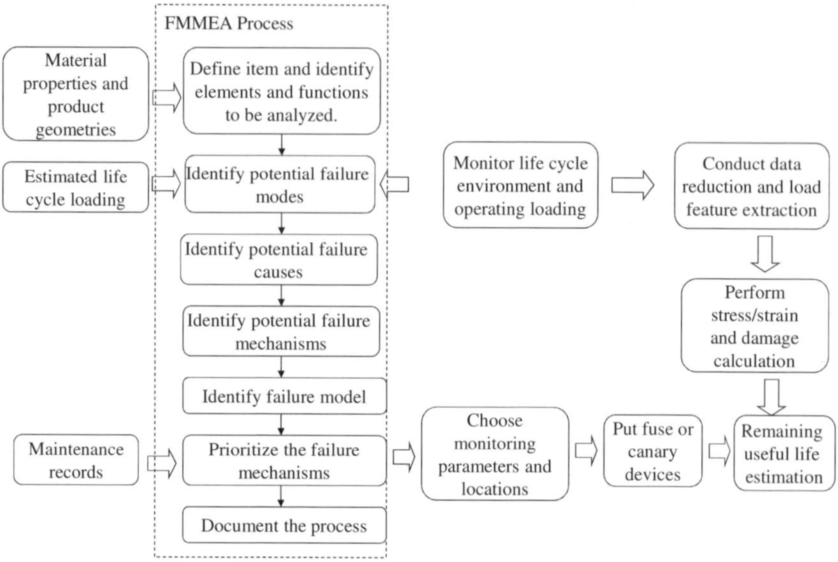

Fig. 4. PoF-based PHM approach [24].

The data-driven approach uses mathematical analysis of current and historical data to detect anomalies and predict failures [17], [27], [28], [29]. This approach begins with a functional consideration that includes an analysis of the system-level priorities, feasibility, and limitations. Then, data acquisition techniques are considered to monitor and collect data on system performance. When the raw data are collected, data feature extraction is conducted to minimize the effect of data noise and scaling. These extracted data features are used to create a product health baseline and to indicate the product performance degradation. The failure

[§] Fuse and canary devices have the same failure mechanisms as the product, but they fail before the product.

criteria must be defined based on the data features, and then the monitored data trends can be used for anomaly detection (diagnostics) and RUL estimation (prognostics) based on the pre-defined failure criteria.

The fusion approach, which combines the PoF and data-driven models to overcome shortfalls in the individual approaches, is shown in Figure 5 [25]. The PoF approach identifies the root causes and failure mechanisms of system failure and then estimates the product RUL under different usage loadings. However, the implementation process for this approach is very complicated for large systems. The data-driven approach can estimate the product RUL based on data trends without understanding the failure mechanisms behind the product degradation and failure, but it is difficult to set a threshold for the data anomaly to indicate failure without knowledge of the failure mechanisms. The fusion approach can compensate for the disadvantages of each individual approach by estimating the RUL of a large system based on the monitored data trends and using the identified failure mechanisms to set a reasonable threshold to warn of impending failure at a certain level of anomalous behavior.

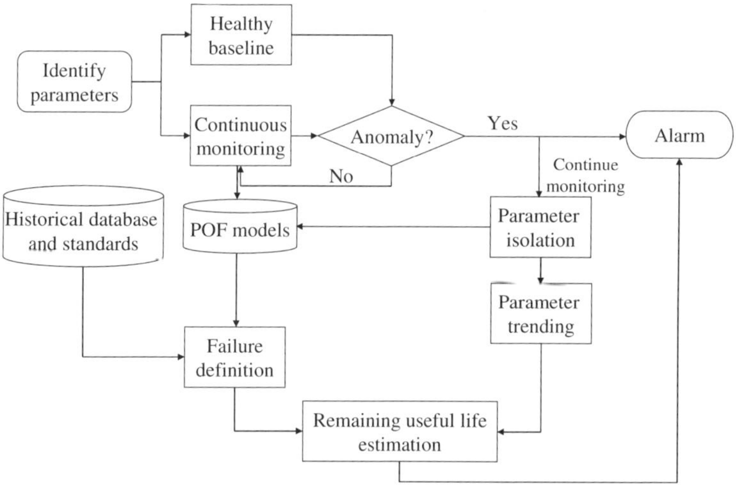

Fig. 5. Flowchart of the fusion approach with PoF and data-driven models [25].

3.2. A Prognostics-based Approach to Assess the Risks in Data Centers

In the present work, a prognostics-based approach is developed to assess and minimize the reliability risks from free air cooling, as shown in Figure 6. This approach starts with system function considerations to analyze which of the system performance parameters indicate the key system functions that need to be monitored. Then, the next step is to identify the product life cycle conditions with the implementation of free air cooling. Based on the identified life cycle condition, a failure modes, mechanisms, and effects analysis (FMMEA) process is conducted to identify the critical failure mechanisms and the associated failure subsystems. As mentioned earlier, several mechanisms may occur at a higher rate in free air cooling conditions if the humidity and temperature are not properly controlled: electrochemical migration, which often occurs in low relative humidity; conductive anodic filament (CAF) formation, which often occurs in high humidity; and creep corrosion, which often occurs at high humidity in the presence of low levels of sulfur-based pollutants. The failure subsystems associated with the critical failure mechanisms are considered as the key subsystems that are the

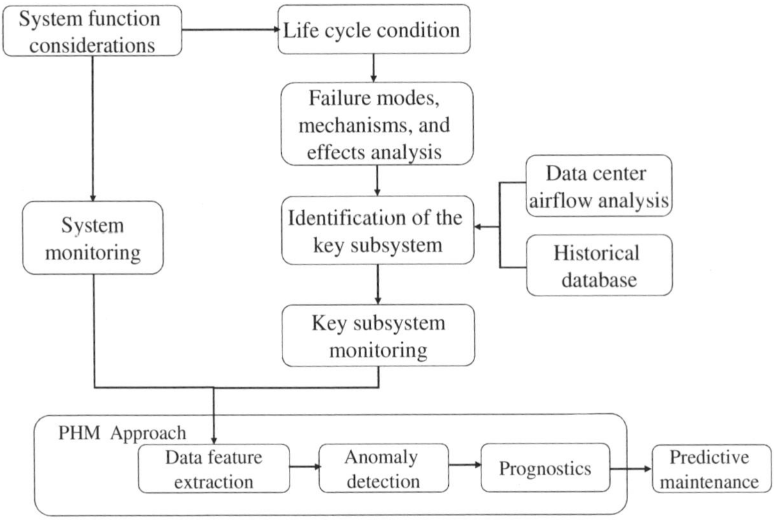

Fig. 1. A prognostics-based approach to mitigating the risks of free air cooling.

most likely to fail first in the system. In addition, the data center airflow analysis and the historical data can be also used to identify the key subsystems.

After the weakest components have been identified, as many parameters as possible (e.g., temperature, voltage, current, and resistance) need to be monitored. System-level parameters must be monitored based on functional considerations. For example, the main function of a power adapter is to provide output voltage to its corresponding telecom equipment, so its output voltage must be monitored, since this parameter may represent the degradation or even failure of the power adapter.

After the monitored data are collected, data feature extractions (e.g., normalization, or mean) are conducted to minimize the data noise. Then, the PHM approach can be implemented for anomaly detection and prognostics (RUL estimation) of the equipment. Predictive maintenance can be scheduled based on the RUL estimation.

3.3. Case Study

For the present study, the power adaptor of a Zonet ZFS 3015P switch was chosen, since this adaptor is widely used in offices and small enterprises. In addition, this power adaptor is a typical switched mode power supply (SMPS), which is usually temperature-sensitive, with failure rates doubling with every 10°C to 15°C temperature rise above 25°C [30]. An image of the power adaptor is shown in Figure 7.

#1—aluminum electrolytic capacitor; #2—output voltage supply

Fig. 2. Power adaptor of a Zonet ZFS 3015P switch.

There were two samples in the test. The power adaptor functions were first considered in order to analyze the system performance parameters that need be monitored. This adaptor provides voltage to the switch, so the output voltage (rated output of 9V) was the system performance parameter monitored. In this case, the power adapter was considered to have failed when its output voltage dropped by more than 10% from the rated value.

With the FMMEA analysis result of SMPS [31] and consideration of the measurement applicability, the voltage across a capacitor, V1 (shown as #1 in Figure 7), was monitored in the experiments. The output voltage, V_{out} (shown as #2 in Figure 7), was also monitored for the power adapter performance trends. The rated operating conditions of the power adapter are 0–40°C and 10%–90% RH. We used a condition of 95°C and 70% RH in the experiments to increase the rate of degradation in order to observe the failure of a new power adaptor within a reasonable time.

The monitored parameter trends for samples 1 and 2 are shown in Figures 8 and 9, respectively. Sample 1 failed at minute 501, and sample 2 failed at minute 389. For both samples it was observed that V_{out} decreased from about 9.3 V to around 3 V, and V1 decreased from about 80 V to about 40 V when the power adaptor failed.

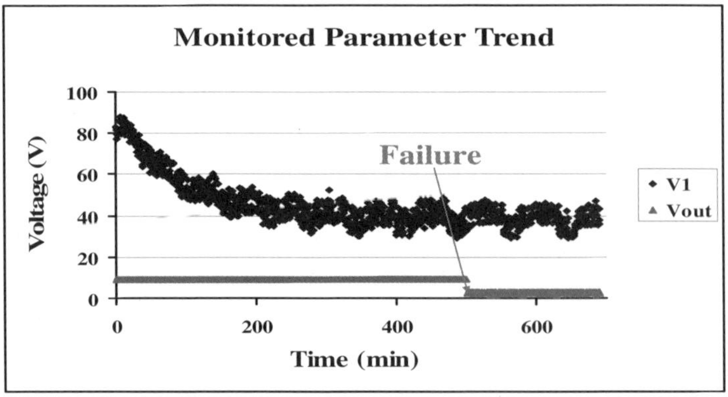

Fig. 8. Sample 1 monitored parameter trends.

The neural network methodology [32] was used to detect the anomaly, since it does not require degradation models and distribution assumptions. The first 20 data points were considered the healthy data and were

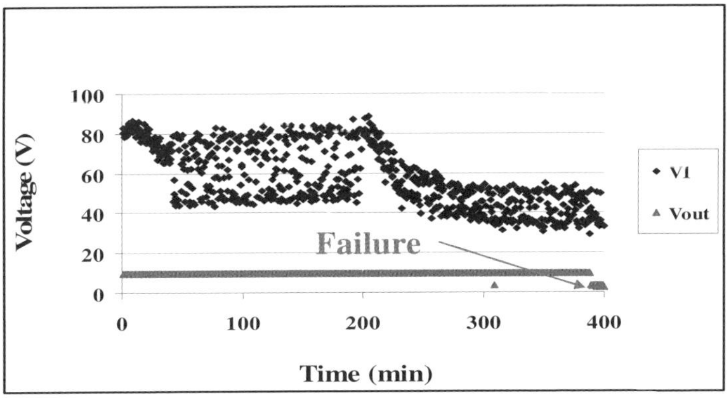

Fig. 9. Sample 2 monitored parameter trends.

selected as the training data to train the neural network. The training was stopped when the error precision between the expected V_{out} calculated by the neural network and the actual V_{out} was equal to or below 10^{-8}. An anomaly was considered to be detected when five consecutive data error precisions went beyond 10^{-4}. When an anomaly was detected, the data for 30 minutes before the anomaly detection (including the anomaly detection time point) were used to predict the failure by the exponential growth model. The anomaly detection and prognostics for samples 1 and 2 are shown in Figure10 and Figure 11, respectively.

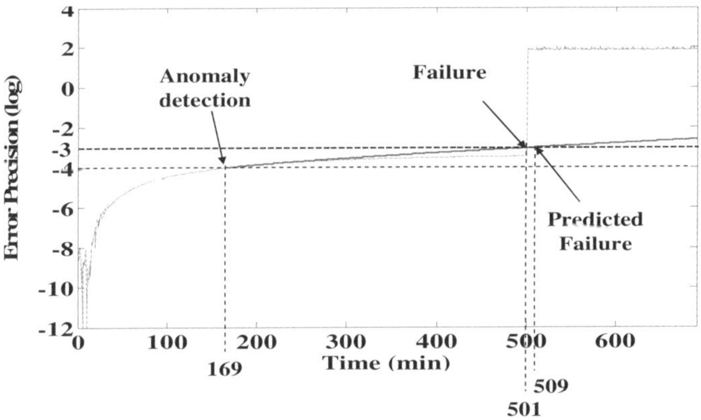

Fig. 10. Sample 1 anomaly detection and prediction with V1.

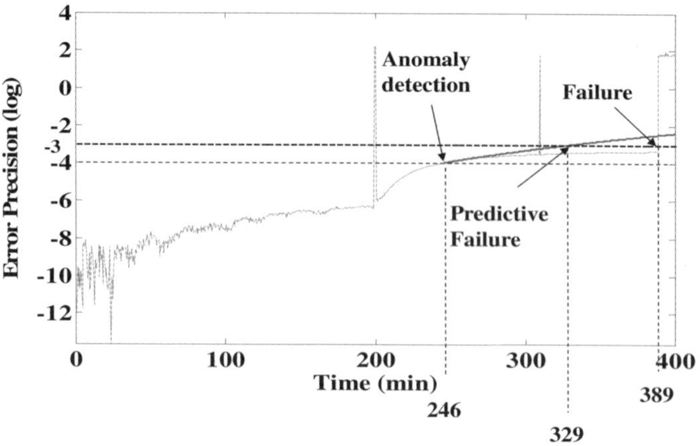

Fig. 11. Sample 2 anomaly detection and prediction with V1.

The anomaly in sample 1 was detected at minute 169, and failure was predicted to occur at minute 509. The anomaly in sample 2 was detected at minute 246, and failure was predicted to occur at minute 329. The actual failures of sample 1 and sample 2 occurred at minutes 501 and 389, respectively. This anomaly detection and prediction method provides early warnings of equipment failure. Additional findings and comprehensive experimental ranges are documented in [33].

4. Summary and Conclusions

Free air cooling is considered to be one of the most effective methods for reducing energy consumption in data centers. However, implementing free air cooling may change the operating environment, including the temperature, humidity, and contamination levels. Such changes in the operating conditions can in turn impact the performance and reliability of the electronics in a data center. This study demonstrated that prognostics-based methods provide a way to assess the health of equipment exposed to conditions such as free air cooling, as well as new equipment, such as that used in the expansion of an existing data center, thus facilitating a pro-active decision making process for reliable and energy-efficient operation of the data center. Prognostics-based assessment can be used as a risk mitigation tool to provide early warning of predicted telecom equipment failure under free air cooling that cannot be otherwise achieved by the traditional methods given in published standards.

In a typical data center with tens or hundreds of thousands of components that are likely to fail, it was indicated that failure modes, mechanisms, and effects analysis (FMMEA) can be used to identify the weakest components, which are the most likely to fail first under the given conditions of the data center. Based on the estimated load conditions and other available product information (e.g., product material properties and configurations), the FMMEA process identifies the potential failure modes and the potential failure causes. The historical data in the field can also contribute to the weakest components identification process. Then, the identified weakest components will be the focus of prognostics-based assessment, since their failures will cause the equipment failures. When an anomaly is detected on one of the weakest components and its remaining useful life is estimated using a suitable algorithm, data centers can schedule maintenance or replacement of equipment to avoid unplanned downtime. This prognostics-based method also identifies whether the equipment failure is intermittent or permanent without having to interrupt data center operation.

Acknowledgements

The authors would like to thank the members of the Prognostics and Health Management Consortium at CALCE for their support of this work.

References:

[1] European Commission, "Code of Conduct on Data Centres Energy Efficiency – Version 2.0," Nov. 2009.

[2] J. G. Koomey, "Growth in Data Center Electricity Use 2005 to 2010," Analytics Press, Oakland, CA, Aug. 2011.

[3] "The Total Cost Ownership of Data Centers," Qpedia Thermal Magazine, Vol. 5 (8), pp. 14–16, Aug. 2011.

[4] Emerson Network Power, "Network Power, "Data Center Users' Group Special Report Energy Efficiency and Capacity Concerns Increase," White Paper, Sep. 2012.

[5] Intel Information Technology, "Reducing Data Center Cost with an Air Economizer," IT@Intel Brief; Computer Manufacturing; Energy Efficiency; Aug. 2008.

[6] R. Miller, "Microsoft's Chiller-less Data Center," Data Center Knowledge, Sep. 2009.

[7] R. Miller, "Google's Chiller-less Data Center," Data Center Knowledge, Jul. 2009.

[8] V.Mulay, "Achieving Efficiency through Concurrent Design of Hardware and Facility," International Microelectronics and Packaging Society (IMAPS), Palo Alto, CA, Nov. 10–11, 2011.

[9] ASHRAE TC 9.9 2004, "Thermal Guidelines for Data Processing Environments," 2004.

[10] ASHRAE Technical Committee (TC) 9.9, "2011 Thermal Guidelines for Data Processing Environments – Expanded Data Center Classes and Usage Guidance," Atlanta, 2011.

[11] Bell Communications Research Inc., Generic Requirements GR-63-CORE, "Network Equipment-Building System (NEBS) Requirements: Physical Protection," Piscataway, NJ, Mar. 2006.

[12] Bell Communications Research Inc., Generic Requirements GR-3028-CORE, "Thermal Management in Telecommunications Central Offices," Piscataway, NJ, Dec. 2001.

[13] H. Bulut, and M. A. Aktacir, "Determination of Free Cooling Potential: A Case Study for Istanbul, Turkey," Applied Energy, Vol. 88 (3), pp. 680–689, Mar. 2011.

[14] A. Lazaro, P. Dolado, J. M. Marín, and B. Zalba, "PCM–air Heat Exchangers for Free-cooling Applications in Buildings: Experimental Results of Two Real-scale Prototypes," Energy Conversion and Management, Vol. 50 (3), pp. 439–443, Mar. 2009.

[15] K. Dovrtel and S. Medved, "Weather-Predicted Control of Building Free Cooling System," Applied Energy, Vol. 88 (9), pp. 3088–3096, Sep. 2011.

[16] J. Dai, D. Das, and M. Pecht, "Risks To Telecommunication Equipment Under Free Air Cooling Conditions And Their Mitigation," MFPT: The Applied Systems Health Management Conference 2011, Virginia Beach, VA, May 10–12, 2011.

[17] J. Dai, D. Das, and M.Pecht, "Prognostics-Based Risk Mitigation for Telecom Equipment under Free Air Cooling Conditions," Applied Energy, In Press, Corrected Proof, Available Online: 2 July 2012.

[18] T. Homorodi and J. Fitch, "Fresh Air Cooling Research," Dell Techcenter, Aug. 2011.

[19] ASHRAE, "2008 ASHRAE Environmental Guidelines for Datacom Equipment," Atlanta, GA, 2008.

[20] D. L. Moss, "Data Center Operating Temperature: The Sweet Spot," Dell Technical White Paper, Jun. 2011.

[21] A. Shehabi, W. Tschudi, and A. Gadgil, "Data Center Economizer Contamination and Humidity Study," Emerging Technologies Program Application Assessment Report to Pacific Gas and Electric Company, Mar. 2007.

[22] X. B. Yu and S. Jin, "Research of Temperature-raising in Base Stations Room," Telecom Engineering Technics and Standardization, Vol. 12, Dec. 2008, ISSN 1008–5599, CN 11-4017/TN, Chinese version.

[23] ASHRAE Technical Committee (TC) 9.9, "Particulate and Gaseous Contamination Guidelines for Data Centers," White Paper, 2009.

[24] M. Pecht, "Prognostics and Health Management of Electronics," Wiley-Interscience, New York, NY, 2008.

[25] R. Jaai, and M. Pecht, "Fusion Prognostics," Proceedings of Sixth DSTO International Conference on Health & Usage Monitoring, Melbourne, Australia, Mar. 2009.

[26] W. Wang, M. Azarian, and M. Pecht, "Qualification for Product Development," 2008 International Conference on Electronic Packaging Technology & High Density Packaging, Shanghai, China, Jul., 2008.

[27] D. Wang, Q. Miao, and R. Kang, "Robust Health Evaluation of Gearbox Subject to Tooth Failure with Wavelet Decomposition," Journal of Sound and Vibration, Vol. 324 (3–5), pp. 1141–1157, Jul. 2009.

[28] W. He, N. Williard, M. Osterman, and M. Pecht, "Prognostics of Lithium-ion Batteries Based on Dempster-Shafer and Bayesian Monte Carlo," Journal of Power Sources, Vol. 196 (23), pp. 10314–10321, 2011.

[29] W. He, Z. Jiang, and K. Feng, "Bearing Fault Detection based on Optimal Wavelet Filter and Sparse Code Shrinkage," Measurement, Vol. 42 (7), pp. 1092–1102, 2009.

[30] K. B. Pressman and T. Morey, Switching Power Supply Design, McGraw Hill, New York, NY, 2009.

[31] S. Mathew, M. Alam, and M. Pecht, "Identification Of Failure Mechanisms To Enhance Prognostic Outcomes," MFPT: The Applied Systems Health Management Conference 2011, Virginia Beach, VA, May 10–12, 2011.

[32] G. Mirchandani and W. Cao, "On Hidden Nodes for Neural Nets," IEEE Transactions on Circuit and Systems, Vol. 36 (5), pp. 661–664, 1989.

[33] J. Dai, "Risk Assessment and Mitigation of Telecommunication Equipment under Free Air Cooling Conditions," Ph.D. Thesis, Mechanical Engineering of University of Maryland, College Park, MD, Aug. 2012.

CHAPTER 15

DAMAGE PRE-CURSORS BASED ASSESSMENT OF ACCRUED THERMO-MECHANICAL DAMAGE AND REMAINING USEFUL LIFE IN FIELD DEPLOYED ELECTRONICS

Pradeep Lall[(1)], Mahendra Harsha[(1)], Kai Goebel[(2)], Jim Jones[(3)]

[(1)]Auburn University
NSF-CAVE3 Electronics Research Center
Auburn, AL 36849
[(2)]NASA Ames Research Center, Moffett Field, CA
[(3)]Oracle Corporation, Sunnyvale, CA
E-mail: lall@auburn.edu

Field deployed electronics may accrue damage due to environmental exposure and usage after finite period of service but may not often have any macro-indicators of failure such as cracks or delamination. A method to interrogate the damage state of field deployed electronics in the pre-failure space may allow insight into the damage initiation, progression, and remaining useful life of the deployed system. Aging has been previously shown to effect the reliability and constitutive behavior of second-level leadfree interconnects. Prognostication of accrued damage and assessment of residual life can provide valuable insight into impending failure. In this paper, field deployed parts have been extracted and prognosticated for accrued damage and remaining useful life in an anticipated future deployment environment. A subset of the field deployed parts have been tested to failure in the anticipated field deployed environment to validate the assessment of remaining useful life. In addition, some parts have been subjected to additional known thermo-mechanical stresses and the incremental damage accrued validated with respect to the amount of additional damage imposed on the assemblies. The presented methodology uses leading indicators of failure based on micro-structural evolution of damage to identify accrued damage in electronic systems subjected to sequential stresses of thermal aging and thermal cycling. Damage equivalency methodologies have been developed to map damage accrued in thermal aging to the reduction in thermo-mechanical cyclic life based on damage proxies. The expected error with interrogation of system state and assessment of residual life has been quantified. Prognostic metrics including α-λ metric, sample standard deviation, mean square error, mean absolute percentage error, average bias, relative accuracy, and cumulative relative accuracy have been used to compare the performance of the damage proxies.

1. Introduction

An electronic component operating in a harsh environment is subjected to both temperature variations as well aging for a finite duration during use-life. Continuous capture of time-temperature history would put immense demands on existing system function. The life prediction models often need well defined accelerated test conditions to provide acceptable accuracy. In addition, the field environment may be often determined by equipment's operating conditions and usage profile resulting in variance in the field usage. Current health management techniques in EPS and Avionic Systems provide very-limited or no-visibility into health of power electronics, and packaging to predict impending failures. [McCann 2005, Marko 1996, Schauz 1996, Shiroishi 1997]. Electronics systems may be subjected to prolonged periods of thermal exposure over wide temperature extremes and long periods of thermal aging at often high ambient temperatures. High-rel systems may need very long operating life times with low downtime during operation. Historical environmental conditions to which the systems may have been subjected may be often unavailable. Prognostics Health Management (PHM) for electronics has wide applicability spanning a number of different areas. Electronics used in critical applications like space, military and avionics require multiple deployments with sequential thermal stresses. Systems may also be subjected to long term dormant storage in uncontrolled thermal environments. Ultra high reliability is needed to ensure predictable operation when needed and avoid any catastrophic damage during the service life.

Previously, Lall [2004[a-d], 2005[a-b], 2006[a-f], 2007[a-e], 2008[a-f], 2009[a-d], 2010[a-j]] has developed leading indicators based prognostic and health management methodologies for residual life computation of electronics subjected to single, multiple and superimposed thermal environments comprising of isothermal aging and thermal cycling. Examples of damage pre-cursors include micro-structural evolution of second level solder interconnects, inter-metallic compound growth, stress and stress gradients. Pre-cursors have been developed for both eutectic 63Sn37Pb and various lead-free alloy compositions like Sn4Ag0.5Cu, Sn3Ag0.5Cu, Sn1Ag0.5Cu, Sn0.3Ag0.7Cu, Sn3Ag0.5Cu-Bi, Sn3Ag0.5Cu-Bi-Ni,

96.5Sn3.5Ag alloy systems on a variety of area-array architectures. Researchers have reported the detrimental effect of thermal aging on the mechanical behavior of leadfree solders used in second-level interconnects of component level packaging. It has been shown that the degradation of the mechanical and creep properties of the solders during aging is caused due to the micro-structural evolution. The aging effects are significant even at room temperature [Zhang 2009]. In electronics assemblies, the built-in-self test (BIST) circuit involving error detection and correction circuits are used to give electronic assemblies the ability to test and diagnose themselves with minimal interaction from external test equipment [Chandramouli 1996, Drees 2004, Hassan 1992, Williams 1983, Zorian 1994]. The results obtained from BIST functions can generate diagnostic information which in turn provides additional confidence in the measurement result and confirms the device availability. BIST helps in minimizing the interaction with external automated test equipment (ATE) as well as provides the advantage of a more robust "at-speed" test of the circuitry, however, the current form of BIST gives little insight about the system level reliability or the remaining useful life of the system. Several studies conducted [Allen 2003, Drees 2004, Gao 2002, Rosenthal 1990] have shown that BIST can be prone to false alarms and can result in unnecessary costly replacement, re-qualification, delayed shipping, and loss of system availability. Fuses and Canaries may be mounted on a part to provide advance warning of failure due to specific wear out failure mechanism. Advanced warning is used to provide a maintenance-window for correction action, after an initial failure or malfunction, to prevent additional or secondary failures [Mishra 2002, Anderson 2004]. However, past efforts have provided limited insight into methods for estimation of remaining useful life.

There is need for tools and techniques which will enable the spot-assessment of the system's health and provide method for estimation of remaining useful life. Prognostication of field-deployed parts after sequential thermo-mechanical damage under thermal aging and thermal cycling based on damage pre-cursors is currently beyond the state of art. In this paper, a PHM method has been developed for the interrogation of system state and the estimation of remaining useful life of systems subjected to sequential stresses. Data on damage pre-cursors has been collected and analyzed for both aging and cycling to develop a technique that can map aging time into number of cycles and thus account for reduction in life due to dormant storage. The methodology involves the use of condition monitoring devices which have been interrogated for aging, cycling and different combinations of aging and cycling for residual life predictions. Developed PHM technique is based on non-linear least-squares method called Levenberg-Marquardt (LM) algorithm which has been developed for two different damage proxies. Prognostic model performance based on standard prognostic metrics has also been evaluated for both the damage proxies to determine which leading indicator of failure can be employed for accurate life prediction. Results of interrogation of system state have been compared with a second set of experimental-matrix to validate the proposed methodology.

2. Test Vehicle

In the present study, the perimeter type ball-grid arrays with traditional eutectic tin-lead solder have been studied under thermo-mechanical loads. Figure 1 shows the PBGA with heat-sink mounted over it and X-ray image showing the array of solder balls in Figure 2. The table below shows I/O counts and other package variables for the test vehicles used in this study. Phase growth and IMC data has been gathered and analyzed using image processing. Components analyzed are perimeter type ball-grid arrays with I/O counts of 456 (Table 1). The boards contain six trace layers to simulate the thermal mass of a true production board, though all functional traces were run on the topmost layer. All pads on the board were non-solder mask defined (NSMD) and had an immersion silver finish. The package parameters of this board are shown in Table 1.

Fig. 1. BGA with Heat Sink.

Table 1. Package Architecture.

Ball Count	456
Package type	26 × 26 Perimeter
Alloy System	Sn63Pb37
Ball Pitch (mm)	1.27
Pad Type	NSMD
Ball Diameter (mm)	0.80
No of parts	11

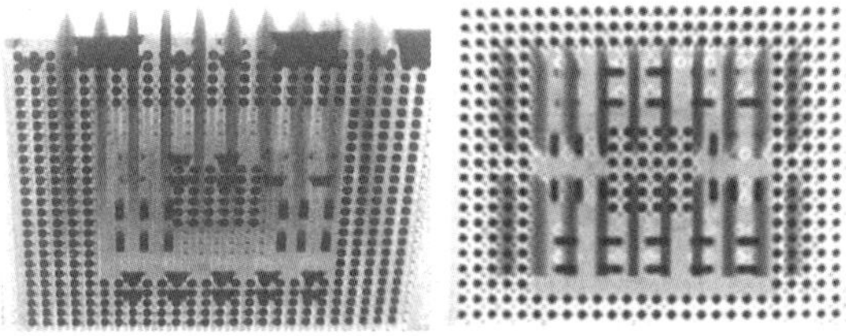

Fig. 2. X-ray images showing 26 × 26 Perimeter Array.

3. Approach for Prognosticating Damage

Field deployed electronic systems may be subjected to different environment profiles and usage profiles during service life. Assemblies may accrue damage at different rates in different field environments. Insight into the pre-failure space is necessary to understand damage initiation and the rate of damage progression. Samples have been withdrawn from the field deployed population to determine the prior accrued damage and assess the remaining useful life. The samples were then subjected to combinations of thermal aging and thermal cycling in the intended application environment and the accrued damage studied using damage precursors. Information is intended to aid decision making regarding assembly preparedness for future deployment. The withdrawn samples have been subjected to two different thermal environments including thermal aging and thermal cycling. Thermal aging is performed at 60°C for varying lengths of time. Assemblies were subjected to thermal cycling between -40ºC and 125ºC at 2.5 hours per cycle. A combined plot for both the environments in terms of damage accrual and remaining-life in terms of number of cycles is shown in Figure 4. Information on progression of damage proxies has been used to estimate prior damage and remaining useful life. Commonality of the damage proxies between thermal aging and thermal cycling has been used to superimpose damage in the two environments. The difference in the slopes between thermal aging and thermal cycling indicates the differing rates of damage accrual in the two environments.

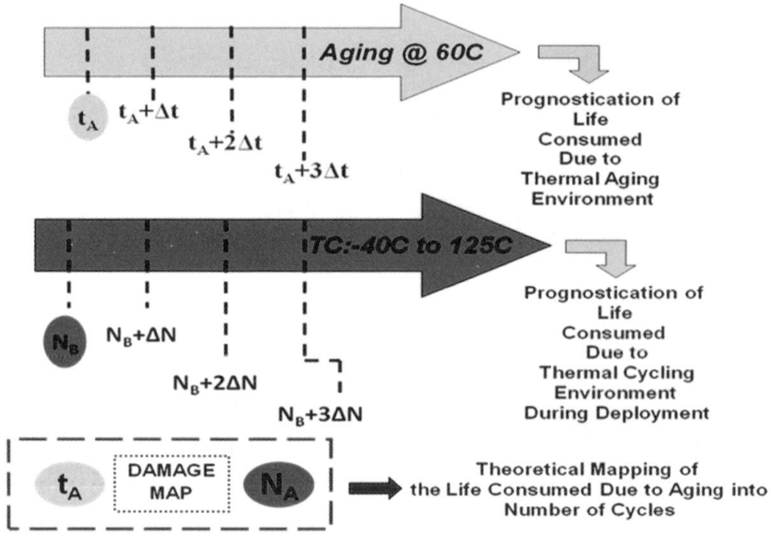

Fig. 3. Stepwise Approach.

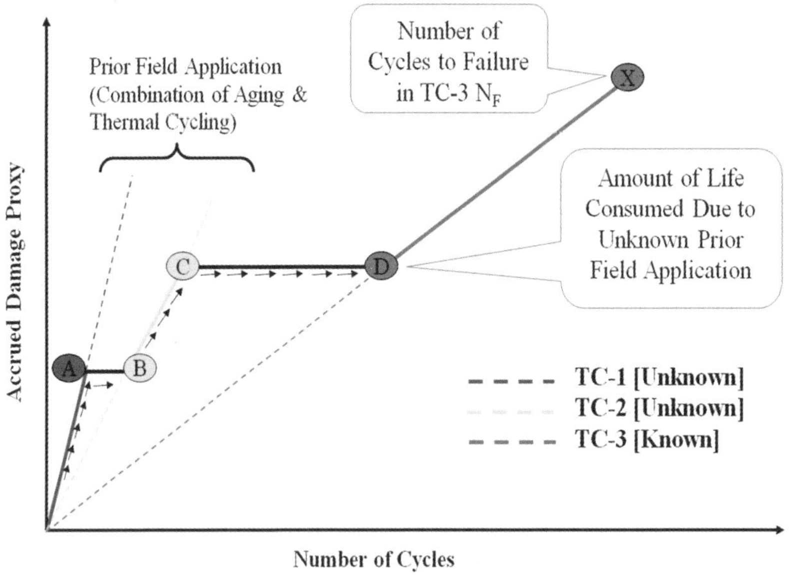

Fig. 4. Life vs Damage Curve for Thermal Environment.

3.1. *Micro-structural Evolution of Damage under Single Stresses*

In this step, board assemblies have been subjected to single stresses of thermal aging and thermal cycling. Samples have been withdrawn periodically and cross-sectioned. Damage proxies studied include the phase-growth parameter and the intermetallic thickness. Previously, it has been shown that the rate of change in phase growth parameter [d(lnS)/d(lnN)] is valid damage proxy for prognostication of thermo-mechanical damage in solder interconnects and assessment of residual life under thermal aging and thermal cycling type of thermo-mechanical stresses [Lall 2004[a-d], 2005[a-b], 2006[a-f], 2007[a-e], 2008[a-f], 2009[a-d]]. The damage proxy [d(lnS)/d(lnN)] is related to the microstructural evolution of damage by the following equation:

$$S = g^4 - g_0^4 = a(N)^b \tag{1}$$

$$\ln S = \ln\left(g^4 - g_0^4\right) = \ln a + b \ln N$$
$$\frac{d(\ln S)}{d(\ln N)} = b \qquad (2)$$

Where, g is the average grain size at time of prognostication, g_0 is the average grain size of solder after reflow, N is the number of thermal cycles, S is the phase growth parameter, parameters a and b are the coefficient and exponent respectively.

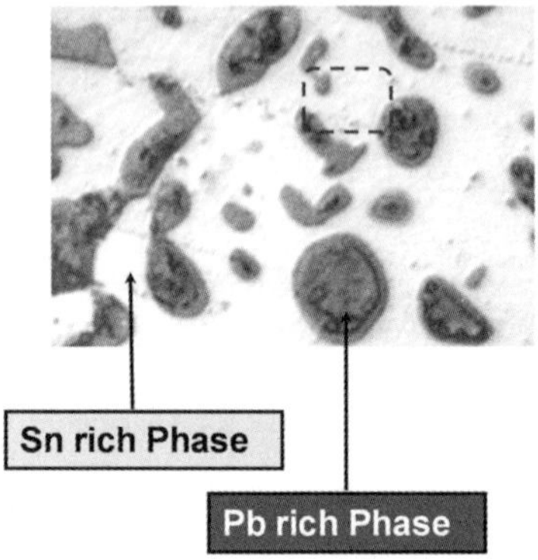

Fig. 5. Phases in Traditional Eutectic Solder.

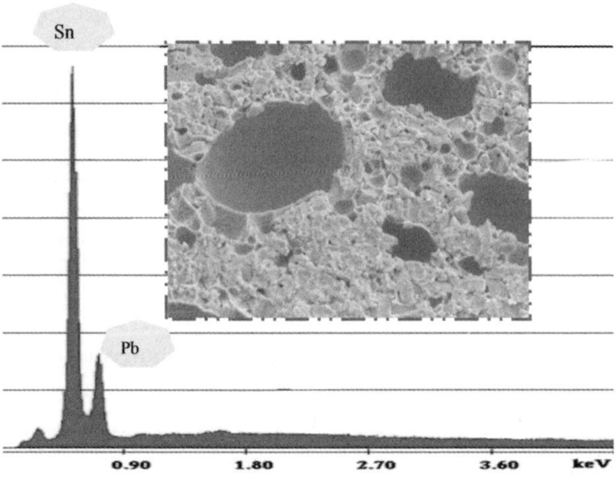

Fig. 6. Microanalysis Spectrum of Phase Growth using Energy Dispersion Spectroscopy (EDS).

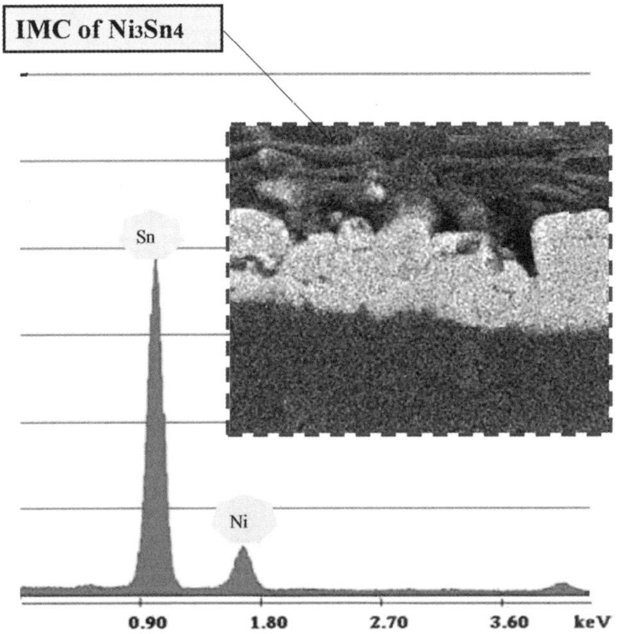

Fig. 7. Microanalysis Spectrum of IMC using Energy Dispersion Spectroscopy (EDS).

The log-plot of the equation provides a straight line relationship between the phase growth parameter and the number of cycles. It is anticipated that the higher temperature cycle magnitude will result in more accrued thermo-mechanical damage and a higher slope of the phase growth parameter versus number of thermal cycle curve. Cross-sectioned/polished samples were then gold sputter coated so as to take SEM images at 750x magnification. Energy Dispersion Spectroscopy (EDS) @ *5000X* to analyze the composition of the solder. Figure 5 shows the Sn-rich and Pb-rich phases in interconnect and Figure 6 shows the Microanalysis Spectrum. Figure 7 shows the composition and spectrum for Intermetallic Compound (IMC). To measure the phase size, first the images were cropped to a size of 60μm × 45μm and then mapped to gray scale image to calculate the average phase size for selected region using imaging software. Figure 8 shows the mapping of actual image to gray scale using imaging software.

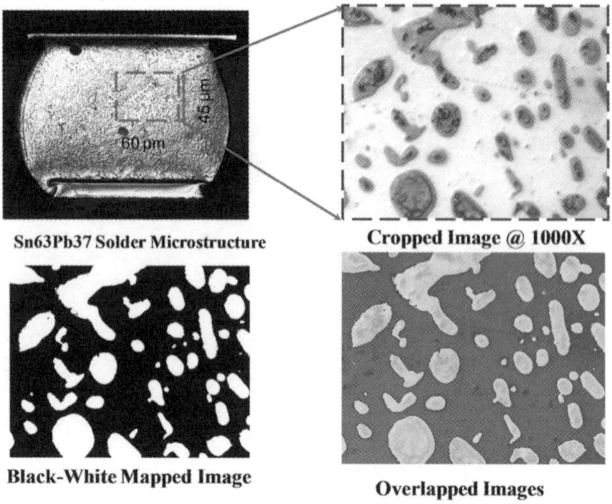

Fig. 8. Micrograph and Gray scale mapping of image using Image analysis software.

Growth of Inter-Metallic thickness during thermal aging has been studied as another leading indicator of failure in bulk solder. From past studies it has been established that growth of Intermetallic thickness is used as a damage precursor for computation of remaining useful life [Lall 2005[a], 2006[c,d], 2007[c,e], 2008[c,d], 2009[c,d]]. The interfacial intermetallic layers are formed between solder and copper, and some precipitates appear near the interface of the IMCs/solder. These intermetallic layers have been identified in SEM micrographs to consist of Cu_3Sn and Cu_6Sn_5 phases [Lall 2005[a]]. In order to investigate the correlation of interfacial intermetallic thickness growth versus thermal aging and thermal cycling, components have been withdrawn and cross sectioned at various intervals of thermal aging. The aged components were cross-sectioned periodically to measure the intermetallic thickness in SEM using 1000x magnification. The mean thickness of intermetallic layers were measured using commercial image processing software on SEM images, as a shown in Figure 5. Trend analysis of intermetallic thickness growth on SEM using image processing software, indicates a square root dependence of IMC thickness versus aging time,

$$y = y_0 + kt^n \tag{3}$$

Where y(t) is IMC growth thickness during aging, y_0 is the initial thickness of intermetallic compounds, k is the coefficient standing for the square root of the diffusivity at aging temperature, and t is test time. The exponent value, n = ½ has been used in Equation (3), which reveals a diffusion-controlled mechanism during aging. The aged components were sliced periodically to measure the Intermetallic thickness in SEM using 1000x magnification. The mean thickness of Intermetallic layers were measured using commercial image processing software on SEM images, as a shown in Figure 9.

Fig. 9. IMC measurement at board side copper pad.

3.2. *Damage Equivalency Relationships for Thermal Aging and Thermal Cycling*

A combined plot for thermal-aging and thermal-cycling in terms of damage accrual proxy and life in terms of number of cycles is shown in Figure 10. For overlapping stresses of thermal aging and thermal cycling, the phase-growth damage proxy has been normalized as follows:

$$S_{N\Delta T} = \left(\frac{g_p}{g_0}\right)^4 - 1 = a_{n\Delta T}(N)^{b_{n\Delta T}} \tag{4}$$

Where g is the phase-growth, subscripts p and 0 indicate point "p" and initial time respectively, $S_{N\Delta T}$ is the normalized phase-growth parameter during thermal cycling, $a_{n\Delta T}$ and $b_{n\Delta T}$ are the normalized phase-growth coefficient and phase growth exponent during thermal cycling. The normalized equation has been used for developing damage equivalency relations between damage accrued in thermal aging and thermal cycling. Phase-coarsening will also occur during thermal aging also. Micro-structural evolution in thermal aging can be represented as follows:

$$S_{Nt} = \left(\frac{g_p}{g_0}\right)^4 - 1 = a_{nt}(t)^{b_{nt}} \tag{5}$$

Where, g is the phase-growth, subscripts p and 0 indicate point "p" and initial time respectively, S_{Nt} is the normalized phase-growth parameter during thermal aging, a_{nt} and b_{nt} are the normalized phase-growth coefficient and phase growth exponent for thermal aging. Since the same value of normalized phase coarsening can be achieved on in either thermal aging or thermal cycling, damage equivalency relationship has been derived by equating the normalized phase-growth parameter:

$$a_{n\Delta T}(N)^{b_{n\Delta T}} = a_{nt}(t)^{b_{nt}} \tag{6}$$

$$N = \left[\left(\frac{a_{nt}}{a_{n\Delta T}}\right)(t)^{b_{nt}}\right]^{\frac{1}{b_{n\Delta T}}} \tag{7}$$

Figure 10 illustrates the solution for damage equivalency at point A and B in thermal cycling and thermal aging respectively. In an operational component, the accrued damage in thermal aging will result in reduction in the thermo-mechanical fatigue life of the component. The normalized phase growth parameter S_A has been measured at point A or at the end of storage and prior to deployment. This damage parameter S_A was then mapped back on the thermal cycling curve (marked by point B) that corresponds to S_B to find the equivalent number of cycles N_B. Point A and point B are points on the thermal aging and thermal cycling curves respectively, corresponding to identical values of the normalized phase-growth parameter, i.e. $S_A = S_B$. Mapped cyclic life, N_B, corresponding to t_A hours of thermal aging represents the reduction in thermal cycle reliability because of pre-conditioning at a high temperature.

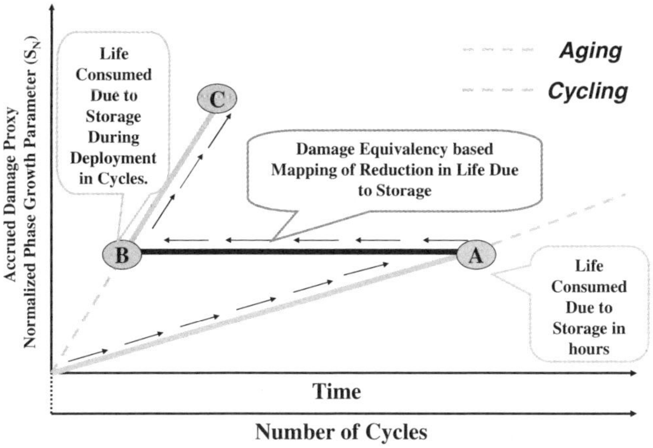

Fig. 10. Concept of Damage Equivalency and Mapping of Life Consumed Due to Storage.

Similar damage equivalency relationships have been derived from the intermetallic growth measurements under thermal cycling and thermal aging. A normalized measure of microstructural evolution has been used in this case.

$$\left(\frac{y_p}{y_0} - 1\right) = k_{N\Delta T} N^{n\Delta T} \tag{8}$$

$$\left(\frac{y_p}{y_0} - 1\right) = k_{Nt} t^{nt} \tag{9}$$

Where y is the intermetallic thickness, subscripts p and 0 indicate point "p" and initial time respectively, $k_{n\Delta T}$ and $n_{n\Delta T}$ are the normalized intermetallic growth coefficient and intermetallic growth exponent during thermal cycling, and k_{nt} and n_{nt} are the normalized intermetallic growth coefficient and intermetallic growth exponent during thermal aging. Convergence between the relationships has been used to validate that damage equivalency from thermal aging to thermal cycling reflects the underlying failure physics.

3.3. *Validation of Damage Equivalency and Prognostication under Overlapping Stresses*

Test assemblies have been subjected to varying periods of thermal aging and separate test assemblies have been subjected to varying periods of thermal cycling. In each case the assemblies were previously field deployed and thus subjected to unknown combinations of thermal aging and thermal cycling. Prior damage in each case has been prognosticated based on the micro-structural evolution of damage. Prognostication involves withdrawal of four-samples from the deployed population of devices at four periodic intervals. The samples were then cross-sectioned and the damage proxies including phase-growth parameter and the intermetallic thickness were measured.

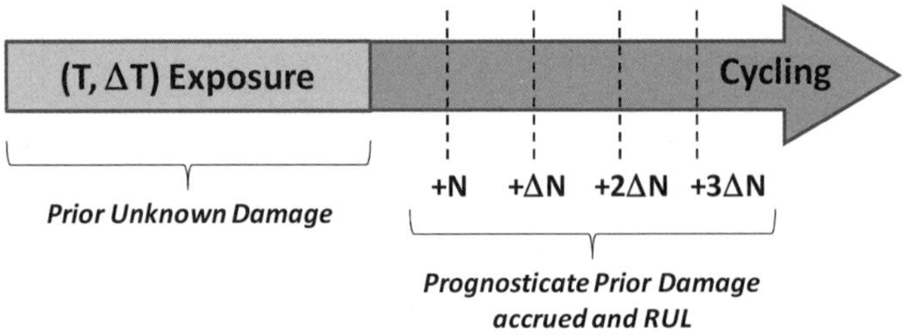

Fig. 11. Prognostication of AGED+THERMAL CYCLED Sample at the same thermal cycle count N_1.

Prior damage accrued was prognosticated. Normalized phase growth parameter was used to compute life consumed due to cycling alone

$$\frac{g_N^4}{g_0^4} - 1 = a_{N\Delta T}(N)^{b\Delta T} \tag{10}$$

$$\frac{g_{N+\Delta N}^4}{g_0^4} - 1 = a_{N\Delta T}(N + \Delta N)^{b_{N\Delta T}} \tag{11}$$

$$\frac{g_{N+2\Delta N}^4}{g_0^4} - 1 = a_{N\Delta T}(N + 2\Delta N)^{b_{N\Delta T}} \tag{12}$$

$$\frac{g_{N+3\Delta N}^{4}}{g_{0}^{4}} = a_{N\Delta T}(N+3\Delta N)^{b_{N\Delta T}} \quad (13)$$

The prognosticated damage was mapped using the damage equivalency relationships. Identical damage can be obtained by several combinations of thermal aging and thermal cycling. However, the user may be interested in particular solution relevant to the system of interest. The solution of the system of equations is thus bounded using trust regions for range of acceptable values. Most operational systems have date-codes indicating the vintage of manufacture and deployment logs. This enables bounding of the analysis. The prognosticated value of accrued damage is represented by N_C. Reduction in cyclic life due to thermal aging is then be mapped to account for the storage time t_A prior to exposure to cycling. The mapped reduction in cyclic life is represented by N_B. The remaining life is then:

$$N_{CYCLIC} = N_F - N_C \quad (14)$$

Where N_{CYCLIC} is the life of the electronic system in thermal cycling. Field-deployed parts were then subjected to thermal cycling to determine the time-to-failure (N_F) and experimentally measure the remaining useful life. The remaining useful life can then be calculated from the prognostic horizon.

4. Levenberg-Marquardt Algorithm

The relationship between the phase growth parameter and time is nonlinear because it contains terms with fourth power. Inverse solution for interrogation of system-state is challenging for damage evolution in such systems. Levenberg-Marquardt (LM) algorithm n iterative technique that computes the minimum of a non-linear function in multi-dimensional variable space has been used for identifying the solution in the prognostication neighborhood [Madsen 2004, Lourakis 2005, Nielsen 1999]. Let f be a assumed functional relation between a measurement vector referred to as prior-damage and the damage parameter vector, p, referred to as predictor variables. The measurement vector is the current values of the leading-indicator of failure and the parameter vector includes the prior system state, and accumulated damage and the damage evolution parameters. An initial parameter estimate p0 and a measured vector x are provided and it is desired to find the parameter vector p, that best satisfies the functional relation *f* i.e. minimizes the squared distance or squared-error, $\varepsilon^T\varepsilon$ with $\varepsilon = x - f(p)$. Assume that g(p) = ε in the squared error. The minimizer parameter vector, p, for the error function has been represented as,

$$F(p) = \frac{1}{2}\sum_{i=1}^{m}(g_i(p))^2 = \frac{1}{2}g(p)^T g(p) \tag{15}$$

$$F'(p) = J(p)^T g(p) \tag{16}$$

$$F''(p) = J(p)^T J(p) + \sum_{i=1}^{m} g_i(x) g_i''(x) \tag{17}$$

Where F(p) represents the objective function for the squared error term $\varepsilon^T\varepsilon$, J(p) is the Jacobian, and F'(p) is the gradient, and F''(p) is the Hessian. The variation of an F-value starting at "p" and with direction "h" is expressed as a Taylor expansion, as follows:

$$F(p + \alpha h) = F(p) + \alpha h^T F'(p) + O(\alpha^2) \tag{18}$$

Where α is the step-length from point "p" in the descent direction, "h". Mathematically, "h" is the descent direction of F(p) if $h^T F'(p) < 0$. If no such "h" exists, then F'(p)=0, showing that in this case the function is stationary. Since the condition for the stationary value of the objective function is that the gradient is zero, i.e. $f'(p + h) = L'(h) = 0$. The descent direction can be computed from the equation,

$$(J^T J)h_{gn} = -J^T g \tag{19}$$

In each step, Newton method uses α = 1, and $p = p + \alpha h_{gn}$, where subscript 'gn' indicates gauss-newton. The value of α is found by line search principle described above. Levenberg-Marquardt algorithm is a hybrid method which utilizes both steepest descent principle as well as the Gauss-Newton method. When the current solution is far from the correct one, the algorithm behaves like a steepest descent method: slow, but guaranteed to converge. When the current solution is close to the correct solution, it becomes a Gauss-Newton method. The LM method actually solves a slight variation of Equation (19), known as the augmented normal equations.

$$(J^T J + \mu I) h = -J^T g \tag{20}$$

The term $\mu$is called as the damping parameter, $\mu>0$ ensures that coefficient matrix is positive definite, and this ensures that h is a descent direction. When the value of μ is very small, then the step size for LM and Gauss-Newton are identical. Algorithm has been modified to take the equations of phase growth and inter-metallic growth under both iso-thermal aging and cycling loads to calculate the unknowns.

5. Case Study: Prognostication of Damage in Leadfree Area-Array Interconnects under Sequential Stresses

Samples subjected to single and sequential stresses have been prognosticated for damage equivalency and assessment of cumulative accrued damage resulting from a combination of thermal aging and thermal cycling. Prognosticated life has been correlated with experimental values. In addition, prognostic metrics have been computed to compare the robustness of the leading indicators and their accuracy in reference to the prognostic horizon.

5.1. *Micro-structural Evolution Under Thermal Aging*

A set of packages was subjected to thermal aging at 60°C and were withdrawn after each 144 hrs increment. The samples were cross-sectioned, polished. Phase growth and intermetallic growth was studied using SEM images at each time intervals.

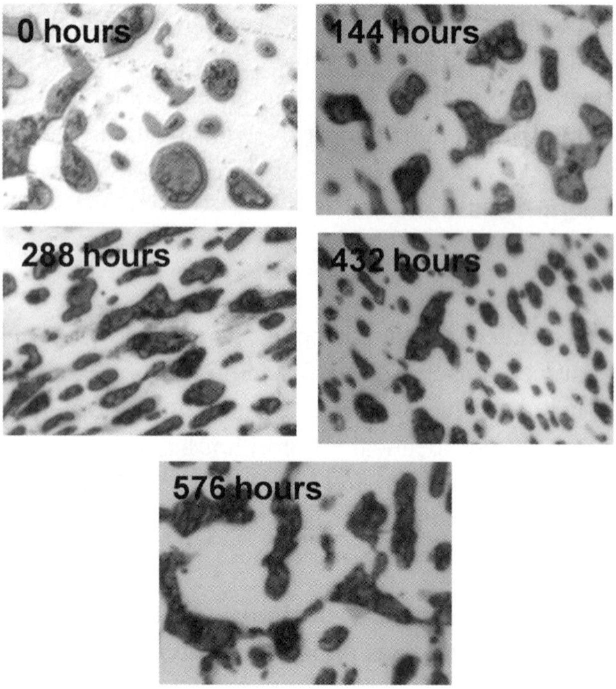

Fig. 12. Microscope images of Phase growth versus different aging time intervals (Thermal aging at 60°C, 456 I/O PBGA, magnification 1000x).

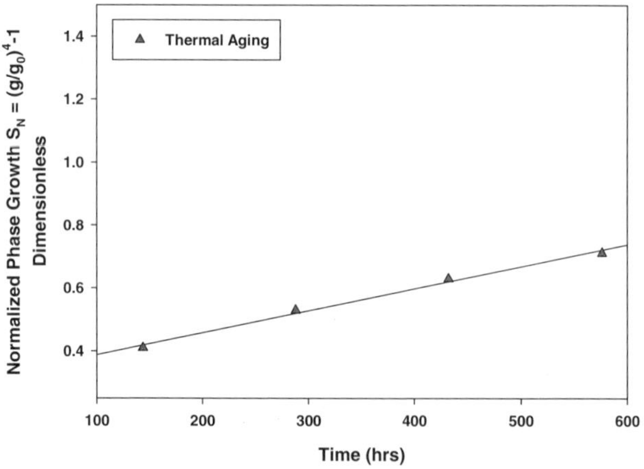

Fig. 13. Plot of Normalized Phase growth versus Thermal Aging Time 456 PBGA, subjected to thermal aging at 60°C.

The image analysis software has been used to measure the average phase size. Figure 12 shows phase size at different aging time intervals and Figure 13 shows the plot of normalized phase growth at various time intervals. The test data has been represented by the following equation:

$$S_{Nt} = \left(\frac{g_p}{g_0}\right)^4 - 1 = a_{nt}\left(t\right)^{b_{nt}} \tag{21}$$

Where, g is the phase-growth, subscripts p and 0 indicate point "p" and initial time respectively, S_{Nt} is the normalized phase-growth parameter during thermal aging, a_{nt} and b_{nt} are the normalized phase-growth coefficient and phase growth exponent for thermal aging.

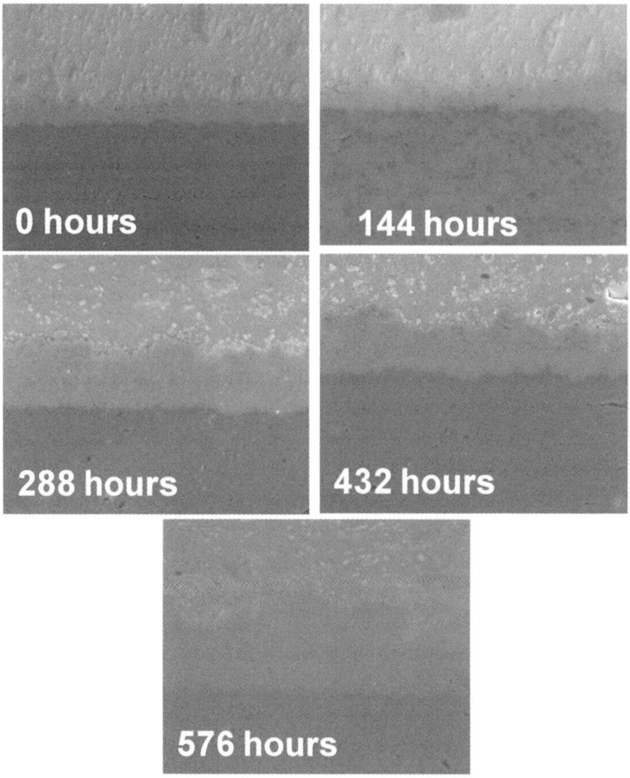

Fig. 14. Microscope images of Intermetallic Thickness versus different aging time intervals (Thermal aging at 60°C, 456 I/O PBGA, magnification 1000x).

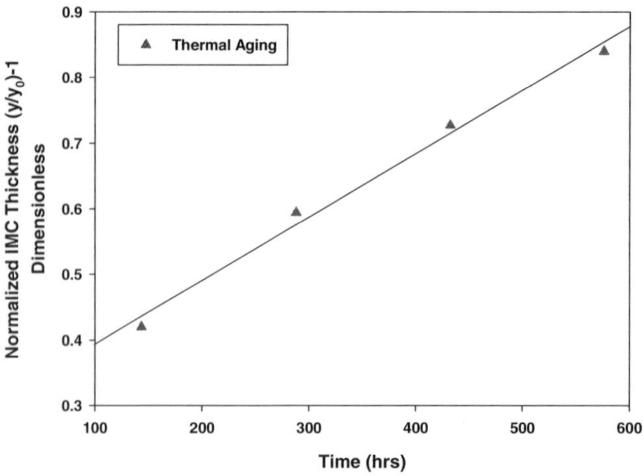

Fig. 15. Plot of Normalized IMC thickness versus Aging Time for 456 PBGA, subjected to thermal aging at 60°C.

The following relationship represents the evolution of phase-growth parameter in thermal aging based on experimental data. The equation parameters have been derived based on experimental measurements of the phase-growth parameters from cross-sections.

$$S_N = 0.056(t)^{0.4} \tag{22}$$

Figure 14 shows the growth of intermetallic thickness at various interval of time when subjected to thermal aging at 60°C. Figure 15 shows a plot of normalized intermetallic growth versus thermal aging time. The data has been fit to an equation for the following form:

$$\left(\frac{y_p}{y_0} - 1\right) = k_{Nt} t^{nt} \tag{23}$$

Where y is the intermetallic thickness, subscripts p and 0 indicate point "p" and initial time respectively, k_{nt} and n_{nt} are the normalized intermetallic growth coefficient and intermetallic growth exponent during thermal aging. The following relationship represents the evolution of intermetallic growth in thermal aging based on experimental data. The equation parameters have been derived based on experimental measurements of the intermetallic growth from cross-sections.

$$\left[\frac{y}{y_0}-1\right]=0.035(t)^{0.5} \tag{24}$$

5.2. *Micro-structural Evolution Under Thermal Cycling*

A different set of assembly was subjected to thermal cycling from -40°C to 125°C and were withdrawn after each 100 cycle increments. The samples were cross-sectioned, polished. Phase growth and intermetallic growth was studied using SEM images at each cyclic intervals.

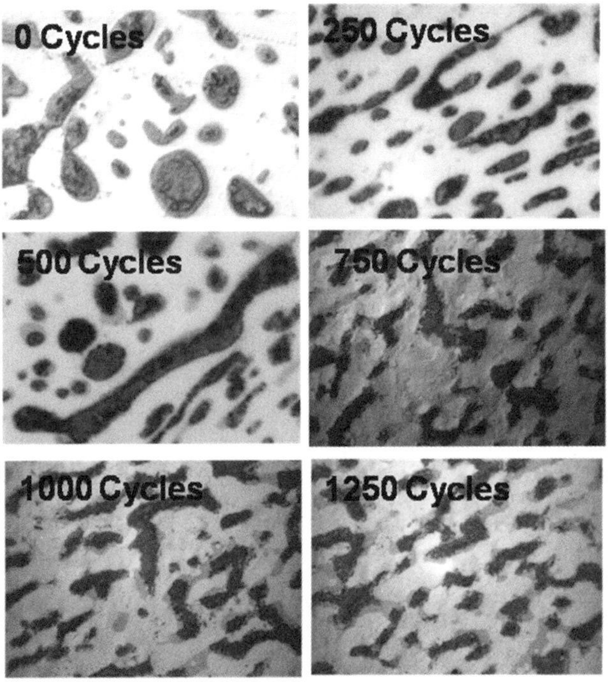

Fig. 16: Microscope image of Phase growth versus different aging time intervals (Thermal cycling -40°C to 125°C, 456 I/O PBGA, magnification 1000x).

The image analysis software has been used to measure the average phase size. Figure 16 shows phase size at different intervals of cycles and Figure 17 shows the plot of normalized phase growth at various intervals of cycles. The test data has been fit to the following equation:

$$S_{N\Delta T} = \left(\frac{g_p}{g_0}\right)^4 - 1 = a_{n\Delta T}(N)^{b_{n\Delta T}} \tag{25}$$

Where g is the phase-growth, subscripts p and 0 indicate point "p" and initial time respectively, $S_{N\Delta T}$ is the normalized phase-growth parameter during thermal cycling, $a_{n\Delta T}$ and $b_{n\Delta T}$ are the normalized phase-growth coefficient and phase growth exponent during thermal cycling. The following relationship represents the evolution of phase-growth parameter in thermal cycling based on experimental data. The equation parameters have been derived based on experimental measurements of the phase-growth parameters from cross-sections.

$$S_N = 0.0128(N)^{0.645} \tag{26}$$

Where S_N is the normalized phase-growth parameter, and N is the number of cycles. Figure 18 shows the growth of intermetallic thickness at various interval of time when subjected to thermal cycling from -40 to 125°C. Figure 19 shows a plot of normalized intermetallic growth versus thermal cycles.

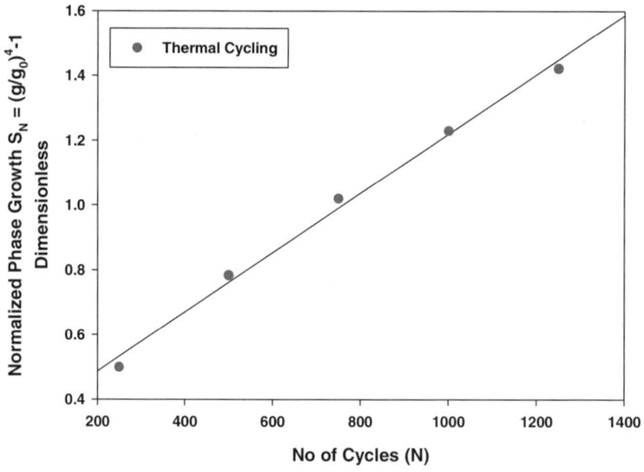

Fig. 17. Plot of Normalized Phase growth versus Aging Time for 456 PBGA, subjected to thermal cycling from -40°C to 125°C.

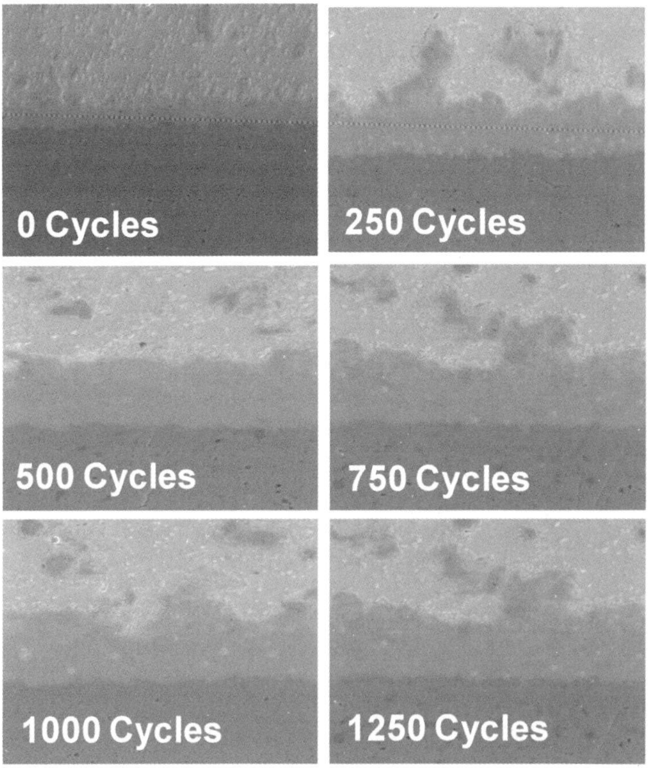

Fig. 18. Microscope image of Intermetallic Thickness versus different aging time intervals (Thermal Cycling at -40 to 125°C, 456 I/O PBGA, magnification 1000x).

The data has been fit to an equation for the following form:

$$\left(\frac{y_p}{y_0}-1\right)=k_{N\Delta T}N^{n\Delta T} \tag{27}$$

Where y is the intermetallic thickness, subscripts p and 0 indicate point "p" and initial time respectively, $k_{n\Delta T}$ and $n_{n\Delta T}$ are the normalized intermetallic growth coefficient and intermetallic growth exponent during thermal cycling. The following relationship represents the evolution of intermetallic growth in thermal cycling based on experimental data. The equation parameters have been derived based on experimental measurements of the intermetllic growth from cross-sections.

$$\left[\frac{y}{y_0}-1\right]=0.0053(N)^{0.82} \tag{28}$$

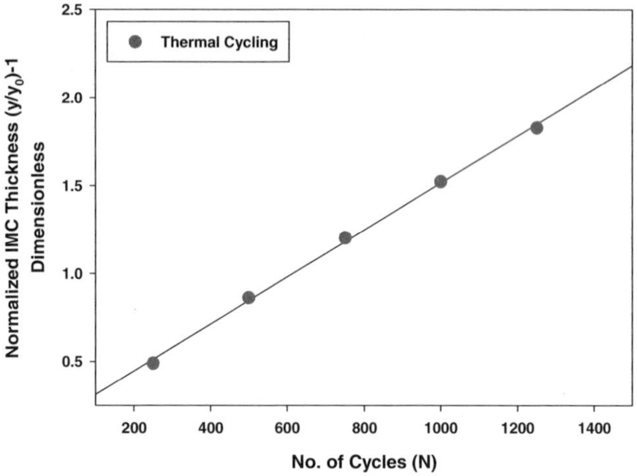

Fig. 19. Plot of Normalized IMC thickness versus No of Cycles for 456 PBGA, subjected to thermal Cycles from -40°C to 125°C.

5.3. *Damage Equivalency Relationships between Thermal Aging and Thermal-Cycling*

Figure 20 shows the combined plot normalized Phase growth versus time and cycles on x-axis. Damage accrued from aging and cycling environments has been equivalenced based on two damage proxies including normalized intermetallic thickness and normalized phase growth. The evolution of normalized phase growth has been plotted versus thermal cycles in -40 to 125°C and thermal aging time at 60°C (Figure 20). A similar value of damage proxy can be obtained by exposure to single stresses of thermal aging and thermal cycling. The exposure length to the environmental stresses however, will be different in each case. This combined plot is helpful for mapping of damage from thermal aging onto thermal cycling.

Phase-growth parameter has been used for damage equivalency using previously derived equations for evolution of S_N versus thermal aging time and thermal cycles. Equations (22) and (26) have been equated to derive the damage equivalence relationships based on phase-growth parameter:

$$0.0128(N)^{0.645} = 0.056(t)^{0.4} \tag{29}$$

$$N = 9.93(t)^{0.62} \tag{30}$$

Where N is the number of thermal cycles at -40 to 125°C and t is the time in hours at 125°C. Damage equivalency has also been studied using the intermetallic growth in second-level interconnects. The evolution of normalized intermetallic growth has been plotted versus thermal cycles in -40 to 125°C and thermal aging time at 60°C (Figure 21). A similar value of intermetallic thickness can be obtained by exposure to single stresses of thermal aging and thermal cycling. The exposure length to the environmental stresses however, will be different in each case.

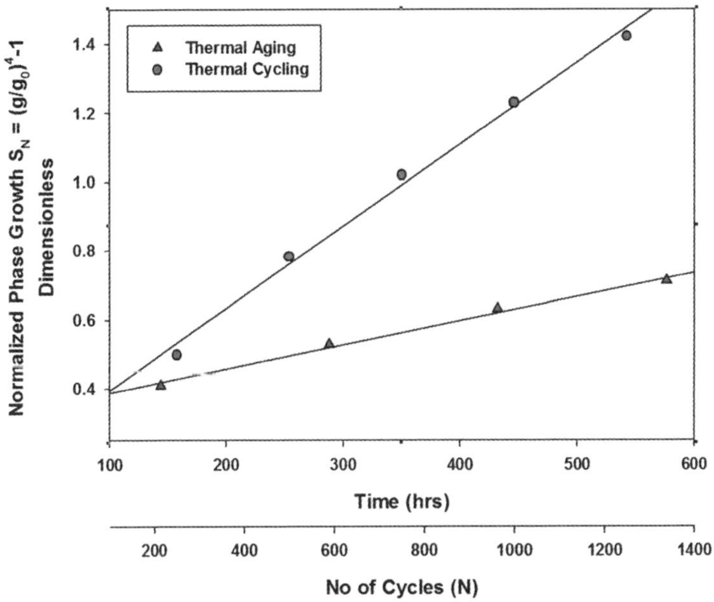

Fig. 20. Combined plot Normalized Phase growth versus thermal aging time/thermal cycling.

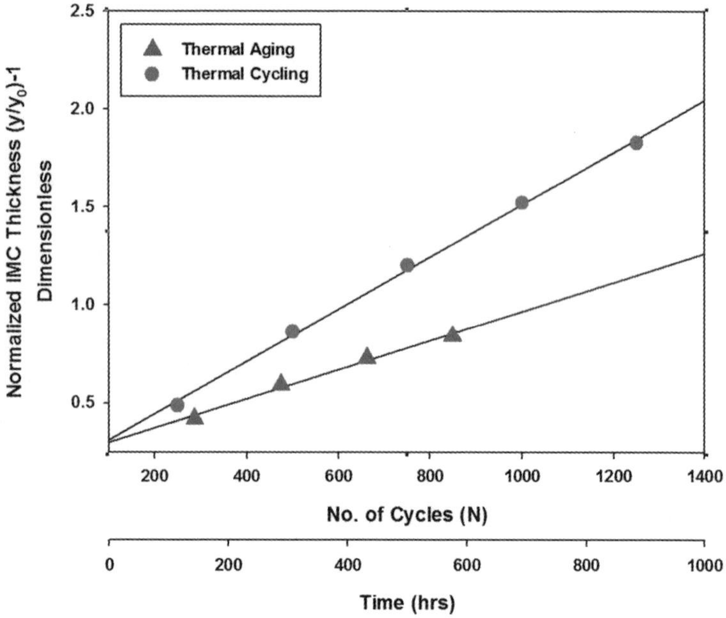

Fig. 21. Combined plot of Normalized IMC thickness versus thermal aging time/thermal cycling.

This combined plot is helpful for mapping of damage from thermal aging onto thermal cycling. Equations (24) and (28) have been equated to develop the damage equivalence relationships based on intermetallic growth:

$$0.0053(N)^{0.82} = 0.035(t)^{0.5} \tag{31}$$

$$N = 9.99(t)^{0.61} \tag{32}$$

Table 2. Damage mapping from leading indicators of failure.

Time in Hrs	Reduction in life from Normalized Phase Parameter	Reduction in life from Normalized IMC Parameter
144	207	217
288	316	333
432	404	428
576	481	511

Comparison of the damage equivalency relationships between thermal aging at 125°C and thermal cycling from -40 to 125°C developed based on the phase-growth parameter shows that both equations have similar coefficients and exponents (Equations (30), (32)). Thc values on the graph and the table have been plotted and shown respectively. The convergence of the damage equivalency parameters from two separate damage proxies bolsters the validity of the correlation. Figure 22 and Table 2 show the reduction in thermo-mechanical fatigue life derived based on the two damage proxies.

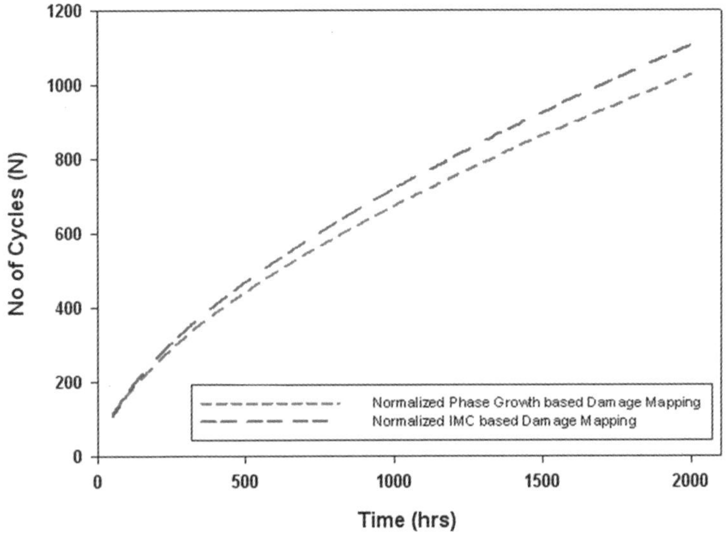

Fig. 22. Combine plot of No of Cycles (N) versus Time in hrs, due to damage mapping from Phase growth and IMC thickness.

6. Prognostication of Damage Under Multiple Stresses

The system state has been interrogated in test assemblies that have been exposed to different stresses of thermal aging at 60°C and thermal cycling from -40 to 125°C using both the damage proxies discussed in this paper. Samples have been withdrawn at periodic intervals and cross-sectioned for measurement of the damage proxies. The LM-Algorithm has been used for interrogation of system-state for accrued damage.

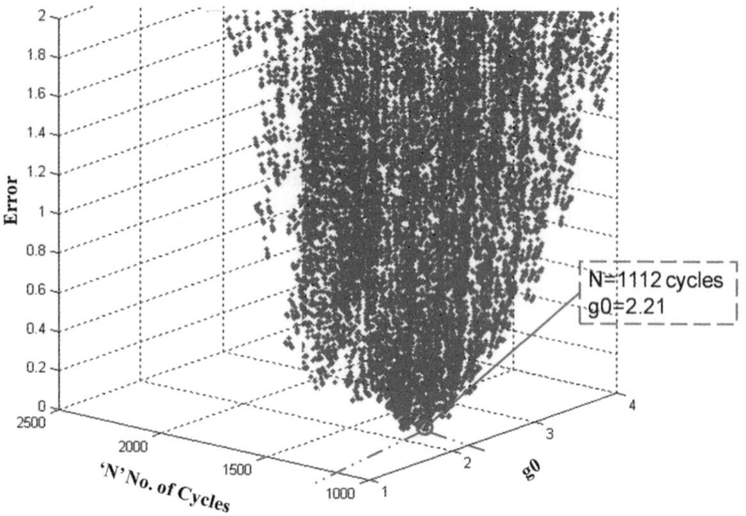

Fig. 23. 3D plot of error versus Number of Cycles, damage incurred due to storage in Thermal Cycling Environment (N).

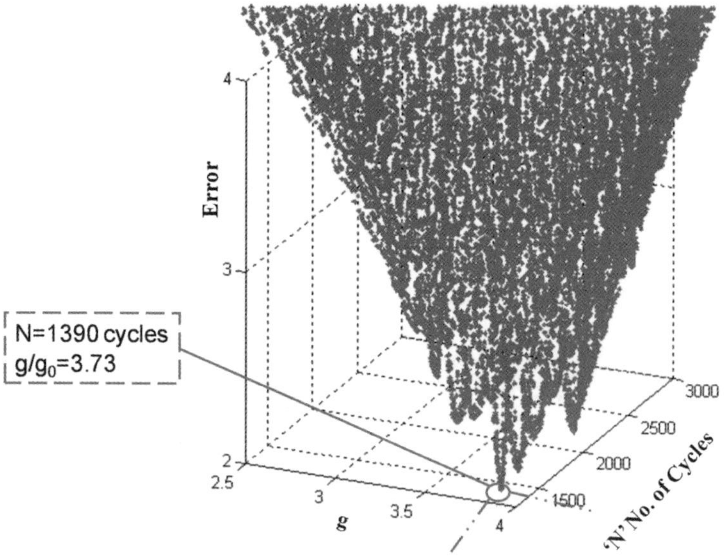

Fig. 24. 3D plot of error versus Number of Cycles, damage incurred due to storage in Thermal Cycling Environment (N+ΔN).

Figure 23 shows the prognosticated value of prior-accrued thermo-mechanical damage in field withdrawn parts without any additional imposed exposure to thermo-mechanical stresses. The prognosticated life has been computed in terms of the number of cycles of accrued damage from -40 to 125°C. The environmental conditions used were identified based on anticipated future deployment environment. Figure 24 shows the prognosticated value of prior accrued damage for field withdrawn parts after they had been subjected to 250 cycles of -40 to 125°C.

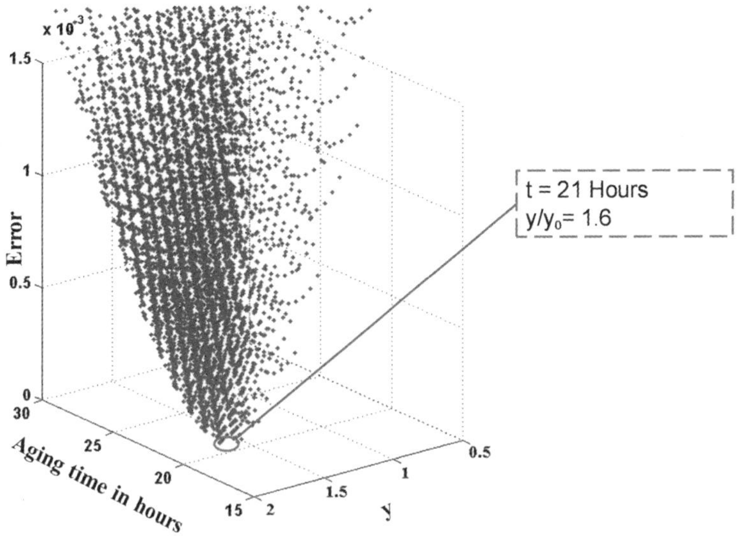

Fig. 25. 3D plot of error versus Aging time damage incurred due to storage in Isothermal Aging Environment (t).

Validation of the prognosticated damage can be assessed by the difference between the field withdrawn parts and the field withdrawn parts with additional 250 cycles of thermo-mechanical damage. Comparison of the prognosticated incremental damage between the two thermal-event waypoints has been compared with experimental incremental damage in Table 3. The prognosticated value of 278 cycles correlates well with the experimental value of 250 cycles. The prior accrued damage under steady-state thermal exposure has also been studied in the field withdrawn parts. A separate batch of field withdrawn parts has also been subjected to additional thermal exposure of 144 hours. Figure 25 shows the prior accrued damage in field withdrawn parts. Figure 26 shows the prior accrued damage in field withdrawn parts with increment thermal exposure.

Table 3. Comparison of the prognosticated and experimental incremental damage.

	Field Withdrawn Parts	Field Withdrawn Parts + 250 Cycles	ΔN Incremental Accrued Damage
Prognosticated Life	1112 Cycles	1390 Cycles	278 Cycles
Actual Life			250 Cycles

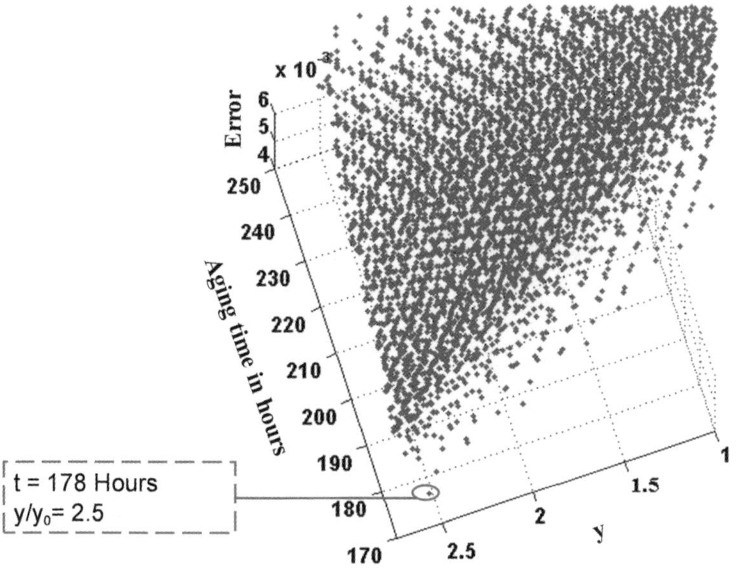

Fig. 26. 3D plot of error versus Aging time damage incurred due to storage in Isothermal Aging Environment (t+144) hours.

Table 4: Comparison of the prognosticated and experimental incremental damage.

	Field Withdrawn Parts	Field Withdrawn Parts + 144 Hours	ΔN Incremental Accrued Damage
Prognosticated Life	21 Hours	178 Hours	157 Hours
Actual Life			144 Hours

The prognosticated value of incremental damage has been compared with the experimental value of incremental damage (Table 4). The prognosticated thermal exposure time-frame has been computed in terms of exposure time at 60°C. The temperature level for prognostication has been identified based on the anticipated temperature of future deployment.

7. Prognostics Performance Metrics

Prognostic metrics have been used to compare the robustness of the prognostication algorithms using these leading indicators of damage for overlapping sequential stresses of thermal aging and thermal cycling. The prognostics metrics used in the paper include: accuracy, precision, Mean Squared Error (MSE), and Mean Absolute Percentage Error (MAPE), α-λ accuracy, relative accuracy (RA) and cumulative relative accuracy (CRA) have been computed to compare the two models [Saxena 2008[a,b]].

7.1. *Average Bias*

The average bias method averages the error in predictions made at all subsequent times after prediction starts for the lth unit under test (UUT). The metric can be extended to average bias over all UUTs to establish overall bias [Saxena 2008[a,b]].

$$B_l = \frac{\sum_{i=P}^{EOP} \left\{ \Delta^l(i) \right\}}{\left(EOP - P + 1 \right)} \tag{33}$$

Where, EOP (End-of-Prediction) is the earliest time index, i, after prediction crosses the failure threshold, P is the time index at which the first prediction is made by the prognostic system, $\Delta^l(i)$ is the error between the predicted and the true RUL at time index *i* for UUT l.

7.2. *Sample Standard Deviation (S)*

The sample standard deviation measures the dispersion of the error with respect to the sample mean of the error. This metric is restricted to the assumption of normal distribution of error [Saxena 2008[a,b]].

$$S(i) = \sqrt{\frac{\sum_{l=1}^{n} \left(\Delta^l(i) - M\right)^2}{n-1}} \tag{34}$$

Where, M is sample mean of the error, and $\Delta^l(i)$ is the error between the predicted and the true RUL at time index *i* for UUT l .

7.3. *Mean Squared Error (MSE)*

The mean squared error averages the square prediction error for multiple UUTs at the same prediction horizon. A derivative of MSE is root mean square error (RMSE) [Saxena 2008[a,b]].

$$\mathrm{MSE(i)} = \frac{1}{\mathrm{L}} \sum_{\mathrm{i=1}}^{\mathrm{L}} \Delta^{\mathrm{l}}(\mathrm{i})^2 \tag{35}$$

Where, $\Delta^l(i)$ is the error between the predicted and the true RUL at time index *i* for UUT l .

7.4. *Mean Absolute Percentage Error (MAPE)*

MAPE averages the absolute error in the predictions of multiple UUTs at the same prediction horizon. Instead of the mean, median can be used to compute Median absolute percentage error (MAPE) in similar fashion [Saxena 2008[a,b]]

$$\mathrm{MAPE(i)} = \frac{1}{\mathrm{L}} \sum_{\mathrm{l=1}}^{\mathrm{L}} \left| \frac{100\Delta^{\mathrm{l}}(\mathrm{i})}{\mathrm{r}_*^{\mathrm{l}}(\mathrm{i})} \right| \tag{36}$$

Where, $r_*^l(i)$ is the true RUL at time t_i given that data is available up to time t_i for the lth UUT.

7.5. *α-λ accuracy*

The α-λ curve has been plotted for both the models as shown in Figure 27 and Figure 28. A comparison α-λ curve is shown in Figure 29. It is a normalized plot of Remaining Useful Life (RUL) Vs Life which is compared against failure data and the error bounds. In this case the experimental data obtained from accelerated testing shown by blue line in the plots and ± 10% error bounds are shown by dotted lines.

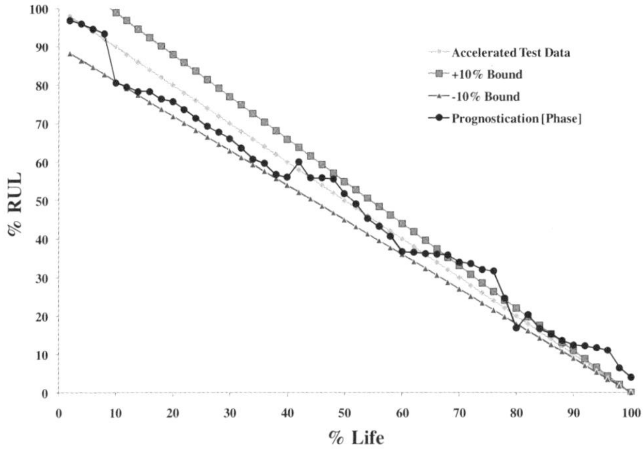

Fig. 27. α-λ curve for Prognostication using phase-growth as the damage proxy.

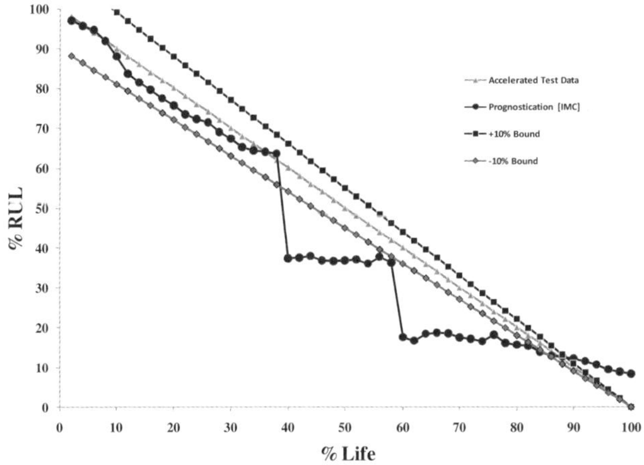

Fig. 28. α-λ curve for Prognostication using IMC as the damage proxy.

It should be noted that the selection of error bounds is application specific and typically tighter bounds are imposed as the criticality of the system increases. In this paper the main idea to plot α-λ curves for the two models was not to find the prognostic horizon point but to compare the relative performance of the models visually. Both leading indicators perform well.

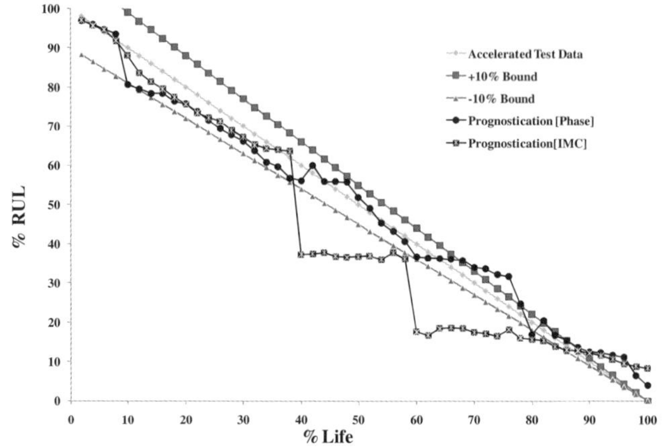

Fig. 29. Combined α-λ curve for Prognostication.

7.6. *Relative Accuracy (RA)*

The relative prediction accuracy has been used to asses whether the predictions fall within a given accuracy levels at a given time instant. The time instant is the fraction of actual remaining useful life from the point when the first prediction is made. An algorithm with higher relative accuracy is desirable [Saxena 2008[a,b]]

$$RA = 1 - \frac{\left| r_*(t_\lambda) - r^1(t_\lambda) \right|}{r_*(t_\lambda)} \tag{37}$$

Where $t_\lambda = P + \lambda\,(EOP - P)$, $r^1(i)$ is the RUL estimate for the lth UUT at time t_i as determined from measurement and analysis, $r_*^1(i)$ is the true RUL at time t_i given that data is available up to time t_i for the lth UUT.

7.7. *Cumulative Relative Accuracy (CRA)*

Relative accuracy has been evaluated at multiple time instances. Cumulative Relative Accuracy has been calculated as the normalized weighted sum of relative prediction accuracies at specific time instances [Saxena 2008[a,b]]

$$\text{CRA} = \frac{1}{\text{EOP} - \text{P} + 1} \sum_{1=\text{P}}^{\text{EOL}} \text{RA} \tag{38}$$

Where EOL represents End-of-Life, *w* is a weight factor as a function of RUL at all time indices. In most cases it is desirable to weigh the relative accuracies higher closer to the EOL.

Table 5. Comparison of Prognostics Metrics for the Two Leading Indicators.

Prognostic Metrics	LM prognostication (Phase)	LM prognostication (IMC)
SSD (S)	61.79280896	195.85
MSE	11225.04082	61228.84
MAPE	0.0157601	0.25
RA (λ=0.5)	0.9624	0.734
CRA	0.016282	0.0141

Table 5 shows the relative accuracy for a point where 50% of the life of system is consumed. Both leading indicator based prognostic algorithms show comparable performance.

8. Residual Life In Multiple Environments

The residual life of the assemblies subjected to multiple thermal environments has been be computed based on the following equation:

$$\text{RUL} = 1 - \left(\frac{\text{N}}{\text{N}_{1\%}} \right) \tag{39}$$

Where, $N_{1\%}$ is the time to one-percent failure of the population, N is the prognosticated prior damage in the field-deployed test assemblies. The $N_{1\%}$ has been measured with two methods. The first method included testing an identical pristine part till failure. The second method included accelerated testing of an identical field extracted part from the same batch till failure. Figure 30 shows the Weibull distribution for an identical pristine part. The time to one-percent failure ($N_{1\%}$) for 456 BGA packages subjected to -40 to 125°C is 1746 cycles (Figure 30). The prior accrued damage (N) was prognosticated to be 1112 cycles.

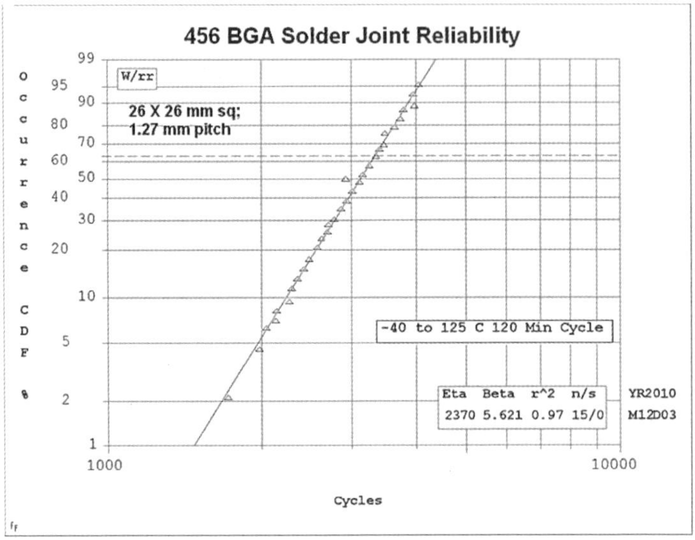

Fig. 30. Weibull plot for pristine 456 BGA packages subjected to thermal cycle -40°C to 125°C.

Fig. 31. Cross-section of identical field extracted part subjected to thermal cycle -40°C to 125°C for additional.

Thus, from above values of LM and Weibull chart, the % life consumed is $(N/N_{1\%}) = 63.7\%$ RUL is $1-(N/N_{1\%}) = 36.3\%$ in the anticipated new environment of -40 to 125°C. The field extracted part has been subjected to thermal cycling at -40 to 125°C for additional 750 cycles. The part was cross-sectioned and the exhibited significant cracking in the solder joints. Figure 31 shows the cross-section of the solder joint with crack on the package side of the solder joint. The total prior accrued damage in the field extracted part based on experimental data in anticipated future environment is (1746 Cycles – 750 Cycles) = 1096 Cycles. The prognosticated value of 1112 Cycles correlates well with the measured value of 1096 Cycles. The prognosticated RUL in -40 to 125°C is (1746 Cycles - 1112 Cycles) = 634 Cycles. The RUL can be any combination of thermal cycling and thermal aging. For example if 50% of the RUL is expected to be spent in thermal cycling and the remaining in thermal aging, then only 50% or (0.5*1112) Cycles = 556 Cycles of damage will be accrued in thermal cycling. The remaining 50% of the damage will be accrued in the form of reduction in fatigue life because of thermal aging. The equivalent thermal aging time can be computed from the damage equivalence relationship developed earlier in this paper:

$$N = 9.93(t)^{0.62} \tag{40}$$

$$\Rightarrow 556 = 9.93(t)^{0.62}$$
$$\Rightarrow t = 660 \text{ hrs}$$

Thus, the 50% duty cycle in the future anticipated environment will involve storage of the part for 660 hours at 60°C and 556 Cycles of -40 to 125°C. A similar procedure can be followed for any other combination of thermal cycling and thermal aging in the operational environment.

9. Summary and Conclusions

A method has been developed for prognostication of accrued prior damage and remaining useful life in field-deployed electronics extracted from service after exposure to overlapping sequential environments of thermal aging and thermal cycling. The presented approach uses the

Levenberg-Marquardt Algorithm in conjunction with microstructural evolution of damage based leading indicator for estimating prior accrued damage. Specific damage proxies examined include the phase-growth indicator and the intermetallic thickness. Damage equivalency relationships between thermal aging during storage life and the resulting reduction in thermo-mechanical reliability in cyclic thermal environments during field deployment has been derived and validated based on two damage proxies. Convergence of the damage mapping to a common solution from data based on the two separate leading indicators has been demonstrated. The viability of the approach has been demonstrated for field-deployed parts extracted from service without any prior knowledge of the stress histories. Percentage of the life consumed has been prognosticated. The remaining useful life has been determined in the anticipated future environment. The prognosticated values have been validated versus experimental data by two methods. The first method involved subjected pristine parts to anticipated future environment. Weibull distribution of failures has been developed for the pristine parts. The second method, involved subjecting a subset of the field extracted parts to the anticipated future environment till failure. The prognosticated prior damage of 1112 Cycles correlates with the experimentally measured value of prior damage of 1056 Cycles from -40 to 125°C. In addition, the prognosticated remaining useful life of 634 cycles correlates well with the measured remaining useful life of 750 cycles. The methodology can be used for prognosticating RUL in any combination of thermal aging and cycling an example calculation of 50% duty cycle in cycling yields a value of 556 thermal cycles from -40 to 125°C and 660 hours of thermal aging at 60°C. The process can be used for any other combination of anticipated thermal cycling and thermal aging. Correlation between the prognosticated damage and the actual accrued damage demonstrates that the proposed approach can be used to assess damage accrued under overlapping thermo-mechanical stresses of thermal aging and thermal cycling. In addition, prognostics metrics have been used to quantitatively evaluate the performance of the prognostic algorithms using both the leading indicators. Results demonstrate that both damage proxies work well in estimating accrued damage and estimating residual life.

Acknowledgments

The research presented in this paper has been supported by NASA-IVHM Program Grant NNA08BA21C from the National Aeronautics and Space Administration.

References

Allen, D., Probabilities Associated with a Built-in-Test System, Focus on False Alarms, Proceedings of AUTOTESTCON, IEEE Systems Readiness Technology Conference, pp. 643–645, September 22–25, 2003.

Anderson, N., and Wilcoxon, R., Framework for Prognostics of Electronic Systems, Proceedings of International Military and Aerospace Avionics COTS Conference, Seattle, WA, Aug 3–5, 2004.

Chandramouli, R., Pateras, S., Testing Systems on a Chip, IEEE Spectrum, Vol. 33, No. 11, pp. 42–47, Nov. 1996.

Drees, R., and Young, N., Role of BIT in Support System Maintenance and Availability, IEEE A&E Systems Magazine, pp. 3–7, August 2004.

Gao, R. X., Suryavanshi, A., BIT for Intelligent System Design and Condition Monitoring, IEEE Transactions on Instrumentation and Measurement, Vol. 51, Issue: 5, pp. 1061–1067, October 2002.

Hassan, A., Agarwal, V. K., Nadeau-Dostie, B., Rajski, J., BIST of PCB Interconnects Using Boundary-Scan Architecture, IEEE Transactions on Computer-Aided Design, Vol. 11, No. 10, pp. 1278–1288, October 1992.

Jarrell, D., Sisk, D., Bond, L., Prognostics and Condition Based Maintenance (CBM) A Scientific Crystal Ball, Pacific Northwest National Laboratory, Richland, WA, International Congress on Advanced Nuclear Power Plants (ICAPP), paper number 194 June 2002.

Lall, P., Islam, N., Rahim, K., Suhling, J., Gale, S., Leading Indicators-of-Failure for Prognosis of Electronic and MEMS Packaging, 54th ECTC, Las Vegas, Nevada, June 1–4, 2004[a].

Lall, P., Islam, N., Shete, T., Evans, J., Suhling, J., Gale, S., Damage Mechanics of Electronics on Metal-Backed Substrates in Harsh Environments, 54th ECTC, Las Vegas, Nevada, June 1 - 4, 2004[b].

Lall, P., N. Islam, J. C. Suhling, and R. Darveaux, Model for BGA and CSP Reliability in Automotive Underhood Applications, IEEE Transactions on Components and Packaging Technologies, Vol 27, No. 3, pp. 585–593, 2004[c].

Lall, P., D. Panchagade, Y. Liu, R. W. Johnson, and J. C. Suhling, Models for Reliability Prediction of Fine-Pitch BGAs and CSPs in Shock and Drop-Impact, 54th ECTC, pp. 1296–1303, Las Vegas, NV, June 1–4, 2004[d].

Lall, P., Islam, N., Choudhary, P., Suhling, J., Prognostication and Health Monitoring of Leaded and Lead Free Electronic and MEMS Packages in Harsh Environments, 55th ECTC, pp. 1–9, Orlando, FL, June 1–3, 2005[a].

Lall, P., Panchagade, D., Choudhary, P., Suhling, J., Gupte, S., Failure-Envelope Approach to Modeling Shock and Vibration Survivability of Electronic and MEMS Packaging, 55th ECTC, Orlando, FL, pp. 480–490, June 1–3, 2005[b].

Lall, P., Gupte, S., Choudhary, P., Suhling, J., Solder-Joint Reliability in Electronics Under Shock and Vibration using Explicit Finite Element Sub-modeling, 56th ECTC, San Diego, California, pp.428–435, May 30-June 2, 2006[a].

Lall, P., Choudhary, P., Gupte, S., Suhling, J., Health Monitoring for Damage Initiation & Progression during Mechanical Shock in Electronic Assemblies, 56th ECTC, San Diego, California, pp.85–94, May 30-June 2, 2006[b].

Lall, P., Hande, M., Singh, N., Suhling, J., Lee, J., Feature Extraction and Damage Data for Prognostication of Leaded and Leadfree Electronics, 56th ECTC, San Diego, California, pp.718–727, May 30-June 2, 2006[c].

Lall, P., Islam, N., Rahim, K., Suhling, J., Gale, S., Prognostics and Health Management of Electronic Packaging, IEEE Transactions on Components and Packaging Technologies, Vol. 29, No. 3, pp. 666–677, September 2006[d].

Lall, P., Islam, N., Shete, T., Evans, J., Suhling, J., Gale, S., Damage Mechanics of Electronics on Metal-Backed Substrates in Harsh Environments, IEEE Transactions on Components and Packaging Technologies, Vol. 29, No. 1, pp. 204–212, March 2006[e].

Lall, P., D. Panchagade, Y. Liu, R. W. Johnson, and J. C. Suhling, Models for Reliability Prediction of Fine-Pitch BGAs and CSPs in Shock and Drop-Impact, IEEE Transactions on Components and Packaging Technologies, Vol. 29, No. 3, pp. 464–474, September 2006[f].

Lall, P., Choudhary, P., Gupte, S., Suhling, J., Hofmeister, J., Statistical Pattern Recognition and Built-In Reliability Test for Feature Extraction and Health Monitoring of Electronics under Shock Loads, 57th ECTC, Reno, Nevada, pp. 1161–1178, May 30–June 1, 2007[a].

Lall, P., Gupte, S., Choudhary, P., Suhling, J., Solder-Joint Reliability in Electronics Under Shock and Vibration using Explicit Finite Element Sub-modeling, IEEE Transactions on Electronic Packaging Manufacturing, Vol. 30, No. 1, pp. 74–83, January 2007[b].

Lall, P., M. Hande, C. Bhat, J. Suhling, Jay Lee, Prognostics Health Monitoring (PHM) for Prior-Damage Assessment in Electronics Equipment under Thermo-Mechanical Loads, 57th ECTC, Reno, Nevada, pp. 1097–1111, May 29 – June 1, 2007[c].

Lall, P. Panchagade, D., Liu, Y., Johnson, W., Suhling, J., Smeared Property Models for Shock-Impact Reliability of Area-Array Packages, ASME Journal of Electronic Packaging, Volume 129, pp. 373–381, December 2007[d].

Lall, P., Hande, M., Bhat, C., Islam, N., Suhling, J., Lee, J., Feature Extraction and Damage-Precursors for Prognostication of Lead-Free Electronics, Microelectronics Reliability, Vol. 47, pp. 1907–1920, December 2007[e].

Lall, P., Choudhary, P., Gupte, S., Suhling, J., Health Monitoring for Damage Initiation and Progression during Mechanical Shock in Electronic Assemblies, IEEE Transactions on Components and Packaging Technologies, Vol. 31, No. 1, pp. 173–183, March 2008[a].

Lall, P., Panchagade, D., Choudhary, P., Gupte, S., Suhling, J., Failure-Envelope Approach to Modeling Shock and Vibration Survivability of Electronic and MEMS Packaging, IEEE Transactions on Components and Packaging Technologies, Vol. 31, No. 1, pp. 104–113, March 2008[b].

Lall, P., Hande, M., Bhat, C., More, V., Vaidya, R., Suhling, J., Algorithms for Prognostication of Prior Damage and Residual Life in Lead-Free Electronics Subjected to Thermo-Mechanical Loads, Proceedings of the 10th ITherm, Orlando, Florida, pp. 638–651, May 28–31, 2008[c].

Lall, P., Bhat, C., Hande, M., More, V., Vaidya, R, Pandher, R., Suhling, J., Goebel, K., Interrogation of System State for Damage Assessment in Lead-free Electronics Subjected to Thermo-Mechanical Loads, 58th ECTC, Orlando, Florida, pp. 918–929, May 27–30, 2008[d].

Lall, P., Iyengar, D., Shantaram, S., S., Gupta, P., Panchagade, D., Suhling, J., KEYNOTE PRESENTATION: Feature Extraction and Health Monitoring using Image Correlation for Survivability of Leadfree Packaging under Shock and Vibration, EuroSIME, Freiburg, Germany, pp. 594–608, April 16–18, 2008[e].

Lall, P., Iyengar, D., Shantaram, S., Pandher, R., Panchagade, D., Suhling, J., Design Envelopes and Optical Feature Extraction Techniques for Survivability of SnAg Leadfree Packaging Architectures under Shock and Vibration, 58th ECTC, Orlando, Florida, pp. 1036–1047, May 27–30, 2008[f].

Lall, P., Shantaran, S., Angral, A., Kulkarni, M., Explicit Submodeling and Digital Image Correlation Based Life-Prediction of Leadfree Electronics under Shock-Impact, 59th ECTC, pp. 542–555, San Diego, California USA, May 25–29, 2009[a].

Lall, P., Gupta, P., Panchagade, D., Angral, A., Fault-Mode Classification for Health Monitoring of Electronics Subjected to Drop and Shock, 59th ECTC, pp. 668–681, San Diego, California USA, May 25–29, 2009[b].

Lall. P., More, V., Vaidya, R., Goebel, K., Prognostication of Latent Damage and Residual Life in Leadfree Electronics Subjected to Multiple Thermal-Environments, 59th ECTC, pp. 1381–1392, San Diego, California USA, May 25–29, 2009[c].

Lall. P., Lowe, R., Goebel, K., Resistance Spectroscopy-based Condition Monitoring for Prognostication of High Reliability Electronics Under Shock-Impact, Proceedings of 59th ECTC, pp. 1245–1255, San Diego, California USA, May 25–29, 2009[d].

Lall. P., Lowe, R., Goebel, K., Prognostics Using Kalman-Filter Models and Metrics for Risk Assessment in BGAs Under Shock and Vibration Loads, Proceedings of 60th ECTC, pp. 889–901, Las Vegas, Nevada, USA, June 2–5, 2010[a].

Lall. P., More, V., Vaidya, R., Goebel, K., Assessment of Residual Damage in Lead-free Electronics Subjected to Multiple Thermal Environments of Thermal Aging and

Thermal Cycling, Proceedings of 60th ECTC, pp. 206–218, Las Vegas, Nevada, USA, June 2–5, 2010[b].

Lall, P., Gupta, P., Panchagade, D., Self-Organized Mapping of Failure Modes in Portable Electronics Subjected to Drop and Shock, Proceedings of 60th ECTC, pp. 1195–1208, Las Vegas, Nevada, USA, June 2–5, 2010[c].

Lall. P., Lowe, R., Goebel, K., Use of Prognostics in Risk-Based Decision Making For BGAs Under Shock And Vibration Loads, Proceedings of ITHERM 2010, Las Vegas, NV, June 2–5, 2010[d].

Lall. P., More, V., Vaidya, R., Goebel, K., PHM-Based Residual Life Computation of Electronics Subjected to a Combination of Multiple Cyclic-Thermal Environments, Proceedings of ITHERM 2010, Las Vegas, NV, June 2–5, 2010[e].

Lall, P., Gupta, P., Panchagade, D., Angral, A., Fault-Isolation in Portable Electronics Subjected to Drop And Shock, Proceedings of ITHERM 2010, Las Vegas, NV, June 2–5, 2010[f].

Lall P., Hinshaw R., Pandher R., Harsha M., Suhling J., Thermo-Mechanical Reliability of SAC Lead-free Alloys Proceedings of ITHERM 2010, Las Vegas, NV, June 2–5, 2010[g].

Lall P., Angral A., Suhling J., Board Trace Fatigue Models and Design Guidelines for Electronics Under Shock-Impact , Proceedings of ITHERM 2010, Las Vegas, NV, June 2–5, 2010[h].

Lall P., Kulkarni M., Angral A., Panchagade D., Suhling J., Digital-Image Correlation and XFEM Based Shock-Reliability Models for Leadfree and Advanced Interconnects Electronic Components and Technology Conference, 2010. ECTC 2010. 60th, 91–105, 2010[i].

Lall P., Shantaram S., Panchagade D.,Peridynamic-Models Using Finite Elements for Shock and Vibration Reliability of Lead-free Electronics Proceedings of ITHERM 2010, Las Vegas, NV, June 2–5, 2010[j].

Lourakis, M., I., A., A brief Description of the Levenberg-Marquardt algorithm implemented by Levmar, Foundation of Research and Technology – Hellas (Forth), Greece, pp. 1- 6, Feb 11, 2005.

Madsen, K., Nielsen, H., B., Tingleff, O., Methods for Non-Linear Least Squares Problems, Technical University of Denmark, Lecture notes, available at http://www.imm.dtu.dk/ courses/02611/ nllsq.pdf, 2nd Edition, pp. 1–30, 2004.

Marko, K.A., J.V. James, T.M. Feldkamp, C.V. Puskorius, J.A. Feldkamp, and D. Roller, Applications of Neural Networks to the Construction of "Virtual" Sensors and Model-Based Diagnostics, Proceedings of ISATA 29th International Symposium on Automotive Technology and Automation, pp.133–138, June 3–6, 1996.

McCann, R. S., L. Spirkovska, Human Factors of Integrated Systems Health Management on Next-Generation Spacecraft, First International Forum on Integrated System Health Engineering and Management in Aerospace, Napa, CA, pp. 1–18, November 7–10, 2005.

Mishra, S., Pecht, M., In-situ Sensors for Product Reliability Monitoring, Proceedings of SPIE, vol. 4755, pp. 10–19, 2002.

Nielsen, H., B, Damping Parameter in Marquardt's Method, Technical Report, IMM-REP-1999–05, Technical University of Denmark, Available at http://www.imm.dtu.dk/˜hbn, pp. 1–16, 1999.

Rosenthal, D., and Wadell, B., Predicting and Eliminating Built-in Test False Alarms, IEEE Transactions on Reliability, Vol. 39, No 4, pp. 500–505, October 1990.

Saxena, A., J. Celaya, B. Saha, S. Saha, and K. Goebel, Evaluating Algorithm Performance Metrics Tailored for Prognostics, IEEE Aerospace Conference, Big Sky, MT, pp. 1–11, March 2008[a].

Saxena, A., J. Celaya, E. Balaban, K. Goebel, B. Saha, S Saha, and M. Schwabacher, Metrics for Evaluating Performance of Prognostic Techniques, Intl. Conf. on Prognostics and Health Management, Denver, Colorado, pp. 1–17, October 2008[b].

Schauz, J. R., Wavelet Neural Networks for EEG Modeling and Classification, PhD Thesis, Georgia Institute of Technology, 1996.

Shiroishi, J., Y. Li, S. Liang, T. Kurfess, and S. Danyluk, Bearing Condition Diagnostics via Vibration and Acoustic Emission Measurements, Mechanical Systems and Signal Processing, Vol.11, No.5, pp.693–705, Sept. 1997.

Williams, T. W., Parker, K. P., Design for Testability- Survey, Proceedings of the IEEE, Vol. 71, No. 1, pp. 98–112, January 1983.

Zhang, Y., Cai, Z., Suhling, J., Lall, P., Bozack M., The Effects of SAC Alloy Composition on Aging Resistance and Reliability, 59th Electronic Component and Technology Conference, San Diego, CA, pp. 370–389, May 26–29, 2009.

Zorian, Y., A Structured Testability Approach for Multi Chip Boards Based on BIST and Boundary Scan, IEEE Transactions on Components, Packaging, and Manufacturing Technology-Part B, Vol. 17, No. 3, pp. 283–290, August 1994.

CHAPTER 16

TOWARDS EMBEDDED COOLING - GEN 3 THERMAL PACKAGING TECHNOLOGY

Avram Bar-Cohen

University of Maryland, Department of Mechanical Engineering, College Park, MD 20742, U.S.A.
abc@umd.edu

From the dawn of the Information Age thermal management technology has played a key role in the continuing miniaturization, performance improvements, and higher reliability of electronic systems. During the past 65 years, thermal packaging has migrated from ventilation and air-conditioning to cabinet cooling, to package cooling with heat sinks and cold plates, and is today addressing on-chip hot spots and near-junction thermal transport. Following a brief history of thermal packaging, attention will turn to a review of emerging micro and nano-technologies for reducing the thermal resistance of defense electronic systems. The asymptotic maturation of current technology and growing thermal management demands in high performance computing and RF systems have led DARPA to initiate efforts in third-generation "embedded" thermal management technology based on intrachip and interchip microfluidic cooling. The motivation, technological thrusts, and promise of this new thermal management paradigm will be discussed.

Introduction

Thermal packaging technology has been a key enabler in the development of today's microelectronic systems, including smart phones, tablet computers, and navigation systems, that have come to define our lives in the 21st Century. Much of the benefit that we derive from miniaturization, higher performance, lower cost and greater reliability of these quintessential 21st Century "widgets," can be traced to improvements in thermal technology, thermal modeling, and the integration of thermal management principles and techniques into electronic product development. However, the user community, as well as many of the engineers developing these components and systems, are generally unaware of the trajectory followed by the evolution in thermal packaging technology over the nearly 70 years of the Information Age. That evolutionary trajectory and the six order-of-magnitude geometric shrink implicit in the arc of "inward migration" of thermal packaging technology – from the floor of the computer room to the silicon chip – is the theme of this essay.

The narrative begins with a brief history of thermal packaging, starting in 1946 with the introduction of the first programmable computer – Mauchly and Eckert's ENIAC (Electronic Numerical Integrator and Computer) at the University of Pennsylvania in 1946 – and covering the seven decades of intervening development. This historical survey will describe the occurrence of two distinct paradigms of thermal packaging technology, extending over five identifiable eras, and tracing the inward migration of cooling techniques from the Gen-1 computer room cooling techniques to the Gen-2 "attached micrcoolers" with engineered thermal paths transferring heat away from the chip to the ambient air or fluid. Attention will be devoted to current thermal management requirements, driven by nano-electronics, which confront packaging engineers with the simultaneous "triple threat" of high-power, "hotspots," and 3D integration. The consequences of the community's ineffectual response to this "triple threat" and the possibility that, in the absence of thermal packaging innovations, we will reach the end of Moore's Law in chip performance, will also be discussed.

Concern about the impact of the end – or even just a weakening of - Moore's Law, led the Defense Department's DARPA (Defense Advanced Research Projects Agency) to initiate a major thermal packaging program some 15 years ago. While the advances in cooling hardware achieved by DARPA's Thermal Management Technologies (TMT) program will be shown to have significantly raised the State-of-the-Art in key cooling components, these developments have been unable to reverse the plateauing of heat removal capability, associated with the current, Gen-2 "attached cooler" paradigm. Thus, the final sections of this essay will deal with the emergence of "embedded cooling," as the Gen-3 thermal packaging paradigm, sequentially examining the limitations of the "attached cooler" technology, the relevant technologies for "embedded cooling," the payoff for the electronic industry in embracing this Gen-3 paradigm, and the challenges to be overcome before embedded cooling can become the preferred choice for the thermal management of high performance electronics.

Thermal Packaging History

The Information Age was born in 1946 with the introduction of Mauchly and Eckert's ENIAC or Electronic Numerical Integrator and Computer, the first programmable computer [1]. The ENIAC, as seen in Figure 1, was housed in 80ft of 8-foot high cabinets, supporting some 17,000 vacuum tubes and dissipating approximately 170 kilowatts, while performing 5000 operations per second (OPS). The ENIAC would be little more than a footnote on a performance map of today's peta-FLOP (10^{12} Floating Point Operations) supercomputers but it did open the door to the symbiotic existence of programmable computing AND thermal packaging. In today's microprocessor-centric world, it is sobering to realize that these 5000 OPS, enabled by 10 W vacuum tubes, required an office-building's worth of electrical power and cooling.

Fig. 1. Electronic Numerical Integrator and Computer, circa 1946.

During the first few decades of the Information Age, thermal management requirements were expressed in terms of maintaining "chilly" but dry computer room air temperatures of 20°C to 25°C. These requirements and the volume of the early computing hardware, defined HVAC (Heating, Ventilation, and Air-Conditioning) technology as the first paradigm and first generation of thermal packaging technology. During this period, thermal management of vacuum tube computers like the ENIAC and IBM's Mark 1, seen in Figure 2, and some of the earliest solid-state computers, like the IBM System/360, shown in Figure 3, along with AT&T's telephone switching equipment, was achieved through better air circulation, removing heated air from the room, the rack, or the cabinet and replacing it with cooler air supplied to the computer facility. While these heat removal requirements, sometimes in excess of 100 kW of heat from a relatively small computer room, drove the expansion of air conditioning technology, during this first era of HVAC thermal packaging, little attention was paid to how the heat was generated by the electronic components or how it made its way to the air circulating through the cabinet and room. Thermal packaging engineers were told to "get that hot air out of here, and we'll be okay."

Fig. 2. IBM Mark 2, circa 1950.

Fig. 3. IBM System 360, circa 1982.

As the Information Age progressed into the 1970's and early 1980's, transistor and package miniaturization resulted in more and more components attached to printed circuit boards (PCB's), ranging in size from just 5 cm or 2in on a side, to a more common 10-15 cm (4"-6") formfactor, but also reaching 50 cm (20") on a side with 60 laminated layers. The prevailing dual-in-line packages, or DIP's, with their spider-like legs or leads, were stuffed into these PCB's, using through-board holes, and wire-wrapped on the backside for the electrical connection to the system. This created relatively high heat dissipation on the populated side of the PCB, but only modest cooling needs on the back of the PCB. Thus, during this second, or "rack cooling," era of the HVAC paradigm, "chimney" type, natural convection cooling of the electronic racks, sometimes assisted by internal cooling coils, was generally adequate to remove the dissipated heat.

However, when surface mount technology (SMT) gained broad acceptance and components could be attached on both sides of the PCB with relatively tight spacings between adjacent solder pads, the "rack cooling" thermal management requirements – and most notably the "cooling density," or heat removal per unit volume – experienced a significant leap. These densely populated printed circuit boards, often in arrays of 5-10 PCB's, were typically assembled in "card cages," as seen in Figure 4, and considerable effort was devoted to maintaining sufficient air flow to keep the maximum air temperature in the same range as the computer room, i.e. less than 25-30°C.

Fig. 4. Typical Printed Circuit Board "Card Cage," circa 1982.

In time it was realized that the DIP's and SMT components on these boards needed to be maintained at a specified "case" temperature. The focus of "rack cooling" thermofluid design then became determining and setting the gap between the PCB's, for both natural and forced convection cooling inside the cabinet, which would yield optimum performance and cooling efficiency [2]. This Gen-1 of thermal packaging, involving implementation of HVAC technology in programmable computers, is thus seen to begin in the 1940's with computer room heat removal and, over the next few decades, migrate into the racks and cabinets containing solid-state electronic components. During the latter years of this paradigm, emphasis was placed on maximizing the cooling density and, hence, the functional performance that could be achieved in the volume available in the standard rack formfactors, while maintaining a specified package surface, or "case," temperature.

In the 1980's the maturation of bipolar technology, principally based on ECL (emitter-coupled logic) and related semiconductor devices, resulted in a rapid escalation of transistor counts and power dissipation in each chip, reaching as high as 10 W-12 W in Motorola's 10,000 transistor MCA chips. To meet functional requirements and compete with early CMOS chips, major computer companies, including IBM, Honeywell, NEC, and Fujitsu mounted such ECL chips to ceramic substrates and created 50-100 chip, field-replaceable multichip modules, whose heat dissipation could exceed 300 W. The escalating thermal challenges in multichip modules, with chip heat flux above 10 Watts/cm^2, a total dissipation approaching 500 W, and module heat fluxes nearing 1 W/cm^2, necessitated a new approach to thermal packaging of high performance electronics.

By the mid-1980's it had become apparent that more aggressive measures were needed to gain control over the "case" temperature of the multichip module and that measures beyond blowing air through the cabinet would be needed to reduce the thermal resistance between the "case" and the ambient while also increasing the cooling density. Moreover, it was argued that, for very high thermal loads, consideration could be given to using refrigerants, which had previously be used to chill the cabinet air, to reduce the local case temperature below the ambient temperature. Thus, in the mid-1980's and despite strong resistance from the electrical designers and computer architects, nearly all the high performance computer platforms adopted liquid and/or refrigerant cooling to provide the required thermal control. The conversion to module- and package-size coldplates defined a new "attached cooler" paradigm and ushered in the second generation of thermal packaging technology [3,4].

The application of such aggressive liquid coldplate, thermal control measures came to a relatively abrupt end, in the late 1980's and early 1990's, when Complementary Metal Oxide Semiconductor (CMOS) technology, with far higher transistor counts but much lower power dissipation per gate, began to replace bipolar devices. Initially, CMOS chips with several million transistors operating in the MHz range dissipated just a few Watts, quite a bit less than the bipolar devices they replaced. These "computers on a chip" enabled high performance mini-computers and desk-top, to be followed by laptop computers. Later CMOS technology penetrated into mainframes and supercomputers, which now no longer required aggressive refrigeration and liquid-cooling approaches. However, the local heat flux on these chips was higher than could be dissipated with direct natural and forced convection, provided by air blowing past these single chip packages and "attached coolers" were needed to reduce the case-to-air thermal resistance.

Fig. 5. Air-Cooled Heat Sink Types – By Manufacturing Technique.

While the first era of the "attached cooler" paradigm involved liquid and refrigeration cooled coldplates, the second era of this paradigm focused on air-cooled heat sinks attached to packages containing single chips or small chip sets. As the CMOS clock speed increased from 1-2MHz to tens of MHz and later hundreds of MHz, and the transistor counts increased to 10's of millions and later 100's of millions of transistors, chip power dissipation increased and began to rival the power dissipation of the older bipolar technology, reaching 10's of watts per chip. The design and optimization of air-cooled heat sinks, or microcoolers, and their associated Thermal Interface Materials, or TIMs, for attachment to the chip packages, became the driver for thermal management technology and led to a proliferation of heat sink designs and manufacturing techniques, as suggested in Figure 5. On-going efforts during this heat sink era, see for example [5] have made it possible to cost-effectively air cool individual chip packages, with power dissipation of several hundred Watts and densely populated boards – as shown for example in Figure 6 – of 1.5 kW to 2 kW.

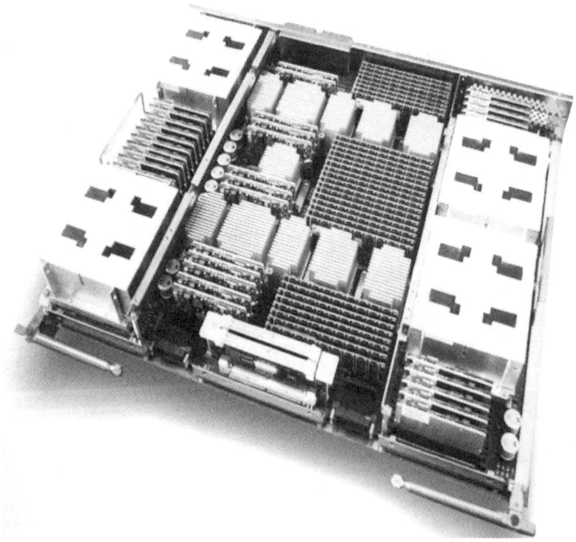

Fig. 6. Fujitsu 8 CPU Air-Cooled Board, circa 1985.

Impact of Nanotechnology

As miniaturization continued throughout the late 1990's and into the 21st century, the advent of nanoelectronics (with device feature sizes below 100 nm) enabled higher and higher transistor densities and switching speeds [6]. At the GHz switching speeds of today, and with billions of transistors on a single silicon chip, chip power dissipation has penetrated into uncharted territory, far above the bipolar chip dissipations of the 1980's and 1990's. While subtle differences exist among the various component roadmaps, following the iNEMI roadmap – shown in Figure 7 - and recognizing some adjustments over time, it can be seen that microprocessors for desktop machines are today designed to operate above 150 watts, while server and computer cluster chips are likely to exceed power dissipations of 500 W per chip. In automotive applications, in which chip sizes are typically smaller than desktop computers and servers, power dissipation is reaching 300 watts with heat fluxes of 200-300 W/cm^2. So, despite the scaling relations implied by Dennard's equation, the introduction of nano-scale electronics has not diminished the need for thermal management. However, providing satisfactory thermal management to address the total chip power dissipation, as well as chip heat flux, continues to be very challenging.

That continuing trend to higher chip power dissipations would be sufficient – in and of itself – to motivate the ongoing efforts to develop enhanced thermal management technology. But, nanotechnology has spawned two complimentary trends – on-chip hot spots and three-dimensional stacking – which, together with rising power dissipation, pose a "triple threat" to successful cooling of advanced electronic components. A recent example of IBM chip technology [7], depicted in Figure 8, provides a representative example of such hotspots, occurring in a 40 mm × 40 mm 900 W multichip module, with four microprocessors, four memories, and a centrally-located switch. For the module described, the resulting average power dissipation of 50 W/cm^2 is within the range of conventional thermal management capability. But, the 2 × 2 mm hot spots dissipate 20% of the 100 watts of each microprocessor, producing a heat flux that is 10 times the average, or

500 W/cm^2, and creating a temperature spike, or hot spot that, as shown in Figure 8, is some 30 degrees above the average temperature of the chip. It is hotspots like these that have become the drivers for the thermal management solution and need to be addressed, often ahead of dealing with the average power dissipation of the chip.

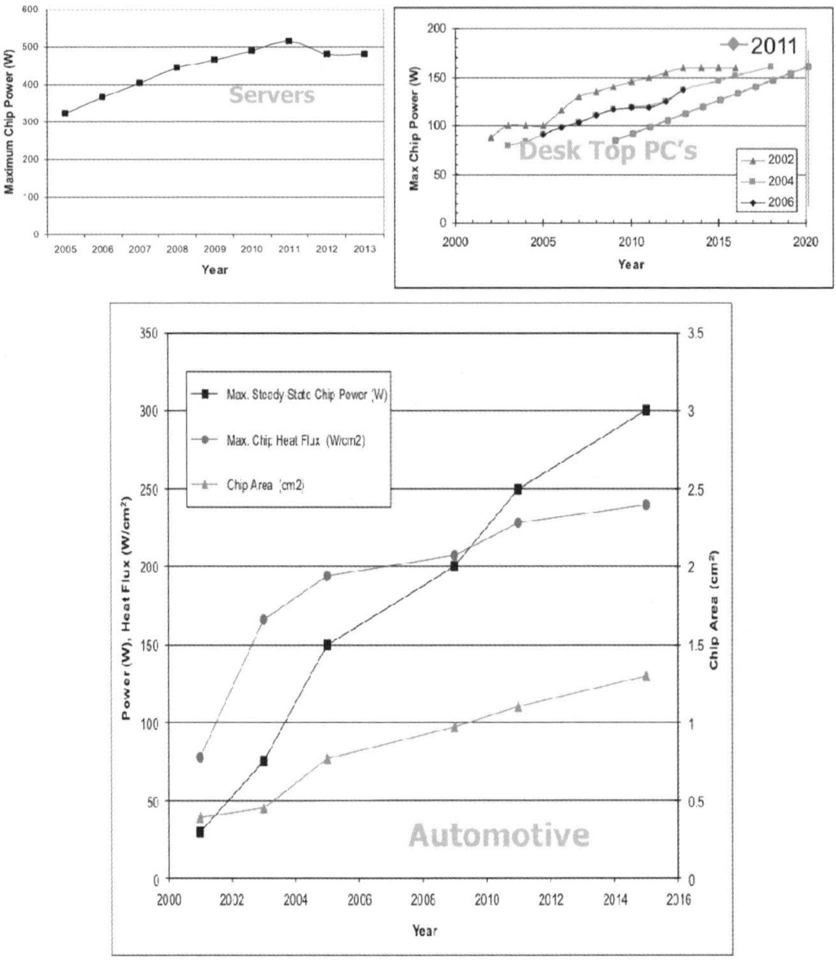

Fig. 7. Chip Power Dissipation - iNEMI Roadmap.

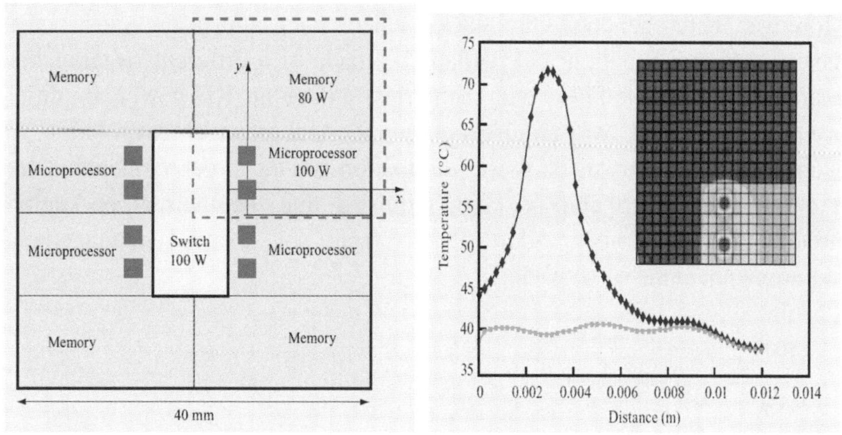

Fig. 8. On-Chip Hot Spots in IBM 3D Chip Stack [7].

To provide a context and perspective on these chip and hotspot heat fluxes, the reader is directed to Figure 9, where it can be seen that the chip hotspots, at values exceeding 1 kW/cm^2, are well beyond heat fluxes encountered in other engineered systems, like nuclear blasts, rocket motor cases, and atmospheric re-entry and are in the range of the heat flux on the surface of the sun. However, while the temperature of the sun's surface is approximately 5600K, the on-chip hot spots need to be maintained at temperatures that are well below 400K (or 130C) and often closer to 350K (80C). Even higher heat flux microprocessor hot spots, in some cases reaching 1.5 kW/cm^2, have been reported and GaN HEMT power amplifiers can produce hot spots in excess of 30 kW/cm^2. Clearly, the kW/cm^2 hot spot heat flux is a formidable thermal management challenge.

Dealing with these hot spots is made even more difficult by the introduction of three-dimensional packaging configurations. Such 3D stacks, with their compressed interconnect lengths, are capable of both improving performance and raising computational energy efficiency by reducing interconnect power dissipation [8]. There is, thus, no doubt that thermal management technology will need to be developed for this formfactor. Examining some representative chip stacks using wire bonding and/or solder bumping, as shown in Figure 10, it is readily apparent that gaining thermal management access to high heat flux hot spots in a stack, especially if they are located on an interior chip, is greatly complicated in such chip stacks.

Moreover, heterogeneous chip stacks, where each chip in the stack could have a different function and perhaps be made of a different material, for example a stack of CPUs and memories, and/or an RF power amplifier inserted into a stack, will require a unique thermal management solution. In such a heterogeneous stack, power dissipation and the target temperature are likely to vary from chip to chip, requiring that the thermal packaging strategy simultaneously addresses a range of thermal management requirements in the same package.

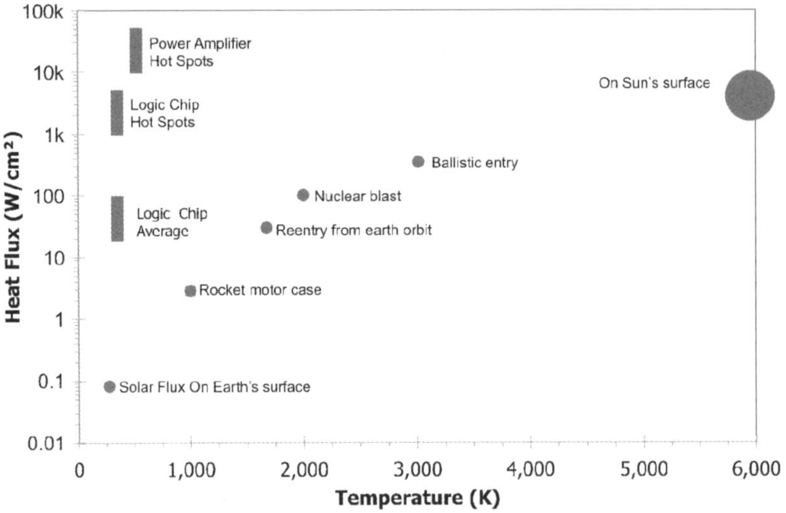

Fig. 9. Typical Heat Fluxes and Operating Temperatures - Semiconductor Chips and Other Thermal Challenges.

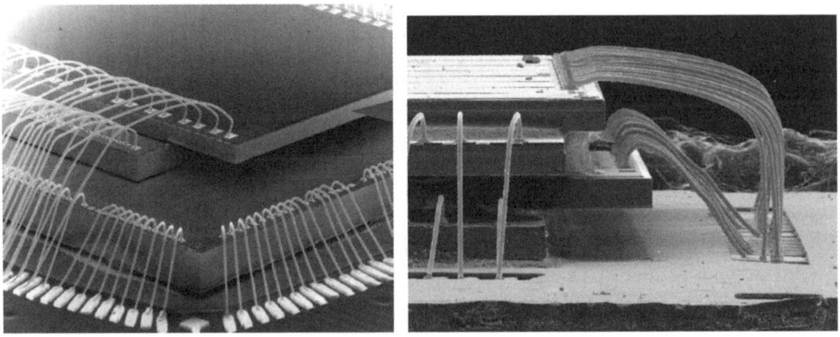

Fig. 10. Typical 3D Chip Stacks, circa 2005.

Returning to the volumetric cooling metric introduced earlier in the discussion, it is daunting to realize that the cooling density for a chip stack can easily exceed 1 kW/cm^3. For example, a stack of ten, 150 W chips, 1 cm × 1 cm in area and 0.5 mm thick (being very generous), and spaced 0.5 mm apart (which is much larger than the actual vertical pitch), would dissipate 1.5 kW in 1 cm^3. These chips could easily dissipate 200 or 300 watts each and the height of the stack could be closer to 5 mm, so that the cooling density could easily reach 5 kW/cm^3. Once again, this thermal management metric can be seen in Figure 9 to fall substantially above the heat density of other, more common thermal management challenges – well above the Cray-3 and the SX3, that are water-cooled computers but only reach 10's of W/cm^3, and interestingly, the human brain dissipating only 0.07 W/cm^3 and operating at even lower temperature.

These figures reveal that thermal management has become far more challenging in the nanoelectronics era, with today's CMOS transistors operating at the 3.5 gigahertz level using features that are typically near 22 nm. As summarized in Figure 11, power dissipation in such chips is increasing and also creating on-chip hotspots. In addition, three-dimensional chip stacks are entering common practice, limiting access to a chip deep in the stack but also violating the traditional allocation of chip "real estate" – providing one side of the chip for the electrical interconnect and the other side of the chip for thermal management. In 3-D stacks there is no "other side," so that the modes of thermal management and the thermal management paradigm clearly have to change to deal with this triple threat.

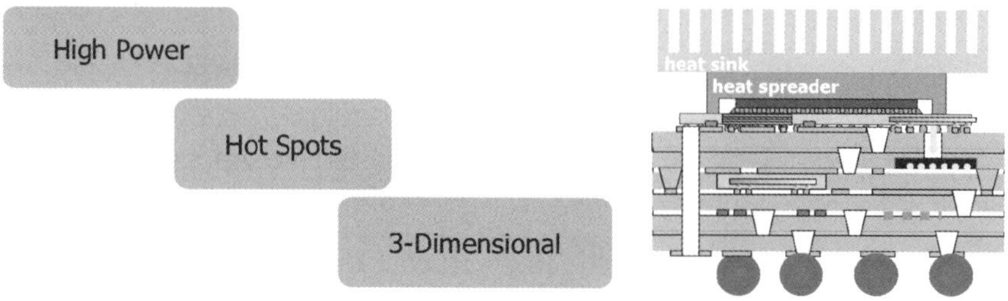

Fig. 11. Nanoelectronic Thermal Packaging - Triple Threat.

Thermal Management and Moore's Law

Moore's Law is easily the best recognized principle in the world of electronic components and systems and, other than Ohm's Law, has probably played the most important role in the success of the electronic industry. The pivotal contribution of thermal packaging to sustaining Moore's Law is less well known but has recently come into focus through the 2012 NRC book, *"The Future of Computing Performance: Game Over or Next Level?"* by Fuller and Millett [9], which posits that power delivery and thermal management are limiting chip and system performance.

Figure 12 provides a graphical depiction of Moore's Law, revealing that in the infancy of the industry in the 1960's, the component counts were doubling every year, then the rate steadied to a somewhat slower rate, doubling every two years, as codified by Gordon Moore in "the Law." While there were occasional departures from Moore's law – and periods of industry-wide angst when component counts and clock frequencies seemed not be increasing at the expected rate – until the last few years, this doubling rate was remarkably well preserved across nearly 4 decades of technology development.

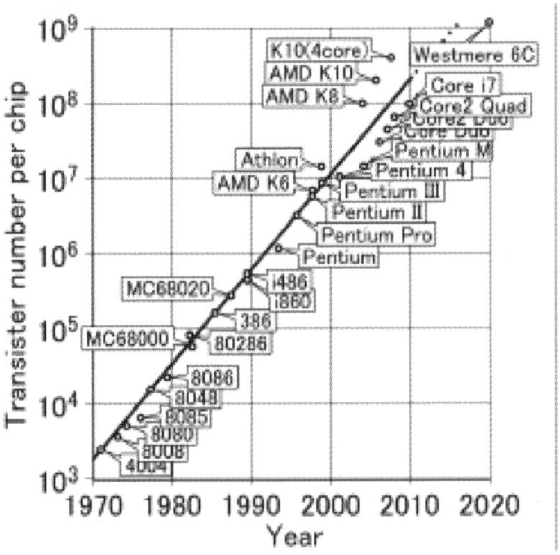

Fig. 12. History of Solid State Transistor Count - Moore's law [8].

The immutability of Moore's Law, as a self-fulfilling prophecy, over those many decades, enabled the growth of an historic symbiosis between the product developers and researchers in the micro-nano electronic industry. The near-certainty, that chip performance, component density, and cost would all follow Moore's law, emboldened product developers and entrepreneurs to stay on an aggressive roadmap, developing ever cheaper, smaller, faster products to capitalize on the expected component improvements and sometimes delaying a product release until it could better intercept the appropriate technology curves. Whenever technology gaps did develop, threatening the Moore's law cadence of progress, investments were quickly committed to the gap-filling research needed to stay on the roadmap. Researchers, developers, and investors – working together to sustain and exploit the technology roadmaps – fueled the explosive global expansion of the electronic industry and underpinned much of the economic growth of the last few decades.

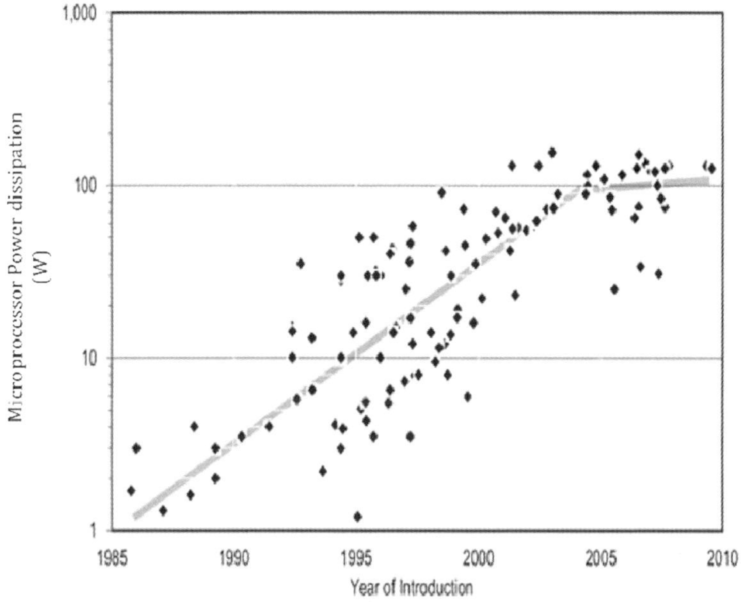

Fig. 13. History of Microprocessor Power Dissipation [9].

The power and durability of Moore's Law, over the past 4 decades, has obscured the vision of many in the electronic industry and hidden the current hiatus in chip performance. But, as seen in Figure 13, and argued by Fuller and Millett (and their authors) in [9], microprocessor power dissipation plateaued around 2005 at approximately 100 W/chip, for an air-cooled microprocessor, capping further improvements in chip performance. Examining Figure 14, it may be seen that along with the plateauing of power dissipation, the clock frequency has stalled at around 3 GHz, driving a single-thread performance stall, as well. And while transistor counts continue to increase, the inability to remove dissipated heat has lead to more and more unused, or so-called "dark," silicon so that the increase in transistor count is not fully exploited. Observing these trends, Fuller and Millett concluded in 2012 that the growth in the performance of computing systems will become limited by their power and thermal requirements within the next decade. It would appear that this has already come to pass.

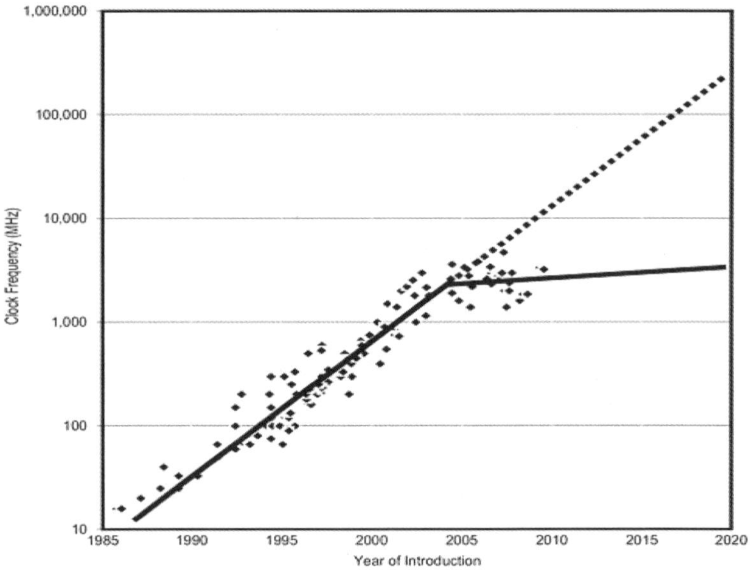

Fig. 14. History of Microprocessor clock frequency [9].

Further support for the existence of a thermal plateau can be seen in Figure 15 which shows package and module heat flux for both bipolar and CMOS technology. The locus for bipolar technology rises rapidly, starting in the 1950's to approximately 12 W/cm^2 in mainframe computers, such as IBM's ES9000, in the 1980s. With the introduction of CMOS, module heat flux drops down to 1-2 W/cm^2 but then begins to rise as CMOS switching speeds increase and, by 2005, reaches the same board/module heat flux of approximately 12-13 W/cm^2.

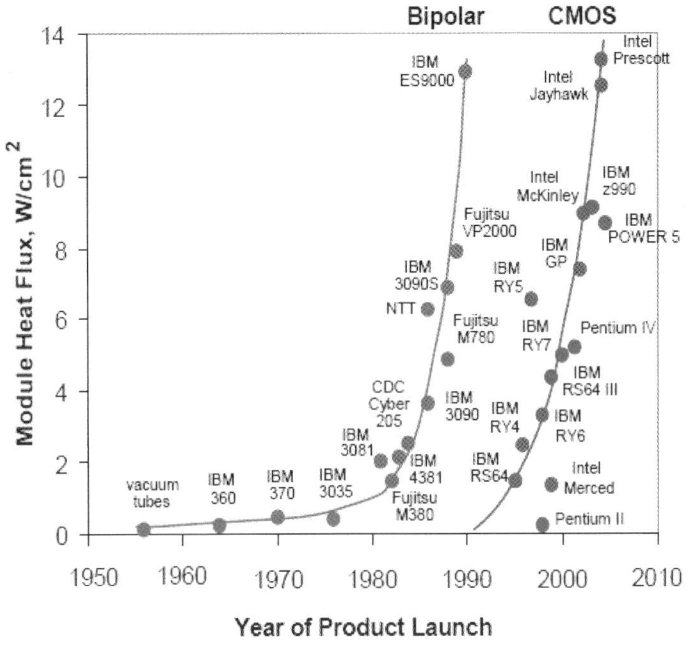

Fig. 15. History of Computer Module Heat Flux – ECL, CMOS [55].

To fully understand the present barriers to further Moore's Law progress in component performance, reference must be made to the so-called Dennard "scaling" that made it possible to progress through the numerous feature shrinks associated with the Moore's Law progression, with only modest increases in power dissipation.

$$P_{density} = N_g C_{load} V^2 f = \text{ power per unit area = heat flux}$$

$$N_g = \text{CMOS gates/unit area}$$

$$C_{load} = \text{capacitive load/CMOS gate}$$

$$V = \text{supply voltage}$$

$$f = \text{clock frequency}$$

$$\mathrm{P}_{density(scaling)} = \left(k^2\right)\left(1/k\right)\left(1/k^2\right)\left(k\right) = 1$$

k = scaling factor, typically 1.4 per geometry shrink

$1/k$ = device feature scaling factor, typically 0.7 per geometry shrink

Dennard's Equation, above, relates power dissipation or heat flux, to the gate count, the capacitive load, the supply voltage, and the clock frequency. In an ideal device, reducing the feature size by a specified "scale factor", k, results in a k^2 increase in gate count, 1/k decrease in capacitive load, $1/k^2$ in supply voltage, and a k increase in frequency. Fortuitously, if the geometric shrink can be implemented perfectly across the chip and the voltage reduced without other complications, the scale changes are "self-cancelling" and power dissipation per unit area, or the heat flux, remains unchanged. For the past 2-3 decades, the scale factor has typically been 1.4, thus reducing the feature size by 0.7, doubling the number of transistors per unit area, and increasing the clock speed by 40% per generation. While theoretically the power dissipation should have remained unchanged, fabrication subtleties and imperfect scaling contributed to modest power increases. Nevertheless, Dennard scaling did maintain a much shallower power dissipation slope (vs time) than the doubling rate of the transistors.

However, with Dennard scaling having driven the supply voltage to 0.9V, it is now far more difficult to implement additional voltage cuts. And, with power dissipation varying as V^2, if voltage scaling ceases, the heat flux will double with every 0.7 shrink. Moreover, with feature sizes already down in the 22 nanometer range, even with significant advances in optics, lithographic and cost considerations may limit how many more feature size "shrinks" can be implemented. Reflecting the voltage and feature size "stall," clock frequency increases can be expected to produce a steep rise in chip power dissipation in the years ahead. If, as suggested by the plateauing of chip heat dissipation shown in Figure 13, such an increase cannot be accommodated, the Moore's Law cadence can not be sustained!

Based on this analysis, Fuller and Mullet concluded that massive concurrency (or on-chip parallelism with a large number of "cores") is the only viable path that can be followed to restoring the Moore's Law progression. The above suggests that better thermal management is another viable approach to this goal and that enhanced cooling could make it possible to restore the chip frequency progression associated with Moore's Law, even if that occurs at the expense of higher power dissipation.

Micro-Nano Materials and Structures for Thermal Management

Recognition of the growing threat to the Moore's Law progression and the implications for electronic product development led the US Defense Advanced Research Projects Agency (DARPA) in the late 1990's to initiate a series of thermal packaging studies in the framework of the Heat Removal by Thermo-Integrated Circuits (HERETIC) program. The HERETIC program covered a broad spectrum of thermal management techniques, and included jet-impingement cooling, thermoelectrics, MEMS-based approaches, and thermoacoustic cooling, among others, see, for example, [10-16]. HERETIC and THREADS, which was focused on device-scale thermal management, were followed by the Thermal Management Technologies Program (TMT), launched in 2008 and targeted at using micro-nano materials and transport processes to reduce the individual, as well as the overall, thermal resistances in the prevailing "attached microcooler" paradigm [17]. The Near Junction Thermal Transport(NJTT) program started in 2012 and the Inter-Intra Chip Enhanced Cooling (ICECool) program in 2013 [18] represent a significant departure from Commercial–Off-the-Shelf technology and will be discussed in a later section.

The TMT program was comprised of 4 thrust areas, each addressing a single element of the thermal resistance chain between the chip and the ambient air. The Thermal Ground Plane (TGP) effort was directed at high-performance heat spreaders, which use phase change processes to replace copper-alloy spreaders in conventional systems. The Microtechnologies for Air-Cooled Exchangers (MACE) required developers to reduce the heat sink-to-ambient thermal resistance by using synthetic jets and microagitators to increase the convective heat transfer coefficients on the heat sink surfaces, reduce the volume of the heat sink blower, and increase the overall system (heat sink and blower) coefficient of performance. Technologies developed under the Nano Thermal Interfaces (NTI) thrust of the TMT program provide significant reductions in the thermal resistance of the thermal interface layer (TIM) between the electronic device and the spreader or a heat sink. The Active Cooling Modules (ACM) thrust was centered on investigating localized active cooling of electronic devices using thermoelectric and Stirling coolers, respectively.

Thermal Ground Plane (TGP): The TGP thrust was focused on developing a highly-conductive, thin, lightweight substrate for electronic systems and multi-chip modules, incorporating device-compatible and micro- and/or nanostructured materials to achieve superior heat spreading and possessing the mechanical properties necessary for hard-mounting integrated circuits. The TGP program was inspired by traditional heat pipes or vapor chambers and the TGPs developed incorporated many of the advantages of that technology, such as high thermal conductivity and high reliability, as well as requiring no moving parts, electrodes, or external power for their operation. However, the difference in thermal expansion coefficients between silicon or ceramic microelectronic components and metallic heat pipes, as well as the need for a planar chip-size attachment surface for these devices, has limited the use of heat pipes in this role. The TGP program aimed to combine the advantages of heat pipes and solid conductors, building thin, flat vapor chambers whose thermal expansion coefficients cover a range typical of electronic devices and which provide high heat flux removal with very low thermal spreading resistance. The primary TGP program goal was a lateral thermal conductivity in the range of 10 kW/m-K–20 kW/m-K, or approximately 25-50 times higher than copper and 10-20 times higher than synthetic diamond.

The effective TGP thermal conductivities achieved by the 8 performers, including GE, Teledyne, UC Santa Barbara, UCLA, University of Colorado, UC Berkeley, NGES, and Raytheon, participating in this program, are shown in Figure 16, along with several analogous commercial off-the-shelf (COTS) products. These prototype TGPs, including more than one formfactor per performer, are seen to provide a range of thermal conductivities all of which exceed copper, nearly all have conductivities better than diamond or thermal pyrolitic graphite, and a few have exceeded the COTS spreaders with conductivities greater than 10,000 W/m-K. Nearly all the performers also succeeded in producing "thin outline" TGP's, just 1 mm thick. Depending on the targeted application for each developer, the TGP's range in area from 4.5 cm^2 to 75 cm^2 and use case materials that include AlN, SiC, Si, Ti, a metal matrix composites, and a metal-clad polymer. In Figure 16 each team's unique wick structure and material is outlined in the figures surrounding the plot. The plot shows the effective conductivity of each TGP in a gravity-neutral, horizontal orientation. Additional details on the characteristics and performance of the TGPs can be found [20-31].

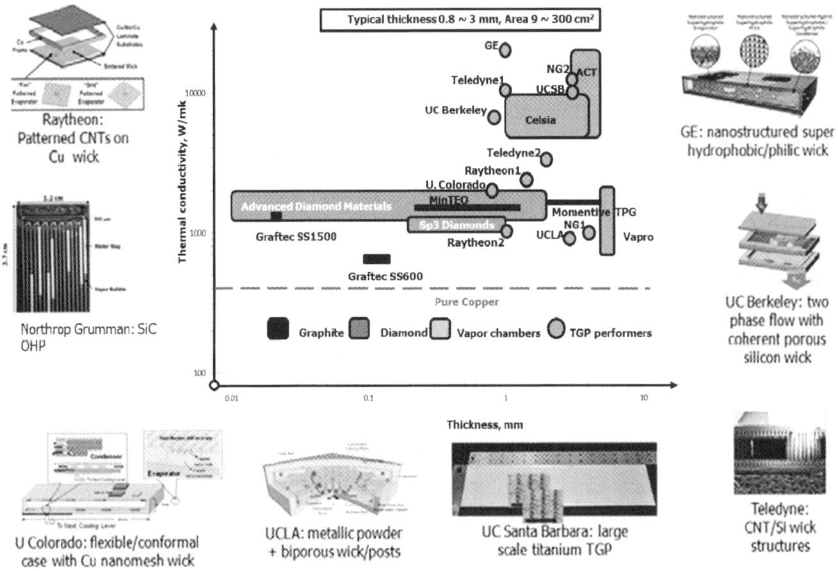

Fig. 16. Effective Lateral Thermal Conductivity of DARPA Thermal Ground Planes (TGP) [19].

Microtechnologies for Air-Cooled Exchangers (MACE): The primary MACE program goal was to reduce the heat sink thermal resistivity to 5 cm^2-K/W while increasing the COP to 20 or above for a combined fan-heat sink volume of 1000 cm^3. Thus, with a MACE heat sink, a 100 W component with a footprint of 10 × 10 cm can be air-cooled with a base-to-air temperature rise of 5°C or less and require no more than 5 W of fan power. The five MACE teams pursued diverse paths to these metrics, including aerodynamically optimized fins and flow paths (UTRC), piezo-driven agitators and synthetic jets with surface microstructured fins, along with a two-tier vapor chamber structure (Thermacore), compressed air micro-jets emerging from thin plenums to drive entrainment (Honeywell), a 3D vapor chamber that combines the pump and condenser in a single unit (MIT), and integrated microchannels and jet impingement (Raytheon). As may be seen in Figure 17, displaying the thermal resistivity of the MACE technologies and several COTS heat sinks, the MACE effort resulted in air-cooled heat sinks with a significantly lower resistivity than typical COTS coolers, with 3 of the MACE designs meeting or improving upon the 5 cm^2-K/W target. A selected bibliography of MACE performer publications can be found in [32-37].

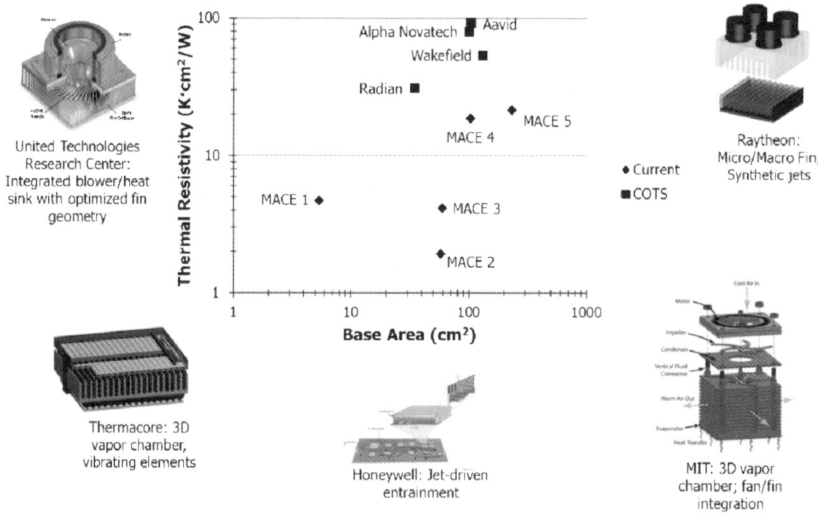

Fig. 17. Thermal Resistivity of DARPA's MACE Air-Cooled Heat Sinks [17].

Nano Thermal Interfaces (NTI): The main objective of the NTI thrust was to develop thermal interface materials (TIMs) based on novel materials and structures that could provide significant reductions in the thermal resistance of the interface layer between an electronic device and a spreader or a heat sink (including the TGP and/or MACE technologies). In many electronic applications, the need for a compliant TIM that can accommodate a large difference in thermal expansion, as would occur in a silicon-to-copper die attach, in-field reworkability, and high reliability have resulted in the widespread use of greases, polymer, and epoxy Thermal Interface Materials (TIMs), whose low thermal conductivity produces a substantial fraction of the junction-to-ambient thermal resistance. While solder TIMs do offer high thermal conductivities and low thermal interface resistances, their mechanical stiffness makes them inappropriate for use across interfaces between materials with large differences in thermal expansion coefficients, due to the risk of delamination and chip cracking. It was, thus, the goal of the NTI thrust to develop TIMs with solder-like thermal resistance, reaching 0.01 cm^2-K/W, along with epoxy-like mechanical compliance and "re-workability" in the field. In addition the NTIs were required to also accommodate non-planarity and roughness on the mating surfaces and survive a minimum of 1000 temperature cycles.

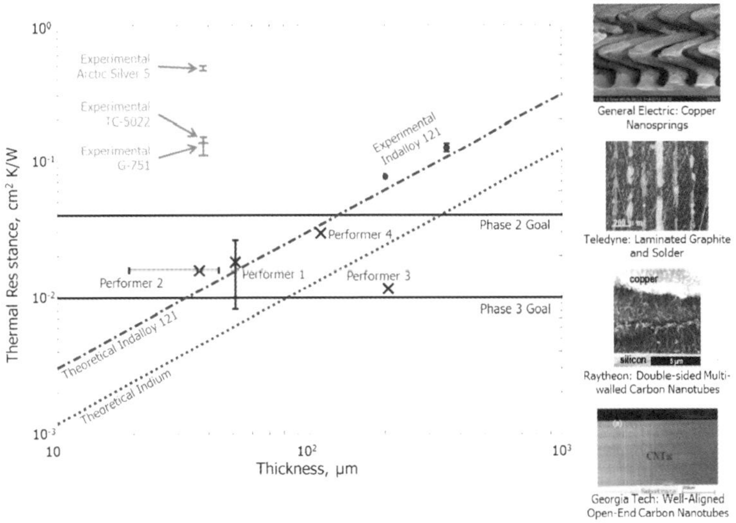

Fig. 18. Thermal Resistivity of DARPA's Nano Thermal Interface (NTI) Structures [17].

Figure 18 displays the thermal resistivity of the 4 NTI devices versus thickness, along with selected COTS thermal greases and epoxies and Indalloy and Indium solder. While the use of thin bond lines of Indium (80 μm) and Indalloy (35 μm) solder is seen to meet the NTI thermal resistance goal, such thin bond lines, if applied with large differential thermal expansion, would be prone to thermal stress and fatigue failure. The NTI technologies include GLAD-applied copper nanosprings (GE), laminated solder and flexible graphite films (Teledyne), multi-walled carbon nanotubes with layered metallic bonding materials (Raytheon), and functionalized-tip carbon nanotubes (Georgia Tech). It is to be noted that all the NTI prototypes improved on the COTS thermal interface materials. The GLAD copper springs (GE) and graphite-solder laminated structures (Teledyne) showed the lowest resistivities and are currently undergoing aging and temperature cycling tests to establish their robustness and durability. A selected bibliography of NTI team publications is included in this paper [37-44].

Active Cooling Modules (ACM): The ACM thrust was centered on the development of thin, high Coefficient of Performance (COP) microcoolers, using novel thermoelectric materials and refrigeration concepts, which can be inserted into the "thermal stack" and produce a local temperature reduction of 15K or more while removing a heat flux of 25 W/cm^2 with a COP of 2 or better. The ACM technologies investigated included near-optimum thicknesses of thin-film thermoelectric superlattices (UTRC), bulk low-conductivity nanostructured materials (GMZ Energy), colloidal high-ZT nanocrystals (Marlow) and two Stirling refrigeration approaches using MEMS-based membranes (Carnegie Mellon University) and miniaturized components (Beck Engineering). Regrettably, none of the ACM approaches has been able to achieve the high heat flux and high COP demands of the ACM program. A selected bibliography of ACM team publications is included in this paper [45-49].

Examining the TMT research thrusts in spreaders (TGP), heat sinks (MACE), and TIMs (NTI) individually or collectively, there is little doubt that the application of micro-nano technology and materials has had notable success in reducing the thermal resistance of individual elements in the conventional thermal resistance chain or "thermal stack." DARPA, together with other DoD agencies and the commercial marketplace has, thus, contributed to significant improvement in thermal management of electronic components with the existing Gen 2, "attached microcooler," paradigm. However, the use of an "attached microcooler" thermal solution forces heat generated in a chip to flow across multiple, high thermal resistance interfaces and spreaders before reaching a "remote" low temperature coolant. It is, thus, incapable of targeting a particular zone or sector on a chip where hot spots may occur, a particularly challenging, high heat flux chip deep inside a three-dimensional stack, or distinct thermal control requirements in two adjacent chips in a stack. As a consequence of this "blunt instrument" approach to thermal management, ongoing miniaturization of the components has not been matched by miniaturization of the thermal solution. On the contrary, as the electronics have shrunk and cooling density has increased, the thermal management hardware has grown.

Consequently, in many of today's high performance electronic systems, the thermal management solution accounts for a very large fraction of the SWAP (size, weight, and power), as well as the cost, and undermines efforts to achieve computational energy efficiency. That large SWAP for the thermal management solution stymies attempts to port advanced systems to smaller form-factors or to reach the SWAP targets of many of the defense-related, as well as the commercial, electronic systems. The continued pursuit of this conventional paradigm does little to address the challenges posed by on-chip hot spots and three-dimensional chip stacks and has failed to yield the extensive enhancements needed to sustain the "Moore's Law" doubling rate in a period when Dennard scaling cannot be counted on to suppress increases in power dissipation. It is, thus, time to consider a new, Generation-3, thermal management paradigm.

Towards a Gen-3 Thermal Management Paradigm

To find a sign post demarking the direction for Gen-3 thermal packaging, one need look no further than the High Performance Computing (HPC) sector. Figure 19 displays three of the leading commercial petaflop-level HPCs in 2012: the (then) top-ranked Japanese K computer, the Milkyway

Chinese computer, and the Jaguar, which was the fastest US supercomputer [50-54]. Several common HPC characteristics may be observed in this Figure: these are large systems, occupying approximately $1000m^3$; they dissipate large amounts of power, in the range of 5-10 MW, and they all use liquid cooling. The Japanese K computer is water cooled. The Chinese Milkyway computer is water cooled and the Jaguar uses refrigerant R134a as the coolant to achieve petaflop performance.

Fig. 19. Leading Commercial High Performance Computers circa 2010: Japanese K Computer (left), Chinese Milkyway Computer (center), and US Jaguar Computer (right) [51-53].

Examining Figure 20, displaying the computational efficiency in megaflops per watt vs the year of launch of various high end computers, it may be seen that in 2012 the K-Computer achieved a computational efficiency of 1 petaflop per kilowatt, representing a considerable improvement over earlier generations of HPC's. Moreover, in the mid-section of the curve, the increase in computational efficiency coincided with the increase in cooling capability and the increase in on-chip performance. Thus, across the decade of 1995-2005, when the SOA in CMOS cooling increased from 1 W/cm^2 to 10 W/cm^2, HPC capability rose from 1 megaflop/watt to 200 megaflops/watt, represented by IBM's Blue Gene 1. It may be argued that in an era, in which interconnect power dissipation rivals and often exceeds the transistor switching power dissipation, this 200x improvement in computational efficiency was facilitated by the use of higher transistor count (and higher power dissipation) chips which enabled a reduction in the signal propagation distance and, in particular, the number of chip crossings required to complete each operation. As more and more transistors and functional capability are provided on a single chip, the energy penalty of communication, of propagating signals off-chip, and from one functional block to another on the same chip, more and more efficient operation can be achieved.

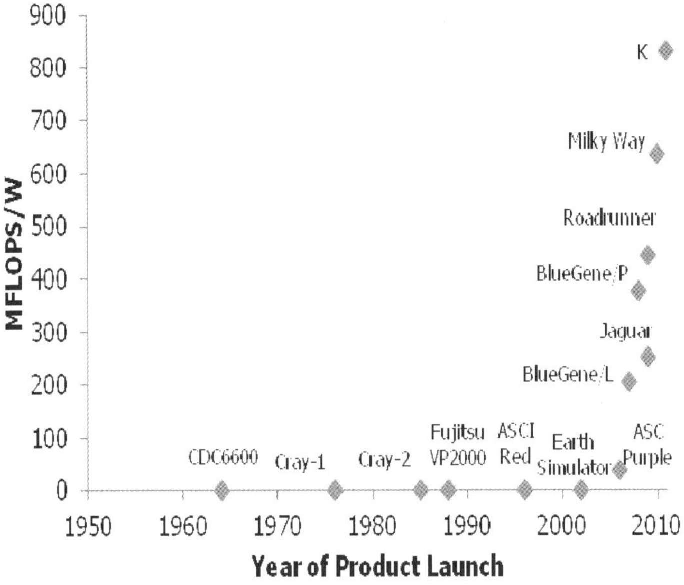

Fig. 20. History of Computational Efficiency.

Further support for the assertion that integral liquid cooling provides a path, if not the best path, for Gen-3 thermal management can be found by comparing two different, 100 terraflop IBM HPC's [55]: the 100-teraflop ASCI Purple Supercomputer operating at Lawrence Livermore Labs in 2005 and entirely air cooled and the IBM Power 775, a more recent HPC delivering approximately the same performance at 93 teraflops, but water cooled. The Table in Figure 21 compares the weight, volume, and system power required to provide this computing capability and reveals that switching from air-cooling to water-cooling allowed designers of the Power 775 to achieve a 52x decrease in weight, 300 times decrease in volume, and a 20 times reduction in system power.

Metric	Remote Cooling (ASCI Purple)	Integral Cooling (Power 775)	Decrease
Weight	178,715 kg	3,410 kg	**52x**
Volume	1800 m^3	6 m^3	**300x**
System Power	3400 kW	175 kW	**20x**

Fig. 21. Size, Weight, and Power Consumption (SWaP) of Two 100 TFLOP IBM Computers [55].

Gen 3 embedded cooling is represented in Figure 22, depicting the move from the remote-cooling, attached microcooler paradigm to intrachip cooling, in which heat is removed at the "source" and extracted by the flowing liquid. Starting with today's state-of-the-art in air-cooled heat sinks, yielding 100 W-130 W per chip, embedded cooling can – over time – reach 10 times that dissipation, i.e. 1 to 1.5 kW by sequentially introducing high-conductivity substrates and vias, dielectric liquids, then directing the liquids better through manifolded structures, by moving from single-phase to evaporative cooling, then by manifolding the distribution of the evaporating fluid and directing the fluid across shorter flow paths, as well as to the hotspot areas, and then completing this sequence with the addition of on-chip thermoelectric devices. It is anticipated that such embedded cooling could be applied to a variety of electronic systems, including microprocessors, power amplifiers, power supplies, and other commercial and defense electronic systems.

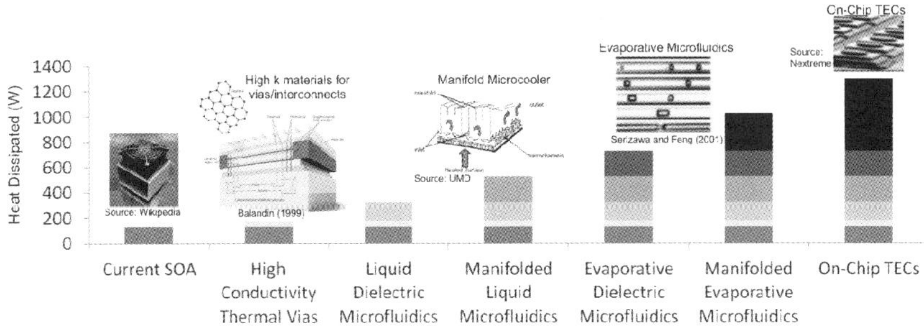

Fig. 22. Gen-3 Embedded Cooling Techniques Under Development.

In addition to improving heat extraction, such an embedded cooling approach can deliver the heat to the external rejection surfaces close to the operating temperature of the electronic devices. Such microfluidic cooling can, thus, provide heat to these surfaces with a large temperature difference relative to the ambient, rather than the diminished temperature driving force experienced in "remote cooling" due to temperature drops at multiple interfaces and spreaders. Consequently, embedded cooling can also reduce the size and weight and/or pumping power required for external heat rejection.

Gen-3 Embedded Cooling – Closure and R&D Challenges

To overcome these limitations and remove a significant barrier to continued Moore's Law progression in electronic components and systems, it is essential to implement aggressive thermal management techniques that directly cool the heat generation sites in the chip, substrate, and/or package. The development and implementation of such "Gen-3" embedded thermal management technology, following on the Gen-1 air-conditioning approaches of the early years and the decades-long commitment to the Gen-2 "remote cooling" paradigm, is the focus of the current DARPA Intra/Inter Chip Enhanced Cooling (ICECool) thermal packaging program. Launched in 2013, ICECool aims to develop and demonstrate "embedded cooling" techniques capable of removing kW/cm^2 chip heat fluxes and kW/cm^3 chip stack heat densities, while suppressing the temperature rise of multi-kW/cm^2 sub-mm hot spots [56,57]. The ICECool program is composed of two thrusts: a 3-year ICECool Fundamentals effort, involving several university teams which are developing embedded cooling building blocks and modeling tools, and a 2.5-year ICECool Applications effort, led by several aerospace performers and culminating in functional electronic demonstration modules. ICECool performers are pursuing the creation of a rich micro/nano grid of thermal interconnects, using high thermal conductivity, as well as thermoelectric, materials to link on-chip hot spots to microfluidically-cooled microchannels. Such intra/inter chip enhanced cooling approaches are required to be compatible with the materials, fabrication procedures, and thermal management needs of homogeneous and heterogeneous integration in 3D chip stacks, 2.5D constructs, and planar arrays. A conceptual ICECool device is shown in Figure 23.

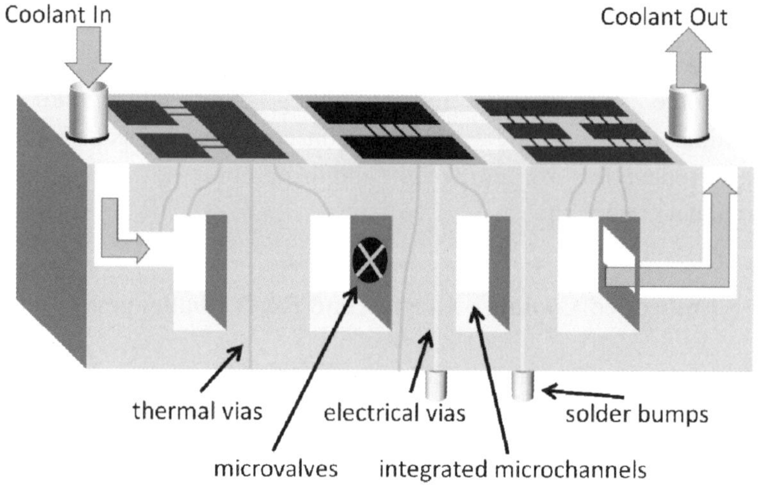

Fig. 23. Intra Chip Enhancd Cooling (ICECool) - Notional Design [56, 57].

An *intrachip* approach would involve fabricating micropores and microchannels directly into the chip while an *interchip* approach would involve utilizing the microgap between chips in three-dimensional stacks, as the cooling channel [58, 59]. In addition to the inclusion of an appropriate grid of passive and/or active thermal interconnects, it is expected that a combination of *intrachip* and *interchip* approaches, linked with thru-silicon and/or "blind" micropores, will confer added thermal management functionality. These microchannels and/or micropores will be integrated into a fluid distribution network, delivering chilled fluid to the chip or package and extracting a mixture of heated liquid and vapor to be transported to the ambiently-cooled radiator.

Some 30 years of thermofluid and microfabrication R&D, driven initially by the publication of the Tuckerman & Pease microchannel cooler paper in 1981 [60] and more recently by compact heat exchanger and biofluidic applications, has created the scientific and engineering foundation for the aggressive implementation of the "embedded cooling" paradigm. Nevertheless, substantial development and modeling challenges must be overcome if Gen-3 techniques are to supplant the current "attcaherd microcooler" paradigm. Successful completion of the DARPA ICECool program requires overcoming multiple microfabrication, thermofluid, and design challenges, including:

- Subtractive and additive **microfabrication** in silicon, silicon carbide, and synthetic diamond of high aspect ratio, thin-walled microchannels and high aspect ratio micropores; low thermal boundary resistance, high thermal conductivity thermal interconnect grids; on-chip, high power factor, high COP thin-film thermoelectric coolers; and hermetic attachment of liquid supply and liquid/vapor removal tubes.

- Convective and evaporative **thermofluid transport** in microchannels and micropores– removal of 1 kW/cm^2 chip heat fluxes with 2-5 kW/cm^2 sub-millimeter "hot spots"; low pumping power liquid-vapor manifolds with Coefficients-of-Performance (CoP) between 20 and 30; high-exit-quality, greater than 90%, evaporative flows without flow instabilities and/or local dryout; and high fidelity thermofluid models for single- and two-phase flow in microchannels, microgaps, and micropores.

- Thermal and electrical **co-design** which moves progressively from passive, thermally-informed designs, which recognize the impact of temperature on functional performance, to active thermal co-design which places functional paths and blocks in the most favorable locations on the chip, to fully-integrated co-design which deals with the impact of microfluidic channels and thermal interconnects on the electrical design and placement of electrical devices and cells, to mature designs that interactively balance the use of resources to optimize layout for energy consumption and functional performance.

- **Physics of Failure** models that address the failure mechanisms and reliability of the Gen-3 thermal management components, including erosion and corrosion in microchannels, microgaps, and micropores; failure modes induced in the electrically active areas of the chip and/or substrate; and the impact of microfabrication and embedded cooling operation on the structural integrity and stress profile of the microchanneled substrate (intrachip) and/or the chip-to-chip bonding (interchip).

Successful development and implementation of this Gen-3 thermal packaging paradigm would place thermal management on an equal footing with functional design and power delivery, transforming electronic system architecture and unleashing the power of nanofeatured device technology, while overcoming the SWaP (size, weight, and power consumption) bottleneck encountered by many advanced electronic systems. After decades of mere "problem solving" with Gen-1 (HVAC) and Gen-2 (spreaders, heat sinks, and TIMs) thermal management technology, it is expected that widespread adoption of Gen-3 "embedded cooling" techniques will provide a significant performance multiplier for advanced electronic components.

References

[1] Burks, A.R. and Burks, A.W., *The First Electronic Computer: The Antanasoff Story*, Michigan: The University of Michigan Press, (1989).

[2] Bar-Cohen, A. and Rohsenow, W.M., "Thermally Optimum Spacing of Vertical Natural Convection Cooled, Parallel Plates," *ASME Journal of Heat Transfer*, 106, pp. 116–123, (1984).

[3] Bar-Cohen, A., "Thermal Management of Air and Liquid-Cooled Multi-Chip Modules," *IEEE Transactions on Components, Hybrids and Manufacturing Technology*, 10(2), pp. 159–175, (1987).

[4] *The Control Data Cyber 205 Computer System*, Control Data Corporation, (1980).

[5] Iyengar, M. and Bar-Cohen, A., "Least-Energy Optimization of Forced Convection Plate-Fin Heat Sinks," *IEEE CPT Transactions*, 26(1), pp. 62–70, (2003).

[6] International Technology Roadmap for Semiconductors (ITRS).

[7] Sri-Jayantha, S.M., McVicker, G., Bernstein, K., Knickerbocker, J.U., "Thermomechanical modeling of 3D electronic packages," *IBM Journal of Research and Development*, 52(6), pp. 623–624, (2008).

[8] Swaminathan, M., Keynote Presentation, *Thermal and Thermomechanical Phenomena in Electronic Systems (ITherm)*, (2012).

[9] Fuller, S.H. and Millett, L.I., *The Future of Computing Performance: Game Over or Next Level?*, Washington, DC: The National Academies Press, (2011).

[10] Amon, C.H., Murthy, J.Y., Yao, S.C., with Narumanchi, S., Wu, C.F., Hsieh, C.C., "MEMs-Enabled Thermal Management of High-Heat-Flux Devices, EDIFICE: Embedded Droplet Impingement for Integrated Cooling of Electronics," *J. Experimental Thermal and Fluid Science,* 25, pp. 231–242, (2001).

[11] Koga, T., Rabin, O., Dresselhaus, M.S., "Thermoelectric Figure of Merit of Bi/Pb1-xEuxTe Superlattices", *Phys. Rev. B*, 62, pp. 16703–16706, (2000).

[12] Zeng, S., Chen, C.H., Mikkelsen, J.C., Jr., Santiago, J.G., "Fabrication and characterization of electrokinetic micro pumps," *Thermal and Thermomechanical Phenomena in Electronic Systems (ITherm)*, 2, pp. 31–36, (2000).

[13] Pettigrew, K., Kirshberg, J., Yerkes, K., Trebotich, D., Liepmann, D., "Performance of a MEMS Based Capillary Pumped Loop For Chip Level Temperature Control," *Proceedings of the 14th IEEE International Conference on Micro Electro Mechanical Systems*, Interlaken, Switzerland, pp. 427–430, (2001).

[14] Travkin, V.S., Hu, K., Rizzi, M., Canino, M., Catton, I., "Revising the Goals and Means for the Base-to-Air Cooling Stage for Semiconductor Heat Removal - Experiments and Their Results", *Proceedings of the 17th Semi-Therm Symposium*, San Jose, California, pp. 85, (2001).

[15] LaBounty, C., Shakouri, A., Bowers, J.E., "Design and Characterization of Thin Film Micro-Coolers," *Journal of Applied Physics*, 89(7), (2001).

[16] Ramaswamy, C., Joshi, Y., Nakayama, W., Johnson, W.B., "Compact Thermosyphons Employing Microfabricated Components," *Microscale Thermophysical Engineering*, 3, pp. 273–282, (1999).

[17] Bloschock, K.P. and A. Bar-Cohen, "Advanced Thermal Management Technologies for Defense Electronics," Proceedings, SPIE 8405, Defense Transformation and Net-Centric Systems, (2012).

[18] Bar-Cohen, A., Maurer, J.J., Felbinger, J.G., "DARPA's Intra/Interchip Enhanced Cooling (ICECool) Program," Proceedings, CS Mantech, pp. 171–174, (2013).

[19] Bar-Cohen, A., Matin, K., Bloschock, K., Jankowski, N., Sharar, D., "Two-Phase Thermal Ground Planes: Technology Development and Parametric Results," Final Report: DARPA Thermal Ground Plane (TGP) Program, DOC# 21479, Defense

Advanced Research Projects Agency (DARPA), Microsystems Technology Office (MTO), 05/13/2013.

[20] Reilly, S., Catton, I., “Utilization of Advanced Working Fluids with Biporous Evaporators,” *ASME Journal of Thermal Engineering and Engineering Applications*, 3(2), (2011).

[21] Oshman, C., Li, Q., Liew, L.A., Yang, R., Lee, Y.C., Bright, V.M., Sharar, D.J., Jankowski, N.R., Morgan, B.C., “Thermal Performance of a Flat Polymer Heat Pipe Heat Spreader Under High Acceleration,” *Journal of Micromechanics and Microengineering*, 22(4), (2012).

[22] Oshman, C., Shi, B., Li, C., Yang, R., Lee, Y.C., Peterson, G.P., and Victor Bright, V., “Development of Thin Polymer-based Flat Heat Pipes,” *ASME/IEEE Journal of Micro-Electro-Mechanical Systems*, 20(2), pp. 410–417, (2011).

[23] Ababneh, M. Gerner, F. Chamarthy, P. De Bock, H. P. J., Chauhan, S, Deng, T, “Thermo-Fluid Model for High Thermal Conductivity Thermal Ground Planes”, *Proceedings of the ASME 2012 3rd Micro/Nanoscale Heat & Mass Transfer International Conference*, Atlanta, GA, (2012).

[24] De Bock, H. P. J., Chamarthy, P., Chauhan, S, Weaver, S. A., Deng, T., “On the Charging and Thermal Characterization of a Micro/Nano Structured Thermal Ground Plane”, *Proceedings of the 12th Intersociety Conference on Thermal and Thermomechanical Phenomena in Electronic Systems*, Las Vegas, Nevada, (2010).

[25] Weibel, J.A., Garimella, S.V., Murthy, J.Y., Altman, D.H., “Design of integrated nanostructured wicks for high-performance vapor chambers,” *IEEE Transactions on Components, Packaging, and Manufacturing Technology*, 1, pp. 859–867, (2011).

[26] Ranjan, R., Garimella, S.V., Murthy, J.Y., “Transport in Passive, High Thermal Conductivity Heat Spreaders,” *ASME Journal of Heat Transfer*, 133, (2011).

[27] Thompson, S.M., Hathaway, A.A., Smoot, C.D., Wilson, C.A., Ma, H.B., Young, R.M., Greenberg, L., Osick, B.R., Van Campen, S., Morgan, B.C., Sharar, D., Jankowski, N., “Robust Thermal Performance of a Flat-Plate Oscillating Heat Pipe During High-Gravity Loading,” *J. Heat Transfer,* 133(10), (2011).

[28] Cai, Q., Chen, B., Tsai, C., “Development and Tests of High Performance Silicon Vapor Chamber,” *Journal of Micromechanics and Microengineering*, 22, (2012).

[29] Cai, Q., Chen, B. C., Tsai, C., Chen, C. L., “Development of Scalable Silicon Heat Spreader for High Power Electronic Devices,” *Journal Thermal Science and Engineering Applications*, 1(4), (2010).

[30] Dussinger, P., Sungtaek Ju, Y., Catton, I., Kaviany, M., ”High Heat Flux, High Power, Low Resistance, Low CTE Two-Phase Thermal Ground Planes for Direct Die Attach Applications,” *Government Microcircuit Applications & Critical Technology Conference*, Las Vegas, Nevada, pp. 257–260, (2012).

[31] Dhillon, N.S., Hogue, C., Chan, M.W., Cheng, J.C., Pisano, A.P., "Integrating Coherent Porous Silicon as a Wicking Structure in the MEMS Based Fabrication

of a Vertically Wicking Micro-columnated Loop Heat Pipe," *ASME 2011 International Mechanical Engineering Congress & Exposition (IMECE)*, Denver, Colorado, (2011).

[32] Zhou, F. and Catton, I., "Numerical Evaluation of Flow and Heat Transfer in Plate-Pin Fin Heat Sinks with Various Pin Cross Sections," *Numerical Heat Transfer, Part A: Applications*, 60(2), pp. 107–128, (2011).

[33] Zhou, F., Hansen, N., Geb, D., Catton, I., "Obtaining Closure for Fin-And-Tube Heat Exchanger Modeling Based on Volume Averaging Theory (VAT)," *Journal of Heat Transfer*, 133(11), (2011).

[34] Allison, J.M., Staats, W.L., McCarthy, M., Jenicek, D., Edoh, A.K., Lang, J.H., Wang, E.N., Brisson, J.G., "Enhancement of convective heat transfer in an air-cooled heat exchanger using interdigitated impeller blades," *International Journal of Heat and Mass Transfer*, 54(21–22), pp. 4549–4559, (2011).

[35] Kariya, H.A., Staats, W.L., Hanks, D.F., Peters, T.B., Cleary, M., Brisson, J.G., Wang, E.N., "Scaling the performance of an air-cooled loop heat pipe with the addition of modular condensers," *Thermal and Thermomechanical Phenomena in Electronic Systems (ITherm)*, pp. 828–833, (2012).

[36] Yeom, T., Zhang, M., Simon, T.W., Cui, T., "High-Frequency, Large-Displacement, and Low-Power-Consumption Piezoelectric Translational Actuators Based on an Oval Loop Shell," *Sensors and Actuators A: Physical*, 176, pp. 99–109, (2012).

[37] Yeom, T., Simon, T. W., Yu, Y., North, M., Cui, T., "Convective Heat Transfer Enhancement on a Channel Wall with a High Frequency, Oscillating Agitator," *ASME 2011 Int'l. Mech. Engr. Congress and Expo*, Denver CO, pp. 875–884, (2011).

[38] Wasniewski, J.R., Altman, D.H., Hodson, S.L., Fisher, T.S., Bulusu, A., Graham, S., Cola, B.A., "Characterization of Metallically Bonded Carbon Nanotube-Based Thermal Interface Materials Using a High Accuracy 1D Steady-State Technique," *J. Electron. Packag.*, 134(2), (2012).

[39] Lin, Z. and Wong, C.P., "Vertically-Aligned Carbon Nanotubes Metallization via Microwave- assisted Polyol Method," *ASE Tech Journal*, (2010).

[40] McNamara, A., Joshi, Y., Zhang, Z., "Characterization of Nanostructured Thermal Interface Materials – A Review," *TMNN 2011*, Antalya, Turkey, (2011).

[41] Shaddock, D.M., Weaver, S., Chasiotis, I., Bahadur, "Development of a Compliant NanoThermal Interface Material," *Government Microcircuit Applications & Critical Technology Conference*, Las Vegas, Nevada, pp. 285–288, (2012).

[42] Shaddock, D.M. Weaver, S. Chasiotis, I. Shah, B., Zhong, D., "Development of a Compliant NanoThermal Interface Material," *ASME 2011 InterPack Conference*, 2, pp. 13–17, (2011).

[43] Zhao, Y, Strauss, D., Liao, T., Chen, Y.C., Chen, C.L., "Development of A High Performance Thermal Interface Material with Vertically Aligned Graphite

Platelets", *Proceedings of the ASME/JSME 2011 8th Thermal Engineering Joint Conference*, Honolulu, Hawaii, (2011).

[44] Zhao, Y, Strauss, D., Liao, T., Chen, Y.C., Chen, C.L., "Experimental Study of a High Performance Aligned Graphite Thermal Interface Material," *Proceedings of the ASME 2012 3rd Micro/Nanoscale Heat & Mass Transfer*, Atlanta, GA, (2012).

[45] Lents, C., Barletta, P., et al., "Advancements in Thin Film Superlattice Thermoelectric Coolers," *Government Microcircuit Applications & Critical Technology Conference*, Las Vegas, Nevada, pp. 245–248, (2012).

[46] Gao, J., Guo, D., McGaughey, A.J.H., Santhanam, S., Yao, S.C., Moran, M., Anderson, W., Wolf, D., Fedder, G.K., "Stirling microcooler array with elemental in-plane flow," *Proceedings of the Government Microcircuit Applications and Critical Technology Conference (GOMACTech 2012),* Las Vegas, Nevada, pp. 289–292, (2012).

[47] Guo, D., Gao, J., McGaughey, A.J.H., Fedder, G.K., Moran, M., Anderson, W., Santhanam, S., Yao, S.C, "Design and evaluation of MEMS-based Stirling cycle micro-refrigeration system," *Proc. ASME 2011 International Mechanical Engineering Congress and Exposition (IMECE 2011)*, Denver, Colorado, (2011).

[48] Ordonez-Miranda, J., Yang, R., Alvarado-Gil, J.J., "A Model for the Effective Thermal Conductivity of Metal-Nonmetal Particulate Composites," *Journal of Applied Physics*, 111, (2012).

[49] Zhou, J., Yang, R., "Quantum and Classical Thermoelectric Transport in Quantum Dot Nanocomposites," *Journal of Applied Physics*, 110, (2011).

[50] S., Anthony, *What do super computers and overclockers have in common?.* Retrieved June 17, 2014, from http://www.extremetech.com/computing/90362-what-do-supercomputers-and-overclockers-have-in-common-water-cooling, (2011).

[51] Fujitsu, *Supercomputer "K computer" Takes First Place in World.* Retrieved June 17, 2014, from http://www.fujitsu.com/global/about/resources/news/press-eleases/2011/0620-02.html, (2011).

[52] Mick, J., *China's New Supercomputer Rejects Intel, AMD CPUs for Homegrown Designs.* Retrieved June 17, 2014, from http://www.dailytech.com/Chinas+New+Supercomputer+Rejects+Intel+AMD+CPUs+For+Homegrown+Designs/article23168.htm, (2011).

[53] Morgan, T.P., *China takes HPC heavyweight title.* Retrieved June 17, 2014, from http://www.theregister.co.uk/2010/10/28/china_tianhe_1a_supercomputer/, (2010).

[54] Bland, B., *Leadership Computing Facility Status and Plans.* Retrieved June 17, 2014, from http://computing.ornl.gov/workshops/peta08/presentations/b_bland.pdf, (2008).

[55] Ellsworth, M.J., "Thermal Design and Implementation of Robust Liquid Cooling Systems for High Performance Computer Systems," *ASME 2011 InterPack Conference*, (2011).

[56] Bar-Cohen, A., "Thermal Packaging: From Problem Solver to Performance Multiplier," Electronic Cooling Magazine, Retrieved December 15, 2013, pp. 8–11, (2013).

[57] Bar-Cohen, A., "Gen 3 Thermal Management Technology: Role of Microchannels and Nanostructures in an Embedded Cooling Paradigm," *ASME Journal of Nanotechnology in Engineering and Medicine*, 4(2), (2013).

[58] Bar-Cohen, A. and Geisler, K.J.L. Cooling the Electronics Brain. In Mechanical Engineering Magazine. Retrieved April 15, 2011, from https://www.asme.org/about-asme/news-media/newsletters, (2011).

[59] Geisler, K.J.L. and Bar-Cohen, A., "Passive Immersion Cooling of 3-D Stacked Dies," *IEEE Transactions on Components and Packaging Technologies*, 32(3), pp. 557–565, (2009).

[60] Tuckerman, D. and Pease, R.F.W., "High-performance heat sinking for VLSI," *IEEE Electron Device Letters*, 2(5), pp. 126–129, (1981).

Index